W0231943

MECHANISMS OF LYMPHOCYTE ACTIVATION AND IMMUNE REGULATION III

Developmental Biology of Lymphocytes

ADVANCES IN EXPERIMENTAL MEDICINE AND BIOLOGY

Editorial Board:

NATHAN BACK, *State University of New York at Buffalo*

IRUN R. COHEN, *The Weizmann Institute of Science*

DAVID KRITCHEVSKY, *Wistar Institute*

ABEL LAJTHA, *N.S. Kline Institute for Psychiatric Research*

RODOLFO PAOLETTI, *University of Milan*

A Continuation Order Plan is available for this series. A continuation order will bring delivery of each new volume immediately upon publication. Volumes are billed only upon actual shipment. For further information please contact the publisher.

MECHANISMS OF LYMPHOCYTE ACTIVATION AND IMMUNE REGULATION III

Developmental Biology of Lymphocytes

Edited by

Sudhir Gupta

University of California
Irvine, California

William E. Paul

National Institute of Allergy and Infectious Diseases
National Institutes of Health
Bethesda, Maryland

Max D. Cooper

The Howard Hughes Medical Institute
University of Alabama
Birmingham, Alabama

and

Ellen V. Rothenberg

California Institute of Technology
Pasadena, California

PLENUM PRESS • NEW YORK AND LONDON

Library of Congress Cataloging in Publication Data

International Conference on Lymphocyte Activation and Immune Regulation (3rd: 1990: Newport Beach, Calif.)
   Mechanisms of lymphocyte activation and immune regulation III: developmental biology of lymphocytes / edited by Sudhir Gupta . . . [et al.].
      p.   cm. — (Advances in experimental medicine and biology; v. 292)
   "Proceedings of the Third International Conference on Lymphocyte Activation and Immune Regulation, held February 16–18, 1990, in Newport Beach, California" — T.p. verso.
   Includes bibliographical references and index.
   ISBN-13: 978-1-4684-5945-6
   1. Lymphocyte transformation — Congresses. 2. Immune response — Regulation — Congresses. I. Gupta, Sudhir. II. Title. III. Title: Mechanisms of lymphocyte activation and immune regulation 3: developmental biology of lymphocytes. IV. Series.
   [DNLM: 1. Lymphocyte Transformation — congresses. 2. Lymphocytes — immunology — congresses. 3. Lymphocytes — physiology — congresses. W1 AD559 v. 292 / WH 200 I572m 1990]
   QR185.8.L9I553   1990
   616.07′9 — dc20
   DNLM/DLC   91-3783
   for Library of Congress   CIP

Proceedings of the Third International Conference on
Lymphocyte Activation and Immune Regulation,
held February 16–18, 1990, in Newport Beach, California

ISBN-13: 978-1-4684-5945-6   e-ISBN-13: 978-1-4684-5943-2
DOI: 10.1007/ 978-1-4684-5943-2

© 1991 Plenum Press, New York
Softcover reprint of the hardcover 1st edition 1991
A Division of Plenum Publishing Corporation
233 Spring Street, New York, N.Y. 10013

All rights reserved

No part of this book may be reproduced, stored in a retrieval system, or transmitted in any form or by any means, electronic, mechanical, photocopying, microfilming, recording, or otherwise, without written permission from the Publisher

PREFACE

Recent advances in the understanding of the major events that shape the immune recognition system have been remarkable. The analysis of immunoglobulin (Ig) gene organization and Ig repertoire diversification in lower vertebrates has provided new insight into this process in mammals. Similarly, the understanding of the early development of lymphocytes and of the acquisition of immunological tolerance has been aided by elegant studies in quail/chicken chimeras, using the power of the distinctive markers of the constitutive cells of these birds. Great strides have been made in understanding the role played by major histocompatibility complex (MHC) molecules in antigen presentation and in repertoire selection within the thymus. The use of transgenic mice expressing specific T-cell receptor (TCR) genes has elucidated the process of both positive and negative selection. In parallel, there has been considerable progress in our understanding of tolerance, based in part on the use of markers for the $V\beta$ genes of T-cell receptors and in part on the analysis of the behavior of long term T-cell lines. This has led to the realization that both clonal deletion and clonal anergy may play critical roles in the maintenance of unresponsiveness to self antigen.

Molecular analysis of the requirements for expression of membrane immunoglobulin molecules has revealed the existence of a complex that appears to be of critical importance in mediating signalling through Ig receptors. In addition, major insights have been obtained into the regulation of expression of genes of immunologic interest. Detailed analyses of the promoter regions of such genes and of the DNA-binding proteins that control their transcription promise to yield important information for the regulation of expression of lymphokines and other lymphocyte products.

Progress in these and related areas was the theme of the Third International Conference on Lymphocyte Activation and Immunoregulation, held in Newport Beach, California on February 16-18, 1990.

The Proceedings is divided into five sections. The first section deals with the phylogeny of lymphocytes. It includes genetic mechanisms influencing the evolution of the immune recognition genes, the ontogeny of thymocytes, and the diversity of antibody repertoire during evolution from cyclostomes to mammals. The second section discusses developmental biology of T lymphocytes. It includes experimental models probing the mechanisms of selection, T cell effector gene programming via signalling, and the use of TCR transgenic mice in the study of MHC restriction and self tolerance at the level of thymus. In addition, origin, development, diversity, repertoire, ligands, and functions of $\gamma\delta$ T cells are discussed. Molecular analysis of the interaction between p56*lck* with CD4 and CD8 antigen, a likely participant in the mechanisms of intrathymic developmental choice, has also been included in this section. The third section includes discussions of multiple mechanisms of induction and maintenance of tolerance to Mls and H2, and induction and maintenance of anergy in mature T cells. Developmental biology of B lymphocytes is featured in Section IV. It includes cloning and characterization of the DNA binding protein that binds to Jκ recombination signal sequence of immunoglobulin gene, B-cell development in fetal liver, the possible role of a "genesearch" retrovirus in gene expression during lymphoid differentiation, and the components of the B-cell antigen receptor complex. Also included is the comparative study of human and murine B-cell development, and characterization of a suppressive stromal cell subclone. The final section deals with the proposed role of somatic hypermutation and gene conversion in antibody diversity, and the molecular control of germline transcription and the role of the latter in immunoglobulin class switching. This book should be of interest to researchers not only in immunology, but also in cell, developmental, and molecular biology.

We wish to thank Miss Nancy Doman for outstanding editorial assistance and tireless preparation of the manuscript.

Sudhir Gupta
William Paul
Max Cooper
Ellen Rothenberg

CONTENTS

PHYLOGENY OF LYMPHOCYTES

DEVELOPMENTAL BIOLOGY OF T LYMPHOCYTES

DEVELOPMENTAL BIOLOGY OF IMMUNOGLOBULINS

IMMUNOGLOBULIN GENES AND B CELL DEVELOPMENT IN AMPHIBIANS

L. Du Pasquier and J. Schwager

Basel Institute for Immunology
Basel, Switzerland

INTRODUCTION

Although all vertebrates can make antibodies, the different classes do not respond to antigenic challenges in exactly the same manner. The differences that one can observe from cyclostomes to mammals can be due to "inside" causes: structure of the Ig molecule, architecture of the Ig loci, properties of the lymphoid cells; or to "outside" causes: difference in selective pressures brought in, for instance, by general physiological differences such as cell cycle properties, developmental differences, etc. Among vertebrates, amphibians (Figure 1) certainly represent a class where the selection pressures are different from those encountered in mammals. Given their mode of development, the individual is exposed to antigen (in most cases) very early (2-4 days after fertilization) which may imply the development of a pressure to develop an immune system early when the total number of lymphocytes is small (orders of magnitude lower than in mammals). Tadpoles indeed produce antibodies but their repertoire is different from that of adults.[1] It is interesting to see how these conditions, unfamiliar to mammals, have influenced the immune system of the frog. From the study of antibody production in anurans and to a certain extent in urodeles, two observations emerged: a restricted heterogeneity of amphibian responses to various antigens (measured by counting isoelectro focusing spectrotypes of the low molecular weight antibodies IgY, see Figure 2) and the poor affinity maturation of the antibodies during the response (10 x at best, versus 10^3 or 10^4 fold in mammals).[1,2,3] Limited protein sequence data of H and L chains were consistent with a low heterogeneity of anti-DNP antibodies.[4] Moreover, studies in genetically identical animals indicated a massive sharing of IEF spectrotypes and idiotypes between cloned individuals and their inheritance from one generation to the other. These facts led to the hypothesis that the low level of antibody heterogeneity in frogs was due to the lack of, or the poor occurrence of, random somatic events during the build-up of their immune system. This situation was thought to be consistent with the properties of the amphibian B cell compartment. It was thought that, when a very small number of lymphocytes was available, random somatic events would be dangerous, implying waste of material in nonproductive rearrangements and deleterious mutations, and that the frog system would be much better off with a limited number of phylogenetically selected germline V genes.[5] One could also have postulated the existence of a simpler Ig locus. The first results on the amphibian Ig loci at the DNA level revealed an architecture similar to that of mammals with a complexity of the heavy chain locus almost as extensive as in mammals, and with even more J_H elements.[6,7] The only factor limiting apparently the diversity of the antibodies was a remarkable sharing of CDR regions within the genes of a given V_H family.[8] It is therefore interesting in the light of these apparently paradoxical results to reconsider the whole issue. This is the purpose of the present review which will first deal with the question: Is the amphibian antibody diversity really lower than in the mouse or man?

Mechanisms of Lymphocyte Activation and Immune Regulation III
Edited by S. Gupta *et al.*, Plenum Press, New York, 1991

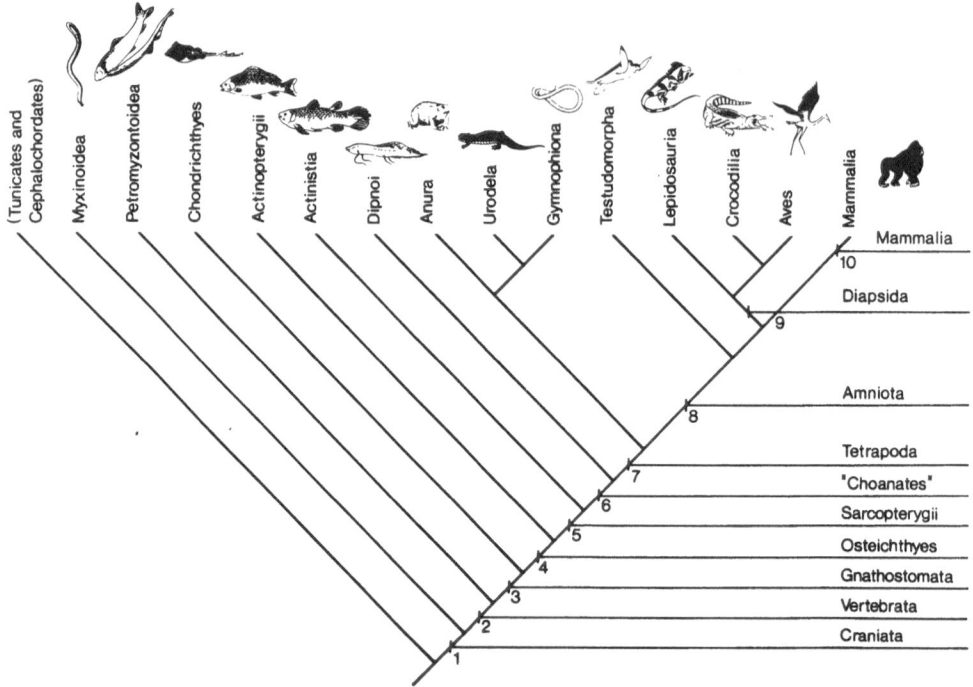

Figure 1. Phylogenetic relationships among living vertebrates (from reference 17, with adaptations). Amphibians represent the first tetrapods that achieved the transition from water to earth about 350 million years ago.

ANTIBODY DIVERSITY IN AMPHIBIANS REVISITED. WHAT ARE THE LIMITING FACTORS?

Antibody Data Revisited: The Repertoire of Xenopus *May be Greater than the IEF Studies Indicate*

1. This is quite possible because the low heterogeneity of the IgY response, the only one that can be studied by this method, does not reflect the heterogeneity of the whole response. In fact, it has been found that IgY heavy chains do not pair with all the possible light chain types, a fact which will lead to a restricted pool of IgY antibodies in terms of the total number of IgY combining sites.[9] This argument does not apply to idiotype studies that were performed on IgM.[4]

2. The apparent low heterogeneity of antibodies may represent, in fact, a large microheterogeneity undetectable by IEF. This is probably true, as seen from sequence data. Figure 3 shows the amino acid sequences derived from two germ-line V_H genes of *Xenopus* that differ at the level of one residue only without a charge difference. It is not difficult to imagine that these two genes could undergo similar rearrangements and that the resulting heavy chain paired with the same light chain would build up two different antibodies with no isoelectric focusing spectrotype difference. The product of these genes paired to an identical V_L product would have created two different antibodies without an isoelectric point difference. One could also argue that the sharing of

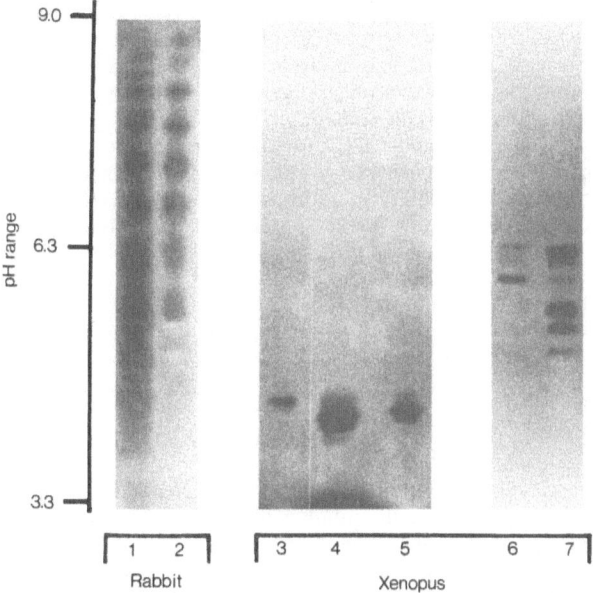

Figure 2. Differences in anti red cell responses between rabbits and *Xenopus*.
1-2: Rabbit IgG anti *Xenopus* red cells isoelectro-focussing pattern revealed as described in reference 18.
3-4-5: *Xenopus* IgY anti-sheep red blood cell antibodies.
6-7: Low heterogeneity of *Xenopus* IgY anti phosphorylcholine antibodies.

CDR or their close relatedness among V_H members would favor this situation in the frog. Yet it is likely to occur also in mammals and this does not entirely explain the differences between frogs and mammals.

DNA Data Revisited: The Potential Repertoire of Frogs May be Bigger Than Guessed From Antibody Data, But What Does Heterogeneity of cDNA Clones Mean in Terms of Repertoire?

    1. *cDNA Clones.* Let us be aware that when dealing with large repertoires and using different methods, one has to compare what is comparable. DNA data will not automatically be predictive of the protein data, and even among DNA studies, total spleen mRNA will give rise to cDNA libraries very different in their representativity from B cell genomic DNA libraries. Analysis of cDNA clones revealed a large heterogeneity among the nonimmune Ig population of *Xenopus*. This was somewhat expected because this kind of approach will automatically reveal the maximum of heterogeneity, irrespectively of its

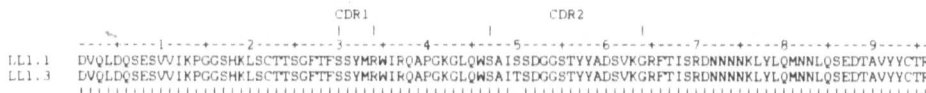

Figure 3. Comparison of two V_H2 segments differing only by one amino acid without charge difference (threonine for serine in CDR2). Data from reference 8.

meaning for the immune system repertoire. cDNA libraries from normal spleen reflect the expressed Ig genes of B cell, plasma cell and those of pre B cell genes that may never turn up a functional molecule. Diversity is also overestimated in such studies because the animals used were outbred and, therefore, polymorphic at the Ig locus.[10,11] The main outcome of these studies is the important somatic diversification in the D_H region. Even in the absence of knowledge on the genomic D_H structure, it is quite obvious that there is somatic diversification in *Xenopus*. If there were to be any differences between mammals and amphibians in somatic events, they will have to be at the level of hypermutations in the CDR_1 and CDR_2 or in the light chains. It was not so suprising in these studies not to find two identical clones since, even if the frog diversity is supposed to be "low", it has always been expected to be at a given time of the order of 10^5 different specificities (in adults). This leads to putative numbers of different V_H (i.e., including somatic variants) of the order of 300; it is therefore quite possible to analyze a few dozen cDNA clones without finding two identical ones provided the libraries contain an even distribution of the gene diversity.

Difficulties in Estimating V_H Gene Numbers from Southern Blots

A common way to calculate the number of genetic elements, knowing their frequency for a given length of DNA, is to count the number of bands in Southern blots and to multiply this by the appropriate factor. Given some possibilities of crosshybridization between certain V_H families, this procedure will lead to overestimation of V_H family members. An example of this can be found in the $V_H 1$ and $V_H 3$ families where some $V_H 3$ members, although clearly $V_H 3$ by their length and their FR3 region, crosshybridize with some $V_H 1 FR_1$ regions. In some cases as well, there is a lack of correlation between restriction fragment length, polymorphism and intrinsic sequence diversity, i.e., different fragments contain identical or nearly identical genes. On the other hand, the number of V_H elements will be underestimated if several tandem repeats give rise to DNA fragments of the same size.

Limiting Factors at the Cellular Level

With allelic exclusion, isotype exclusion and clonal selection operating in frogs like in mammals,[12] the commitment of lymphocytes appears to be a strong prerequisite to the functioning of the vertebrate immune system. This clearly implies that the number of cells at a given time will be a limiting factor. Assuming an average clone size of 10 cells with 10^5 B cells per tadpole cannot have more than 10^4 clones, probably no more than 10^2 V_H and 10^2 V_L will be expressed if random pairing of H and L chains take place. Cellular limitations are exacerbated in *Xenopus* with the existence of polyploid species such as *X. ruwenzoriensis* (108 chromosomes instead of 36 in *X. laevis*) which have fewer lymphocytes than *X. laevis*. The clone size in this species is probably smaller. In other words, the size of the pool of lymphocytes must impose a limit on the expression of newly incorporated genotypes of variable regions.[13] These limitations are not so important in mammals, probably since the system has a profusion of cells and that there is apparently no pressure to have the system working against the outside world when it contains as few cells as a frog tadpole.

In conclusion, differences remain between amphibians and mammals. The next section will now bring some new data dealing with the developmental aspect of gene expression and structural aspects peculiar to the frog that may contribute to our understanding of the above mentioned difference. Maybe we will then find out whether the frog immune system is something other than a mere mouse Ig locus to which only 10^4 clones of lymphocytes have been given, as has been implied in the "protecton" hypothesis.[14] In other words, we hope to find out whether special coadaptations have appeared in evolution, such as lack of hypermutation, lack of wastage in rearrangements, special usage of Ig genes, limited and distinct waves of rearrangement that could fit the observed data and their theoretical consequences. A consequence of the recently discovered greater simplicity between frogs and mouse potential may be to reconsider the actual size of vertebrate antibody repertoires in the context of cell numbers. To a certain extent, cell numbers must also be limiting in mammals in the sense that not all the potential repertoire can be expressed. In fact, it may be that the repertoires expressed early in mice and frogs are closer to each other than previously thought. The former would therefore also have to be smaller than previously thought.

REARRANGEMENT AND V_H USAGE DURING ONTOGENY

Given the limitation of one or the other method (DNA versus protein studies), we decided to complement the already existing data by two studies, one involving the study of B cell DNA in tadpoles and adults, the other involving the comparison of cDNA clones between tadpoles and adults. Both analyses were done within a single diploid heterozygous genome, that of the LG_{15} hybrid. B cell DNA libraries in EMBL3 were made after panning tadpoles or adult B cells with anti-Ig antibodies. Many J_H positive clones were isolated and the nonrearranged clones were sorted out with a probe immediately 5' to the J_H cluster. In parallel, LG_{15} cDNA clones were isolated from cDNA libraries made from larval and post metamorphic spleen mRNA. The analysis of the data led to a couple of interesting observations.

All Kinds of Rearrangement Can Take Place. We thought that in order to avoid wastage, the frog rearrangement mechanisms could have been different from mammals. This is not apparent from the study of rearranged genes where already three of the four possible types have been sequenced: functional rearrangement of a functional gene, abortive rearrangement (with frameshift mutations in the D_H region) of a functional gene, correct rearrangement to a pseudo gene (a V_H3 member with an exceptionally long leader and no promoter). This is consistent with the observations made earlier on Southern blots suggesting that rearrangements occur on the two chromosomes.[6] From the sequence of 8 rearranged genes (tadpoles and adults), 4 were abortive. In both larval and adult cases, the representation of one given abortive rearrangement was overwhelming. In adults, a V_H3 pseudo gene rearranged correctly to D8 and J_H3 represented half of the clones (although each clone was different from one another). Out of 39 larval clones, 25 corresponded to another V_H3 rearranged abortively (frameshift in the D region) to J_H5. Whether this reflects a library artefact or the over representation of a certain load of DNA in the spleen will have to be examined further. It is likely that some wastage occurs at the time of rearrangement (occurrance of double abortive arrangements) and this, in view of the small number of cells engaged, may restrict considerably the repertoire. Therefore, if there is limitation in wastage it has to be at the level of somatic hypermutations. The present results on rearrangement fulfill the prediction that

Table 1. Rearrangements of V_H Genes in *Xenopus*

cDNA clones from	LG_{15} tadpoles	Pool of adults
Associated with V_H1	2	8
V_H2	1	15
V_H3	1	4
V_H4	1	3
others	2	2

B cell rearranged genes from	LG_{15} tadpoles	LG_{15} adults
Associated with V_H1	1	1
V_H2	3	2
V_H3	25 (all identical)	6 (all identical)
V_H4	1	-

evolving a strictly regulated pathway implied the unlikely development of a set of complex stepwise switches of which all the intermediary steps would have to be selected.

The Low Heterogeneity of the Tadpole Response and its Differences from the Adult is Not Due to a Different Usage of V_H Families. Like the adult, tadpoles use mainly $V_H 1$ to $V_H 4$ families (Table 1). This observation, put in the context of V_H locus archi tecture and cell economy, "makes sense" from the immune system point of view. An ordered usage[15] of one member of a family after the other would be bad for the functional diversity of the repertoire, especially in frogs where the members of a single family do not differ enormously from each other in the CDR1 and CDR2. A better repertoire would be available if at least one or two members of each V_H family would be expressed. This prediction seems to be fulfilled by the observations of several V_H families being used in tadpoles. At this point, we do not know whether this is done due to V_H usage through differential promoter activity, a possibility in *Xenopus* given the special organizations of its V_H promoters, or to an interspersion of V_H genes, which in the eventuality of an ordered usage would permit the expression of many families. Interspersion of $V_H 1$ with $V_H 2$ and $V_H 2$ with $V_H 3$ families has been indeed noticed.[8]

STRUCTURAL LIMITATION

The comparison of many (21) cDNA and rearranged B cell DNA clones showed a constant difference between mammals and *Xenopus* in V_H CDR_3 regions. The length of this CDR measured between the last V_H residue and the tryptophan of J_H does not exceed 11 aa in *Xenopus* versus up to 20 in mouse and in chicken. A rearranged B cell gene contained a CDR3 of 17 residues but there is no evidence that it is a functional gene. Assuming a similar usage of amino acids in both classes of vertebrates, this length constraint will limit the frog repertoire. It has been argued that only H chain with an appropriate length in the DJ region could pair with the light chain and form a functional receptor.[14] Consequently, a limited diversity of the Ig light chains may impose a limited length of CDR3 and thus further reduce the diversity of functional molecules. It is thus possible that in *Xenopus* wastage would be reduced by producing DJ CDR3 region always within the appropriate length.

```
A    C40        5' TATGGGGTG 3'
                   |||||||||
     141        3' ATACCCCAC 5'


     C8         5' ACGCTAGCGGGTACAGCT 3'
                   |||||||||||||| |||
     R4         3' TGCGATCGCCCAT--GCT 5'


     J12        5' GGGGGTGG    3'
                   ||||||||
     T11        3' CCCCCACCGTA 5'


                     T  L  A  G  T  A
B    C8         5'   ACGCTAGCGGGTACAGCT    3'
                     ||||||||||||||
     C14        5' GAGTACGCTAGCGGGTAC      3'
                   E  Y  A  S  G  Y
```

Figure 4. Various examples of D_H sequences implying inversions (somatic or in the germ line?) of the D element (A), and usage of different reading frames (B). Data from references 6, 8, 11 and unpublished observations.

Table 2. Factors Affecting the Antibody Repertoire in *Xenopus*

Heavy chain properties	Possible Consequences
V_H higher multiplicity, poor diversity	Microheterogeneity
J_H higher diversity (9_H)	More combinatorial possibilities (compensation for a putative lack of somatic hypermutations in frogs)
CDR_3 length limitation	Reduced heterogeneity Better efficiency in HL pairing = reduced wastage
Rearrangement not $\neq$ Mammals	No $\neq$, some wastage occurs
HL pairing problems in switch to IgY or selected switch	Reduction of diversity
Limitation in rearrangement waves?	Reduction of diversity
Long B cell generation + Small B cell number	Limited wastage possibilities + Reduction of expressed diversity
No hypermutations?	Reduction of diversity
Usage of D_H in several reading frames, and usage of inverted D_H	Maximum diversity potential in the absence of hypermutations

Lack of Somatic Mutations in CDR1 and 2

   In the single case where the use of a specific germline V_H could be identified in a rearranged B cell DNA clone, there was no difference in CDR_1 or $_2$. Similarly, in the few cases where sharing of CDR could be monitored between some germline genes and cDNA clones, these regions were not modified. However, the significance of this observation is not compelling, since in all cases, these genes were rearranged to $c\mu$ and could correspond to the germline repertoire where the lowest incidence of somatic mutations is expected.

Table 3. Comparison of Mouse IgH and TCR Genes (reference 19) with *Xenopus* IgH Genes

	Mouse Ig heavy chain	Mouse T-cell receptor				Xenopus Ig heavy chain
		α	β	γ	δ	
Variable segments	250 - 1000	100	25	7	10	100
Diversity segments	10	0	2	0	2	>8
Ds read in several frames	rarely	-	often	-	often	often
Inverted Ds	?	-	?	-	?	often
N-region addition	V.D, D.J	V.J	V.D.D.J	-	$V.D_1,D_1 D_2,D_2.J$	$V.D_1,D_1 D_2,D_2J$
Joining segments	4	50	12	2	2	8 - 9
Hypermutation	yes	no	no	no	no	no?

The analysis of many cDNA clones and of putatively functional rearranged gene in B cells has revealed in several cases that, unlike in mammals, D_H elements could be used in different reading frames (Figure 4) and in complementary orientation (inverted D). At this point it is difficult to distinguish whether *Xenopus* uses its D_H elements in a special manner (somatic inversion at a high frequency, or D_H segments inverted in the germ-line?) or whether simply because of the lack of somatic hypermutations, all the possible usages of D_H segments are clearly visible in the frog, whereas in mammals they would be unevenly diluted out by the enormous somatic variation. Whichever reason is correct, it again points out a low incidence of somatic events during the diversification of the amphibian antibody repertoire.

DISCUSSION AND CONCLUSION

The evolutionary processes that led the present coadaptations in the amphibian immune system have been multiple with some contradicting each other. For instance, the multiplicity of J_H elements can be taken as an important increase in potential diversity. This would be true only if everything else in the frog immune system behaves like that in mammals. In the opposite case, the increase in J_H elements may appear only as a compensation to the lack of hypermutation mechanisms. The J_H situation in *Xenopus* is actually, to a certain extent, reminiscent of the T cell receptor of mammals where no hypermutations take place and where the number of J_H has been estimated to be about 50 in the mouse and 58 in man.[16] Let us take another example, the apparent paradox of having a large potential with a restricted expression. This indeed raises interesting evolutionary questions about the maintenance of a large pool of genes. Maybe *Xenopus* is now in a phase of expansion and its actual complexity is the product of a recent series of duplications. If indeed the frog immune system does not make extensive use of somatic hypermutations to diversify its antibody response potential, it would actually be useful to periodically duplicate some sets of V genes, allowing them to mutate, thereby providing new possibilities of selection, followed probably by elimination. With respect to the development of the lymphocyte compartment, wastage, that was thought to be so undesirable that it would impose better regulation of certain gene expression, does occur to a certain extent during the rearrangement phase, and thus is not entirely eliminated (See Tables 2 and 3). Indeed, data from polyploid species show that *X. laevis*, in which most studies have been performed, retains some flexibility at the cellular level: In *X. ruwenzoriensis* (108 chr.) antibody diversity is larger than in 36 chr. species, and this species seems to have taken advantage of the duplications of the V genes even though their lymphocyte number is smaller. This suggests that the lower diversity of *X. laevis* is not in fact limited in an absolute manner by cell numbers. *X. laevis* could indeed express as many genes as *X. ruwenzoriensis* does.

In summary, the low diversity that we observe in amphibians is certainly due to more than one cause, which need not be the causes that would explain lower diversity in other species. The multiplicity of causes acting in one direction clearly underlines the fact that this low diversity is a necessary compromise linked to the properties of the cell development in this species, and not merely an evolutionary accident. The structural characteristics of the *Xenopus* Ig locus appear more and more as the evolutionary consequence of the lack of, or poor contribution of, somatic hypermutations in the diversification of the antibody repertoire. This situation represents the best evolutionary compromise for a system where B cells are limiting and where the immune system is exposed extremely early to antigenic challenge.

REFERENCES

1. L. Du Pasquier, J. Schwager, and M. F. Flajnik, The immune system of *Xenopus, Ann. Rev. Immunol.* 7:251 (1989).
2. L. Du Pasquier, Evolution of the immune system, *in*: "Fundamental Immunology" (second edition), W. E. Paul, ed., Raven Press Ltd., New York (1989).

3. J. Charlemagne, Antibody diversity in amphibians. Non-inbred axolotls use the same unique heavy chain and a limited number of light chains for their anti-DNP antibody response, *Eur. J. Immunol.* 17:621 (1987).

4. D. Brandt, M. Griessen, L. Du Pasquier, and J. C. Jaton, Antibody diversity in amphibians: evidence for the inheritance of idiotypic specificates in isogenic *Xenopus*, *Eur. J. Immunol.* 10:731 (1980).

5. L. Du Pasquier, Antibody diversity in lower vertebrates--why is it so restricted? *Nature* 296:311 (1982).

6. J. Schwager, D. Grossberger, and L. Du Pasquier, Organization and rearrangement of immunoglobulin M genes in the amphibian *Xenopus*, *EMBO J.* 7:2409 (1988).

7. L. Du Pasquier and J. Schwager, Evolutions of the immune system, *in*: "Progress in Immunology VII," F. Melchers et al., eds., Springer Verlag, New York, Berlin, Heidelberg (1989).

8. J. Schwager, N. Bürckert, M. Courtet, and L. Du Pasquier, Genetic basis of the antibody repertoire in *Xenopus*: analysis of the V_H diversity, *EMBO J.* :2989 (1989).

9. E. Hsu and L. Du Pasquier. Studies on *Xenopus* immunoglobulins using monoclonal antibodies, *Mol. Immunol.* 21:257 (1984).

10. G. W. Litman, M. J. Shamblott, R. Haire, C. Amemiya, H. Nishikata, K. Hinds, F. Harding, R. Litman, and J. Varner, Evolution of immunoglobulin gene complexity, *in*: "Progress in Immunology VII," F. Melchers et al., eds., Springer-Verlag, New York, Berlin, Heidelberg (1989).

11. E. Hsu, J. Schwager, and F. W. Alt, Evolution of immunoglobulin genes: V_H families in the amphibian *Xenopus*, *Proc. Natl. Acad. Sci. USA* 86:8010 (1989).

12. L. Du Pasquier and E. Hsu, Immunoglobulin expression in diploid and polyploid inter-species hybrids of *Xenopus*: evidence of allelic exclusion, *Eur. J. Immunol.* 13:585 (1983).

13. H. R. Kobel and L. Du Pasquier, Genetics of polyploid *Xenopus*, *Trends Genet.* 2:310 (1986).

14. R. E. Langman and M. Cohn. The E.T. (elephant tadpole) paradox necessitates the concept of a unit of B cell function: the protecton, *Mol. Immunol.* 24:675 (1987).

15. F. W. Alt, G. D. Yancopoulos, T. K. Blackwell, C. W. Wood, E. Thomas, M. Boss, R. Coffman, N. Rosenberg, S. Tonegawa, and D. Baltimore, Ordered rearrangement of immunoglobulin heavy chain variable region segments, *EMBO J.* 3:1209 (1984).

16. E. Lai, R. K. Wilson, and L. E. Hood, Physical maps of the mouse and human immuno-globulin-like loci, *Adv. Immunol.* 46:1 (1989).

17. F. H. Pough, J. B. Heiser, and W. N. McFarland *in* "Vertebrate Life" (Third edition), McMillan Publishing Company, New York (1989).

18. L. Du Pasquier and M. R. Wabl. Antibody diversity in amphibians: inheritance of isoelectric focusing antibody patterns in isogenic frogs, *Eur. J. Immunol.* 8:428 (1978).

19. M. M. Davis and P. Bjorkman, T-cell antigen receptor genes and T-cell recognition, *Nature* 334:395 (1988).

EVOLUTIONARY DEVELOPMENT OF IMMUNOGLOBULIN GENE DIVERSITY

G. W. Litman, C. T. Amemiya, F. A. Harding, R. N. Haire, K. R. Hinds,
R. T. Litman, Y. Ohta, M. J. Shamblott, and J. A. Varner

Tampa Bay Research Institute
St. Petersburg, Florida

THE STRUCTURE OF IMMUNOGLOBULINS IS PRESERVED THROUGHOUT THE VERTEBRATES[1]

In all jawed vertebrate species, antibody diversity is specified by an immunoglobulin monomeric structure consisting of two heavy chains and two light chains. Different immunoglobulin classes are associated with various polymeric configurations of the basic monomer. Immunoglobulin-like heterodimers also occur in the jawless vetebrates (cyclostomes) of which the only extant representatives are the lampreys and hagfishes (Varner and Litman, unpublished observations);[1-2] primary structure data that would firmly establish these as antibodies, however, are not available presently. During the past several years, our laboratory has identified immunoglobulin genes in species that are considered to represent significant departure points in vertebrate phylogeny.[3] Taken together with the descriptions of immunoglobulin genes in avians[4,5] and mammals,[6] a reasonably complete picture of the overall evolution of immunoglobulin gene structure and diversity is emerging. In all of these species, the most distinctive features of the immunoglobulin gene system, segmental organization and selective rearrangement in somatic tissues, are preserved. The structures of individual heavy chain variable (V_H), diversity (D_H), joining (J_H) and certain constant region (C_H) segmental elements are also remarkably similar. Two additional features of higher vertebrate immunoglobulin genomic organization, the split leader and recombination signal sequences (RSSs), also are found in lower vertebrates. Furthermore, the exon-intron organization of the constant region found in *Heterodontus* (horned shark), the most phylogenetically primitive vertebrate studied thus far, is equivalent to that described in mammals.[7] To a certain degree, some of these findings are not unexpected since in many cases the genes in lower vetebrates were detected by cross-hybridization with a particular mammalian immunoglobulin heavy chain gene probe.[8,9] Any gene detected using this procedure would have to be at least 60% related at the nucleotide sequence level. Although it would be interesting to compare the sequences of these genes to one another, the V_H genes are members of extensively diversified multigene families and there are few guidelines that could be applied to determining whether similarities and differences in gene sequence reflect vertical

----

[1]In this review, we distinguish jawed and jawless vertebrates. Jawed vertebrates include all of the extant sharks, skates and rays, as well as all of the subsequent vertebrate radiations. Jawless vertebrates (and craniates) include lampreys and hagfishes which are separated by a considerable evolutionary distance and exhibit major morphologic differences that are consistent with their classification in separate groups.

Mechanisms of Lymphocyte Activation and Immune Regulation III
Edited by S. Gupta *et al.*, Plenum Press, New York, 1991

11

evolutionary relationships. When weighed against the dramatic differences in immunoglobulin gene organization that have occurred during vertebrate phylogeny, such comparisons most likely could afford relatively little insight into the principal mechanisms of evolutionary change.

UNIQUE PATTERNS OF GENE ORGANIZATION ARE ASSOCIATED WITH MAJOR PATHWAYS OF EVOLUTION

As indicated above, the individual segmental elements that constitute immunoglobulin genes are remarkably related at the DNA sequence level, even in cases where the segmental elements, e.g., D_H and J_H, were not selected by cross-hybridization directly but rather have been detected by sequencing of an adjacent DNA segment(s) (i.e., the gene segment[s] is in close chromosomal proximity to a gene that was localized by cross-hybridization and/or has been recovered in a cDNA).[10-12] The genomic organization of these genes in different vertebrates, however, varies markedly. In *Heterodontus*, immunoglobulin heavy chain gene segmental elements (regions) are found in 16,000 bp linkage units consisting of a single variable (V_H), two diversity (D_H), one joining (J_H) and one constant (C_H) region.[7,11] Hundreds of these clusters are present in the genome of an individual and contrast markedly with the organization of mammalian immunoglobulin genes. In *Heterodontus*, antibody gene diversity presumably is generated by the same rearrangement mechanism(s) that is employed by higher vertebrate species, but recombination between segmental elements found in different clusters may not occur. First detected in *Heterodontus*, this form of gene organization has been found in representatives of a different elasmobranch order as well as in a chimera (*Hydrolagus*, ratfish), a species that represents an independent evolutionary line of the chondrichthyes (cartilaginous fishes) (Amemiya and Litman, unpublished observations). With the exception of species-specific differences in intron length, there appears to be relatively little deviation from the basic "cluster-type" organization throughout the entire radiation of the chondrichthyes. These findings suggest that during one major period in vertebrate phylogeny, an entirely unique form of immunoglobulin gene arrangement is found which has not been observed in any other vertebrate species thus far examined.

A second unusual feature of immunoglobulin genes in cartilaginous fishes involves the joining of segmental elements in the germline of approximately one-half of the clusters.[11,13] In all cases examined thus far, fully joined (VDJ) segments maintain a correct reading frame (Hinds and Litman, unpublished observations);[7] if joining were random, only one sequence in three would be expected to correctly encode the J_H segment. For VD-J-joined genes in *Heterodontus*[11] and VD_1-D_2J-joined genes in *Raja erinacea* (little skate) (see below[13]), the RSSs, found at the 3' of D in *Heterodontus* and at the 3' of D_1 and 5' of D_2 in *Raja*, are similar to those found in the non-joined clusters that have been shown to be expressed functionally. The V_H coding segments of joined genes are not related appreciably more to one another than to non-joined "typical" genes.[11] Although "joined" genes appear to be potentially functional, it has not been possible to detect stable transcripts arising from joined germline genes using a variety of conventional oligonucleotide-based probing technologies, mRNA blot analyses or PCR amplification. Different forms of "germline joining" occur in different lineages of the elasmobranch. For example, in the selachians (e.g. *Heterodontus*), the predominant form of joined genes is VD-J;[11] whereas, in the batoids (ex. *Raja* [skate]), VD_1-D_2J germline joining is encountered most frequently.[13] A number of additional probing strategies and other assays for low level transcription in both spleen and other lymphoid tissues are being employed in a continuing effort to understand the functional significance of germline joining.

A third unique feature of the organization of heavy chain genes in the cartilaginous fishes involves the absence of the regulatory octamer (ATGCAAAT or ATTTGCAT) of the B cell-specific immunoglobulin promoter,[11] an invariant component of immunoglobulin gene expression in higher vertebrates (see below). The absence of this sequence motif in cartilaginous fishes suggests that the manner in which these genes are regulated in a tissue-specific fashion may be unique. It should be noted, however, that *Heterodontus* light chain genes possess the regulatory octamer in a 5' position that corresponds to the location of this regulatory sequence in higher vertebrates,[14] whereas in *Raja* the light chain genes lack an octamer (Shamblott and Litman, unpublished observations). In summary, there are at least three major differences in the organization and regulation of immunoglobulin heavy chain genes in the most phylogenetically primitive jawed vertebrates.

THE MODERN FORM OF GENE ORGANIZATION

Immunoglobulin gene organization in more phylogenetically advanced vertebrates is typified by extended tandem arrays of related segmental elements, i.e. $V_1V_2V_3V_n-D_1D_2D_3D_n-J_1J_2J_3J_n-C_n$ where n varies from four to several hundred, depending on the particular type of segmental element.[6] In even the most primitive (least derived) members of the Osteichthyes, immunoglobulin gene organization is similar and in some ways identical with that found in higher vertebrate species. Specifically, in *Elops*, a primitive teleost, clusters of V_H segments are located within 100 kb of a single constant region gene.[12] For *Elops*, there appears to be only a single constant region gene which is in marked contrast to the large numbers of C_H genes in the cartilaginous fish. J segments are located approximately 1.5 kb upstream of the constant region segments[12] (Amemiya and Litman, unpublished observations) and D segments, presumably, occur between the V_H and J_H elements. Large numbers of V_H segments have been characterized and significant numbers of pseudogenes, encountered only infrequently in cartilaginous fish, are present. Joining (or fusion) of germline segmental elements, a common feature for the immunoglobulin heavy chain genes of all cartilaginous fish (see above), has not been observed. In the bony fish and all higher vertebrate forms, a typical regulatory octamer is located upstream of the transcriptional start site. In summary, a major difference in overall gene organization and regulation is found in the more phylogenetically advanced forms. The contrasts between the organization and relative complexities of immunoglobulin genes in cartilaginous vs. bony fishes, their relative complexity and the associations between gene organization and the generation of antibody diversity are of considerable interest and significance in understanding the evolution of this multigene family.

THE V_H REPERTOIRE IN CARTILAGINOUS FISH IS RESTRICTED

Estimates of the size of the V_H repertoire in lower vertebrate species are based on comparisons of both germline and rearranged, expressed (cDNA) immunoglobulin genes. Our initial analyses of V_H gene sequences in *Heterodontus*, including genes that exhibit germline-joining, indicate that all V_H genes are members of a single family, i.e., the genes possess $\geq 70\%$ absolute nucleotide identity. The genes are considerably more related to one another than the 70% cut-off for establishing family identity.[11] In order to extend these analyses, large numbers (~ 700) of individual phage that were selected with either C_H or J_H probes (which should detect all immunoglobulin genes regardless of V_H family) were screened with a V_H probe. Under the hybridization conditions employed, members of a unique V_H gene family (see below) would be expected to be V_H^-, J_H^+ or C_H^+. Wherever a lack of concordance between C_H and V_H hybridization was observed, C_H elements could be mapped to a λ arm, i.e., V_H was not integrated during the cloning. In a similar experiment in which V_H probes were used to detect possible V_H^+, J_H^- or C_H^- variants, only two clones were detected; both contained multiple V_H inserts, of which some segments were overt pseudogenes. Although some of the clones detected in these experiments are variants of the "typical" cluster, e.g., the genes are $V_H-J_H-C_H$ (D_H is deleted), these do not exceed 1% of the total and do not deviate from the basic "cluster-type" organization (Hinds and Litman, unpublished observations).

Using *Heterodontus* V_H probes, gene homologs comprising the same V_H family have been isolated from the skate (*Raja*), a member of a separate order of cartilaginous fish.[13] Although most of the V_H genes in *Raja* belong to the same V_H gene family as in *Heterodontus*, a small number of cDNA clones that hybridize weakly with the homologous *Raja* V_H probe were detected and shown to exhibit ~ 61% nucleotide identity. These V_H genes are associated with a different C_H isotype exhibiting only a limited degree of nucleotide identity with the predominant, "μ-type" C_H. The heavy chains of a second skate immunoglobulin class designated as IgR may be encoded by this newly identified gene.[20] Thus, independent V_H and C_H families have evolved within a different order of cartilaginous fish. *Raja* genes are distinguished further from those of *Heterodontus* in that their J_H segments exhibit more sequence variation.[13]

Although the diversity of V_H genes in *Heterodontus* is limited by the presence of only a single V_H family and a high degree of sequence similarity in D_H and J_H segments, N-region, junctional and complementarity determining region (CDR) diversity are extensive. Furthermore, somatic mutation in V_H coding regions has been detected using an oligonucleotide-based approach that identifies the germline clusters which give rise to a specific rearrangement(s).[19] Thus, *Heterodontus* exhibits two significant restrictions in the immune repertoire: 1)V_H, D_H and J_H diversity is considerably lower than in higher vertebrates and 2) combinatorial rearrangement may not be present. In view of the marked restriction in the D_H and J_H repertoire, this latter effect may be insignificant, i.e., no selective advantage would be realized in recombination between clusters.

V_H GENES IN BONY FISH

In order to examine V_H family diversity in a more phylogenetically recent form, an *Elops* spleen cDNA library was screened with a homologous C_H probe as well as with a probe specifying the primary V_H gene family (V_HI).[12] V_HI$^-$, C_H^+ cDNAs were selected and characterized. A second distinct immunoglobulin V_H family (V_HII) exhibiting < 60% relatedness to V_HI at both the nucleotide and inferred amino acid level also was detected. V_HI- and V_HII-specific probes produce different, complex hybridization patterns in Southern blot analyses of genomic DNA. The number of components hybridizing with the V_HI- and V_HII-specific probes are comparable and presumably some of this complexity is due to large numbers of pseudogenes. It is possible that other V_H families would be detected if additional rounds of screening/selection were carried out. The primary significance of these findings is that V_H family diversity expanded in the bony fishes. In addition, these species and their more recent descendants appear to utilize combinatorial joining as a means for expanding the antibody repertoire.

THE V_H REPERTOIRE IN AMPHIBIANS IS NOT RESTRICTED

As indicated above, V_H family diversity is present in species that are found at or above the evolutionary level of the Osteichthyes. As part of our efforts to examine developmental regulation of immunoglobulin gene expression in *Xenopus laevis* (African clawed frog), V_H-, CDR-, D_H-, as well as junctional-type diversity have been analyzed.[15] Immunoglobulin constant region probes that are specific for the three *Xenopus* immunoglobulin constant region isotypes (C_μ,[16] C_χ,[17] C_ν[18] as well as a J_H-specific probe were used to identify immunoglobulin cDNAs. Approximately 200 individual (J_H^+ and/or C_H^+) clones were identified by an iterative screening procedure from a cDNA library. Initially using an available V_H-specific probe (V_HI), non V_H-hybridizing clones were classified in an unassigned group and a randomly selected member of this group (designated V_HII) was used to screen the remaining clones. By repeating this procedure, eleven distinct V_H gene families were detected. The assignment of these as independent genes was confirmed by DNA sequence analyses; the difference between families actually exceeds that described for different murine V_H families. In addition, D_H and J_H segments have been shown to be complex as have the CDR segments of the various V_H genes. The immunoglobulin gene system of *Xenopus* appears to be diversified as extensively as that of higher vertebrate systems. It is reasonable to assume that the light chain genes of *Xenopus* also may be highly complex.[15,19] Somatic mutation of *Xenopus* immunoglobulin genes is likely, although the extent and relative frequency of this process as well as its role as an antigen-driven process may differ from that seen in mammals.

THE INDEPENDENT EVOLUTION OF IMMUNOGLOBULIN AND T CELL ANTIGEN RECEPTOR SYSTEMS

The evolution of immunoglobulin and T cell antigen receptor (TCR) as independent genetic systems is of considerable interest; however, at present, there is relatively little experimental data that can resolve whether or not T cell immunity preceded B cell immunity.

Typically, it is held that T cell immunity is the more primitive since allograft rejections have been well documented in protochordates[20] and more primitive metazoans.[21] Rejection of allografts, which in higher vertebrates is based on MHC recognition, however, does not necessarily imply that a T cell (or its more primitive evolutionary counterpart) mediates the recognition. To date there has been no conclusive (molecular) evidence for MHC-like molecules in any species below the phylogenetic level of the amphibia, although progress has been made recently towards defining MHC-like molecules in teleost fish (K. Hashimoto and Y. Kurosawa, personal communication). TCR-like molecules have not been described at the molecular genetic level in any species below the phylogenetic level of mammals, although protein molecules resembling components of the TCR complex have been detected in avians.[22] Recently, progress has been made towards identifying TCR genes in an avian (M. Cooper, personal communication).

The *Heterodontus* immunoglobulin gene system is the most primitive immune system (B or T cell) characterized to date. *Heterodontus* "immunoglobulin" genes share several properties with higher vertebrate immunoglobulin and TCR genes. Their V_H (D_H and J_H) segments are the most homologous to those of immunoglobulin,[10,11] their constant regions possess exon/intron organization of μ-type genes[7] and the differential splicing of their secretory vs. transmembrane segments is unequivocally that of immunoglobulin.[7] As indicated above, somatic mutation occurs in rearranged *Heterodontus* immunoglobulin genes but not in the genes encoding mammalian TCRs. *Heterodontus* immunoglobulin genes differ from those of higher vertebrate immunoglobulin in that the regulatory octamer is absent,[11] the organization is cluster-like[10,11] and multiple D segments are present.[11] These properties are also associated with some TCR genes. Furthermore, the decamer/nonamer sequence that is associated with the regulation of β TCR has been detected upstream of the initiation codon in *Heterodontus* immunoglobulin genes (E. Davidson, personal communication and unpublished observation).[11]

CONCLUSIONS

Based on these observations and additional data from our laboratory, hypotheses about the principal events that gave rise to the immunoglobulin gene systems of contemporary vertebrates can be made. The first event in the evolution of this rearranging-gene system presumably involved the segmentation of a single exon encoding a protein that possessed the basic conformation of an immunoglobulin domain. Either the "invasion" of the target sequence by a transposon containing an RSS-like sequence or intragenic recombination could have been the next critical step. In functional terms, rearrangement of the segmented gene would introduce sequence variation at the joining junction analogous to junctional and N-type diversity seen in higher vertebrates. This split exon evolved into a V-J-C- or V-D-J-C-type structure through duplication and/or recombination. Further development of this system involved unit duplication of these V-J-C or V-D-J-C clusters. Sequence diversification in CDRs of the V segments would expand the range of combining site specificities. An alternative pathway of gene evolution resulted in duplication (and diversification) of the individual V, D and J segments in a single cluster. Significant chromosomal distances between the individual elements would promote nonrestrictive rearrangement of segmental elements, facilitating combinatorial diversity. Further diversification of V_H framework regions resulted in the separate V_H gene families found in contemporary vertebrates and duplication and diversification of C_H exons expanded constant region diversity. Somatic mutation (and hypermutation) localized to the immunoglobulin locus became an increasingly important factor in the generation of antibody diversity.

Some tenets of these hypotheses already have been established putatively through these investigations. It is likely that the continuing study of genes in species occupying distant positions in vertebrate phylogenetic development will provide additional insight into the fundamental genetic mechanisms that have influenced the evolution of the immune recognition genes.

REFERENCES

1. R. L. Raison, C. J. Hull, and Hildemann, W. H., Characterization of immunoglobulin from the Pacific hagfish, a primitive vertebrate, *Proc. Natl. Acad. Sci. U. S. A.* 75:5679 (1978).

2. K. Kobayashi, S. Tomonaga, and K. Hagiwara, Isolation and characterization of immunoglobulin of hagfish, *Eptatretus burgeri*, a primitive vertebrate, *Mol. Immunol.* 22:1091 (1985).

3. G. W. Litman, M. J. Shamblott, R. Haire, C. Amemiya, H. Nishikata, K. Hinds, F. Harding, R. Litman, and J. Varner, Evolution of immunoglobulin gene complexity, *in*: "Progress in Immunology VII," F. Melchers, ed., Springer-Verlag, Berlin (1989).

4. C.-A. Reynaud, V. Anquez, H. Grimal, and J.-C. Weill, A hyperconversion mechanism generates the chicken light chain preimmune repertoire, *Cell* 48:379 (1987).

5. C.-A. Reynaud, A. Anquez, and J.-C. Weill, Somatic hyperconversion diversifies the single V_H gene of the chicken with a high incidence in the D region, *Cell* 59:171 (1989).

6. T. K. Blackwell and F. W. Alt, Immunoglobulin genes, *in* "Molecular Immunology," B. A. Hames and D. M. Glover, eds., IRL Press Ltd., Oxford (1988).

7. F. Kokubu, K. Hinds, R. Litman, M. J. Shamblott, and G. W. Litman, Complete structure and organization of immunoglobulin heavy chain constant region genes in a phylogenetically primitive vertebrate, *EMBO J.* 7:1979 (1988).

8. G. W. Litman, L. Berger, K. Murphy, R. T. Litman, K. R. Hinds, C. L. Jahn, and B. W. Erickson, Complete nucleotide sequence of an immunoglobulin V_H gene homologue from *Caiman*, a phylogenetically ancient reptile, *Nature* 303:349 (1983).

9. G. W. Litman, L. Berger, K. Murphy, R. Litman, K. R. Hinds, and B. W. Erickson, Immunoglobulin V_H gene structure and diversity in *Heterodontus*, a phylogenetically primitive shark, *Proc. Natl. Acad. Sci. U. S. A.* 82:2082 (1985).

10. K. R. Hinds and G. W. Litman, Major reorganization of immunoglobulin V_H segmental elements during vertebrate evolution, *Nature* 320:546 (1986).

11. F. Kokubu, R. Litman, M. J. Shamblott, K. Hinds, and G. W. Litman, Diverse organization of immunoglobulin V gene loci in a primitive vertebrate, *EMBO J.*, 7:3413 (1988).

12. C. T. Amemiya and G. W. Litman, The complete nucleotide sequence of an immunoglobulin heavy chain gene and analysis of immunoglobulin gene organization in a primitive teleost species, *Proc. Natl. Acad. Sci. U. S. A.* 87:811 (1990).

13. F. A. Harding, N. Cohen, and G. W. Litman, Immunoglobulin heavy chain gene organization and complexity in the skate, *Raja erinacea, Nuc. Acids Res.* 18:1015 (1990).

14. M. J. Shamblott and G. W. Litman, Genomic organization and sequences of immunoglobulin light chain genes in a primitive vertebrate suggest coevolution of immunoglobulin gene organization, *EMBO J.* 8:3733 (1989).

15. R. N. Haire, C. T. Amemiya, D. Suzuki, and G. W. Litman, Eleven distinct V_H gene families and additional patterns of sequence variation suggest a high degree of immunoglobulin V_H gene complexity in a lower vertebrate: *Xenopus laevis, J. Exp. Med.* 171:1721 (1990).

16. J. Schwager, C. A. Mikoryak, and L. A. Steiner, Amino acid sequence of heavy chain from *Xenopus laevis* IgM deduced from cDNA sequence: Implications for evolution of immunoglobulin domains, *Proc. Natl. Acad. Sci. U. S. A.* 85:2245 (1988).

17. R. N. Haire, M. J. Shamblott, C. T. Amemiya, and G. W. Litman, A second *Xenopus* immunoglobulin heavy chain constant region isotype gene, *Nucleic Acids Res.*, 17:1776 (1989).

18. C. T. Amemiya, R. N. Haire, and G. W. Litman, Nucleotide sequence of a cDNA encoding a third distinct *Xenopus* immunoglobulin heavy chain isotype, *Nuc. Acids Res.* 17:5388 (1989).

19. E. Hsu, J. Schwager, and F. W. Alt, Evolution of immunoglobulin genes: Families in the amphibian *Xenopus, Proc. Natl. Acad. Sci. USA* 86:8010 (1989).

20. V. L. Scofield, J. M. Schlumpberger, L. A. West, and I. L. Weissman, Protochordate allorecognition is controlled by a MHC-like gene system, *Nature* 294:499 (1982).
21. W. H. Hildemann, I. S. Johnson, and P. L. Jokiel, Immunocompetence in the lowest metazoan phylum: Transplantation immunity in sponges, *Science* 204:420 (1979).
22. C.-L. Chen, L. L. Ager, G. L. Gartland, and M. D. Cooper, Identification of a T3/T cell receptor complex in chickens, *J. Exp. Med.* 164:375 (1986).

STUDIES ON THE ONTOGENY OF THE IMMUNE FUNCTION IN BIRDS

Nicole Le Douarin

Institut d'Embryologie Cellulaire et Moléculaire du CNRS
et du Collège de France
Nogent-sur-Marne Cédex, France

Although formal genetics of birds is poorly developed if compared to that of certain mammalian species, the avian model presents certain advantages for immunological studies. The fact that birds possess a specialized organ for B cell differentiation, the bursa of Fabricius, was at the origin of the distinction of the two lymphocytes lineages which differ by their differentiation site and their role in the immune function.[1-5] Moreover, the hematogenic hypothesis of the origin of the blood-forming organs was proposed as a result of investigations carried out in birds.[6,7] By taking advantage of the possibility of operating on avian embryos during the critical period of organogenesis, it was thereafter demonstrated that all the hematopoietic organs have to be seeded by stem cells of extrinsic origin during embryogenesis to differentiate and become functional.[8-10] Thus, the decisive steps of thymic ontogenesis have been precisely defined in two closely related species of birds, the chick (*Gallus gallus*) and the Japanese quail (*Coturnix coturnix japonica*), both of which belong to the same family (the *Phasianidae*). The interest of working on these particular species resides in the cell marking system based on the unique structure of the interphase nucleus of all embryonic and adult cell types of the quail.[11,12] Furthermore, development of the avian embryo allows foreign tissues to be grafted in the embryo *in ovo*, either between quail and chick or between chickens differing by the major histocompatibility complex (MHC), prior to the onset of immune system differentiation. One can, by this means, construct viable birds that are able to hatch and study the immunological status of the graft after birth when the host's immune function has reached maturity.

I will review in this article some of the data that resulted from these experiments and also the work done in the mouse with a view to extending these observations to mammals.

THE QUAIL-CHICK MARKER SYSTEM APPLIED TO THE STUDY OF THYMUS ONTOGENY

The quail-chick marker system, although primarily based on the conspicuous differences existing between the nuclear structure in these two species of birds, now also relies on species-specific monoclonal antibodies (Mabs) identifying surface molecules that characterize particular cell types of the endothelial, hemopoietic and immune systems (Table 1). The α-MB1 and QH1 Mabs recognize surface glycoproteins carried by endothelial and blood cells (except erythrocytes) of the quail.[13,14] Public antigenic determinants of class II antigens of quail and chick can be specifically recognized by the TAC1 and TAP1 Mabs respectively.[15,16] Moreover, Mabs specific for the chicken TCRs and

Mechanisms of Lymphocyte Activation and Immune Regulation III
Edited by S. Gupta *et al.*, Plenum Press, New York, 1991

19

Table 1 SURFACE MARKERS FOR QUAIL AND CHICK CELLS

Species specificity / Antibody		Chick	Quail	
Hemangioblastic lineage	**MB1** (Peault et al , 1983, PNAS)		All blood cells (except erythrocytes) All endothelial cells	Hemangioblastic lineage
	QH1 (Pardanaud et al , 1987, Development)			
MHC	**TAP1**	MHC class II (BLc) α,β heterodimers		
	TAC1		MHC class II (BLq) $\alpha\,\beta$ heterodimers	
	TAC2 (Le Douarin et al , 1983, Prog Immunol) (Guillemot et al , 1984 J Exp Med)	MHC class II (BLc) " "	MHC class II (BLq) " "	
T cell markers	**TCR1** (Chen et al , 1988, Eur J Immunol)	$\gamma\delta$ TCR		
	TCR2 (Sowder et al , 1988, J Exp Med)	$\alpha\beta$ TCR		
	CT3 (Chen et al , 1988, J Exp Med ,)	CD3 complex		
	CT4 (Chan et al , 1988, J Immunol)	CD4 antigen		
	CT8 (Chan et al , 1988, J Immunol)	CD8 antigen		

associated cell surface molecules (CD3, CD4, CD8) have been developed recently.[17-24]

Thymus development has been investigated by constructing chimeric organs in which either the epithelial or mesenchymal derivatives (or both), or the hemopoietic cells (HC), i.e. the lymphocytes and the macrophages and dendritic cells of the thymic medulla, can, at the will of the investigator, be either of the quail or of the chick species.[8,9;15,16] From such experiments, the precise timing of colonization of the endodermal epithelium thymic anlage arising from the third and fourth pharyngeal pouches could be established.

It turned out that, in both the quail and chick species, HC seeding of the thymus takes place according to a cyclic periodicity in which short phases of colonization, during which the thymus is receptive for HC, alternate with periods of non-receptivity lasting 5 days in the quail and 4 days in the chick. Three waves of HC could be evidenced in both species during embryonic and early postnatal life.[25,26] As shown in Figure 1, the thymic epithelial rudiment removed from the quail embryo before the end of the fifth day of incubation is devoid of cells of the hemopoietic lineage and, if grafted into a chick, is the site of differentiation of lymphocytes and medullary dendritic cells belonging to the chick species.

THE EARLY STEPS OF THYMIC ONTOGENY IN THE MOUSE EMBRYO

In the mouse, thymus development proceeds from the endoderm of the third branchial pouch, which starts to grow and be colonized by HC from E11 onward. The fact that the presumptive thymic rudiment is devoid of HC and unable to develop into a lymphoid thymus at E10 if it is cultured in the absence of a source of HC has been fully documented in the past (see [27] and the references therein).

Chimeric thymuses can be constructed between strains of mice differing either in their Thy1 alleles or in their MHC haplotype by grafting the third branchial arch region from E10 embryos into the coelomic cavity of E17 normal fetuses *in utero*,[28] under the kidney capsule of 2 month mice[29] or under the skin in the dorsal aspect of the neck of newborn nude mice (Khazaal et al., in preparation).[27,30]

The exclusively host origin of both the lymphoid and macrophage/dendritic cell populations that developed in the grafted thymuses could be demonstrated in all cases.

POST-NATAL IMMUNOLOGICAL STATUS OF TISSUE GRAFTS INTRODUCED INTO THE EMBRYO

Heterospecific Grafting of Quail Organ Rudiments into Chick Embryos

The anterior limb bud and the epithelio-mesenchymal rudiments of the bursa of Fabricius and of the thymus were implanted from quail into chick embryos in an isotopic and isochronic manner at E4 to E4,5, i.e. before the onset of HC colonization of both primary lymphoid organs (see Figure 1).

The bursal, thymic and limb rudiments developed normally during embryogenesis, the two former organs being colonized by HC and becoming lymphoid according to the developmental timing characteristic of the quail. However, after birth the bursa of Fabricius and the limb underwent acute immune rejection, the latter during the first or second week of life and the former within the 3 first postnatal weeks. The thymus was the only organ not to be rejected, an observation already reported in allogeneic transplants of T cell-depleted fetal thymuses in mice.[31,32]

The Thymic Epithelio-Mesenchymal Rudiment Tolerizes for Embryonic Tissue Grafts Across Species Barrier

This result prompted us to try and induce tolerance of the quail wing and bursa of Fabricius by implanting simultaneously the thymic epithelio-mesenchymal rudiments taken from the same quail donor. The graft was done *in situ* after partial or complete removal of the thymic primordium of the chick host.

At the time of operation, neither the quail nor the chick thymic epithelium had been seeded by HC. Therefore, the lymphocytes and macrophages/dendritic cells that the quail thymus subsequently contained were of chick host origin.

Of 291 embryos receiving the double wing-thymus transplants at E4.-/-4.5., 16 hatched and survived in a healthy state. In all these chimeras the wing remained in good condition for a longer period than in chickens in which no thymus had been implanted. In two birds (ADC19, ADC68), signs of rejection appeared at P28 and P39 respectively. In the others, the wing remained healthy for more than 2-3 months and even more than a year (cf ADC60, which was allowed to survive for all this time). The other birds were sacrificed and their thymuses examined for chimerism by using either the anti-class II (BL antigens in birds) antibodies (TAC1 and TAP1) that allow identification of the species origin of the two classes of thymic cell types expressing the BL antigens, i.e., the epithelial stroma and the HC derived medullary dendritic cells (Figure 2). In all cases where tolerance of the wing was induced, at least one third of the thymic lobes of the double chimeras were derived from the quail graft.[33,34]

Similar results were obtained in the case of grafts of the quail bursa of Fabricius. No immune attack of the quail bursal stroma occurred when the thymic rudiments from the same donor were partly or totally substituted for their chick counterparts.[35]

If one excludes the unlikely possibility that T cells that differentiate in chick thymic lobes can recirculate in quail lobes, it must be assumed that tissue graft tolerance can exist even if only a subset of T cells have been subjected to quail antigens during the intrathymic phase of differentiation and then tolerized. These tolerant quail T cells are apparently able to prevent the chicken thymus-derived lymphocytes from being activated by the peripheral quail antigens and to reject the limb or the bursa.

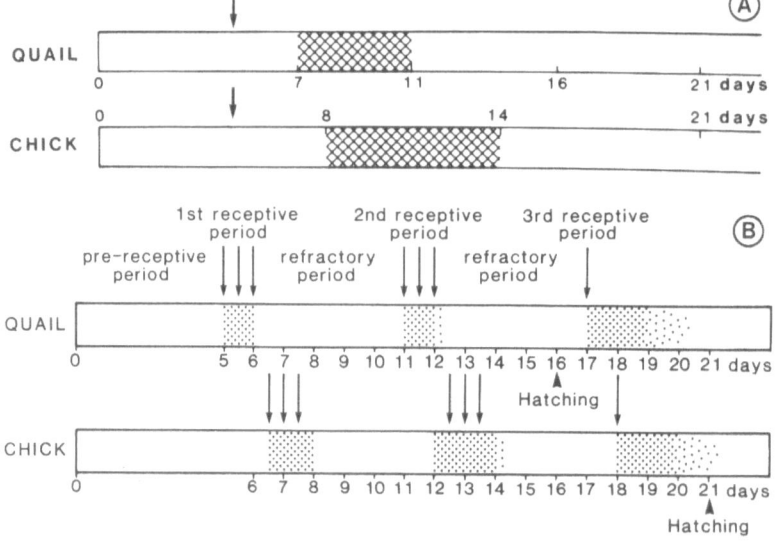

Figure 1. Colonization of bursal (A) and thymic (B) epithelio-mesenchymal rudiments by HC in quail and chick embryos as deduced from hetero-specific transplantations of the bursa and thymus organ rudiments between quail and chick embryo. Note that the bursa is the site of a single influx of HC while the thymus received HC by waves occurring according to a cyclic periodicity.

Moreover, it is clear that even when grafted into a chick, the quail thymus undergoes physiological regression 5 months earlier than its chick counterpart. Therefore, in thymic quail-to-chick chimeras, the quail thymic lobes are already regressed, whereas the chick lobes are still in a functional state. During this postnatal period, however, tolerance of the quail organ is still maintained.

Allogeneic Grafts of Embryonic Limb Bud Primordia in the Chicken Species

Tissue grafts from a histoincompatible donor of the same developmental stage were introduced into an early chick embryo host, in order to probe the immune response to the graft after birth, when the host has reached immune maturity.

Limb buds from B4 or B12 chicken strains were grafted *in situ* on (B15/B21) F1 recipients that were allowed to hatch. The grafted wing grew normally and was tolerated almost perfectly during the host's lifetime, although reversible rejection crises affected more or less severely the fundamentally healthy state of the grafted tissues. Skin grafts of the same MHC haplotype as the wing were performed on the adult wing-chimera and were permanently tolerated. In contrast, host peripheral blood lymphocytes maintained their capacity to proliferate against donor cells in MLR.

These results, while showing that *in vitro* and *in vivo* tolerance are separable phenomena, suggest the existence of a peripheral mechanism inducing tolerance to self that complements the elimination of self-reactive clones by the thymus.

The Role of Thymic Epithelium in Tissue Tolerance Induction in the Mouse

The mouse embryo is not accessible for microsurgery during the days immediately following implantation in the uterus when the decisive events of organogenesis take place. Therefore, the thymic epithelial rudiments cannot be removed or engrafted at E10 before it has been seeded by HC. In order to design an experimental paradigm as close as possible to the one we applied to the avian embryo we turned to the athymic nude mouse

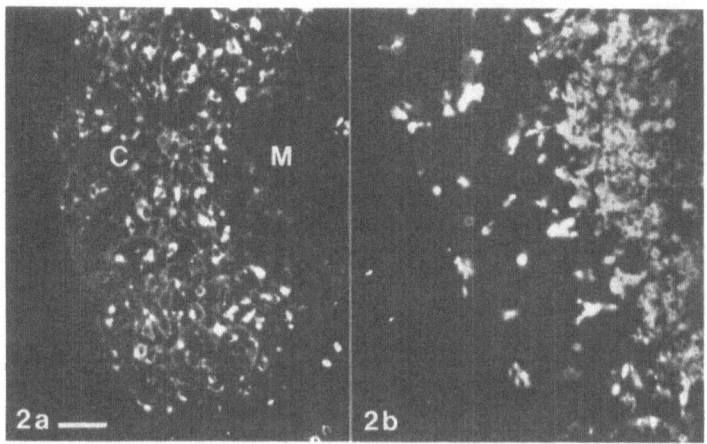

Figure 2. Immunocytochemical analysis of chimerism in the grafted thymus from a double chimera (wing-thymus) at 48 days. Two consecutive sections were stained with, in a) TAC1 Mab stains the epithelial cells of the cortex that are derived from the quail transplant, in b) The TAP1 Mab identifies chick macrophages and dendritic cells densely packed in the medulla and for some of them dispersed in the cortex. Bar = 50 μm. C: cortex; M: medulla.

in which we could transplant foreign thymic epithelial grafts as we had done in the pre-thymic chick embryo.

BALB/c (H-2^d) nude mice were grafted between the day of birth and P13 with 10-15 third branchial arch regions removed from E10 mice of either C3H (H-2^k) or BALB/c euthymic embryos, according to the experimental series (Figure 3).

Restoration of T-cell function was observed in iso- and allogeneically grafted mice. Of 50 newborn mice grafted with BALB/c E10 thymic rudiments, 12 were reconstituted successfully. This was readily apparent at about 3 months of age by the fact that the mice did not lose weight and remained healthy, in contrast to their unreconstituted littermates.

At 4-5 months of age, these animals received a syngeneic and an allogeneic skin graft. Rejection of the allogeneic skin occurred between 15 to 28 days after implantation (mean value: 18 ± 1.4 days), but the syngeneic skin was accepted in all cases. In normal euthymic BALB/c mice of our colony, C3H or C57BL/6 skin grafts were rejected at times ranging from 12 to 15 days.

The 12 mice were sacrificed at 7-10 months of age for T-cell function tests. The neck region was dissected and transplants were removed and treated for histological examination. Microscopic observation showed well-differentiated thymuses and, in some cases, various tissues that had developed from branchial arch mesenchyme and endoderm (cartilage, parathyroid). Spleen and lymph nodes were aseptically removed for in vitro tests. Lymphocytes from the engrafted nude mice were able to react against allogeneic stimulator cells, whereas non-transplanted nude mice were unresponsive.

In the mixed lymphocyte reaction (MLR), the stimulation index varied from 1.53 to 10.23 for spleen or lymph node cells, with a mean value for the 12 mice tested of 3.55 ± 0.63 (for spleen cells). Con A stimulation was significantly higher than in nu/nu mice that had not been grafted.

Cytotoxicity against C3H and C57BL/6 splenocytes was significant in all cases. Particularly high responses were obtained in three of the six mice tested. Non-grafted nude mice failed to generate specific alloreactive cytotoxic T cells. The percentage of Thy-1-2

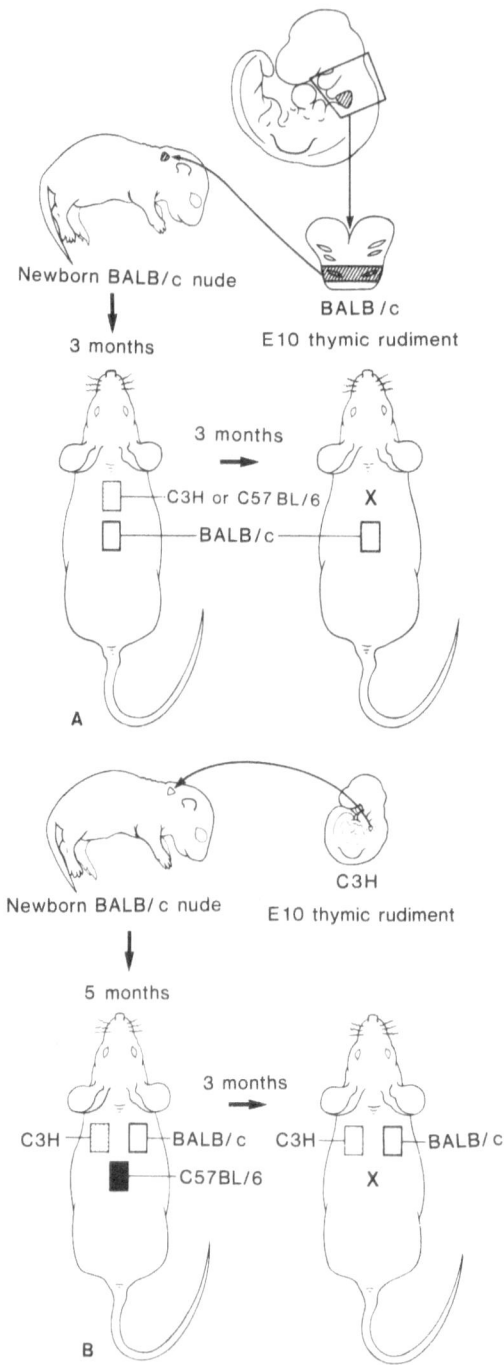

Figure 3. Summary of the graft experiments. The 3rd branchial pouch
removed from a 10-day euthymic BALB/c embryo is grafted under the
skin of a newborn BALB/c nude. A: At 3 months, this nude recipient received two skin grafts: one from an euthymic adult BALB/c
and another from an euthymic allogeneic adult C3H or C57BL/c. B:
At 3 months, this nude recipient received three skin grafts: one from
an euthymic adult BALB/c and the other two from euthymic allogeneic
adults of the C3H and C57BL/6 strains.

splenocytes varied from 27 to 33% in the 12 reconstituted mice, a number close to that found in euthymic BALB/c mice (35-38%) and, in any case, much higher than in non-grafted nude BALB/c mice (3-10%).[30]

The fully allogeneic thymic epithelium has the same potential to reconstitute the T-cell compartment of athymic nude mice and to induce tolerance for skin grafts of the thymic MHC haplotype. Eighty-four BALB/c (H-2^d) nude mice were grafted with thymic rudiments from C3H (H-2^k) E10 embryos. Sixty-seven of these mice died before 5 months, which corresponds to the lifespan of non-grafted nu/nu control mice, but the remaining 17 animals were in good health. Tail skin grafts of recipient (BALB/c), donor (C3H), and third-party (C57BL/6) types were performed on 15 of these mice. Rejection of the C57BL/6 skin occurred after 9-30 days, in all of them. Syngeneic BALB/c skin grafts and skin grafts of thymic epithelium donor haplotype were tolerated in all cases.

Thirteen of these mice were sacrificed 3-5 months after grafting the skin. They were from 9 to 12 months old and in good health. Well-developed thymuses were recovered, as in the case of syngeneic grafts.

We compared the ability of splenic T lymphocytes of nude mice reconstituted with syn geneic or allogeneic thymus epithelium with normal euthymic BALB/c mice to develop *in vitro* responses against BALB/c, C3H, and third-party MHC mismatched targets. The majority (9/13) of tolerant nude mice showed a readily detectable response, indicating that donor H-2-reactive T cells are not deleted in these animals, and are functionally competent to divide (Table 2). In the same way, they showed a killing activity against C3H cells comparable to that they displayed against C57BL/6 cells. Four others were fully tolerant.[27]

CONCLUSIONS AND DISCUSSION

The work reported in this review was initiated to answer a series of questions on the ontogeny of the immune system. In our first studies on the avian embryo, the problem was to see whether the cell marker provided by quail-chick chimeras could help in finding the origin of the lymphocytes that differentiate in the primary lymphoid organs and subsequently in deciphering the relationships existing between the HC and the bursal and thymic epithelial stroma. This led to the definitive demonstration that all the lymphocytes of the bursa and thymus as well as the dendritic cells of the thymic medulla were of hemopoietic origin. In the two species studied (i.e. the quail and the chick), the thymus was shown to be colonized by waves of HC of relatively short duration (24-36 hours according to the species considered) separated by longer periods during which the thymus is not receptive for HC. The bursa of Fabricius, in contrast, turned out to be the site of a single colonization by HC that lasts from 4 to 6 days.[10] I proposed that the mechanisms underlying HC seeding of the primary lymphoid organs involve a chemotactic phenomenon.[36] In this view, the chemoattractant being produced by the thymic (or bursic) epithelial stromal cells would be responsible for the invasion of the epithelial rudiments of thymus and bursa by circulating HC. The involvement of such a mechanism in thymic development was subsequently confirmed by a body of experimental data.[37-42]

Later, embryonic grafting experiments of selected regions of the neural primordium between quail and chick embryos raised the question of the immunological status after birth of embryonic grafts from either xeno- or allogeneic donors. We found that chickens in which parts of the neural epithelium had been replaced by their quail counterpart were able to hatch and had a sensorimotor behavior compatible with survival in a healthy state for a time ranging from several weeks to a few months.[43-45] However, after a while, peripheral and central nervous tissues arising from the quail graft were subjected to the host's immune rejection. The neural chimeras developed lesions similar to those of experimental allergic encephalomyelitis (EAE) and neuritis (EAN) that, in certain animals, extended to the recipient's nervous tissues. The privileged immunological status of neural tissues due to the absence or very weak expression of MHC antigens by neural tissue, along with the presence of a blood-brain barrier, were proposed to be responsible for long-term tolerance of the neural derivatives in this system. This is why grafts of other somatic tissues were carried out, either between quail and chick or between MHC-

Table 2

Typical responses		Responder cells	Stimulator cells BALB/c	C3H	C57BL/6
	Non-Tolerant 9/13	BALB/c Nu + Tep 1	8,733	35,269*	35,219*
		BALB/c	4,521	48,487	32,356
		BALB/c Nu + Tep 2	3,116	17,270*	16,236*
		BALB/c	8,587	30,267	26,882
	Tolerant 4/13	BALB/c Nu + Tep 3	4,875	6,889°	11,236°
		BALB/c	5,478	17,191	16,051

Incorporation of 3HTdR (mean cpm for 3 replicates)
* Significant at 99% (F-test)
° Not significant (F-test)

mismatched chickens to test the effect of embryonic grafts on tolerance induction. The main point was that the grafts were devoid of hemopoietic cells and therefore these experiments were radically different in principle from those of Billingham et al.[46]

The main results that came out of the experiments that have been described above concern the role of the thymic epithelium in tolerance induction. Although it is unanimously recognized that the intrathymic phase of differentiation of T lymphocytes involves both positive and negative selection of self-MHC-reactive T cells, the role of the different thymic stromal components on this process is not fully understood. Clonal deletion of autoreactive clones whose T cell receptors have a high affinity for self MHC + self peptide complexes is postulated to be the fundamental mechanism ensuring tolerance. The task of presenting self antigens for negative selection has been assigned to those MHC-expressing cells of the thymus that are of hemopoietic origin, i.e., the medullary cells of the macrophage/dendritic lineage. On the other hand, positive selection responsible for restriction in the ability of T cells to kill the individual's own virus-infected cells is supposed to be carried out by epithelial cells.

It is interesting to underline that in the avian experimental model, host and donor thymuses are at the same developmental stage and therefore differentiate simultaneously. This developmental synchrony could be an important parameter in tissue tolerance induction.

In any case, donor thymus does not have to substitute completely for that of the host in order to induce tissue tolerance, since only a third of the mass of quail thymic lobes is sufficient for inducing long term organ tolerance.

In the mouse, attempts to induce skin graft tolerance in adult life by implanting an allogeneic E10 epithelial thymic rudiment into the E17 fetus *in utero* have been done in our laboratory. Although in most cases the thymus becomes colonized by host HC and undergoes an apparently normal lymphoid differentiation,[28] no tolerance to skin grafts of the thymus H2 haplotype has ever been induced (unpublished results).

In this system, the host's thymus is ready to release a first wave of T cells to the periphery around the time of birth, whereas the grafted thymus is much less mature. Since operating on the embryo *in utero* is not feasible at E10, we thought of using nude mice as recipients in order to devise a model as close as possible to the one used in birds. Engrafting of the E10 epitheliomesenchymal thymic anlage from either C3H or C57BL/6 into BALB/c nude mice resulted, in all cases, in tolerance of skin grafts from either $H2^k$ or $H2^b$ and rejection of a third party skin. Consequently, in contrast to the assumption that tolerance to self is dependent only upon hemopoietically derived cells of the thymus, our experiments show that both the medullary dendritic and the epithelial cells of the thymus must play a decisive role in self/non-self discrimination resulting in tolerance to tissue grafts.

The fact that T cells, able to proliferate if confronted with the H2 haplotype of the donor thymic epithelium in MLR, are nearly always present in the host suggests that clonal deletion, if it occurs in contact with the thymic epithelium, is far from reaching the extent observed normally when both epithelial and HC cells of the thymus are of the same MHC haplotype. This indicates that tissue tolerance does not exclusively depend upon deletion of autoreactive clones. This was even more obvious in the avian system, where tolerance of skin grafts is induced by limb bud implantation at E4 in allogeneic combinations.[47]

When limb bud grafts are carried out early in embryonic life between MHC-mismatched chickens, not only is the wing tolerated after birth but the continuous presence of the foreign tissue during development also prevents rejection of adult skin from the same donors.

It is noteworthy that no hemopoietic cells are included in the grafted limb, which at the time of transplantation merely contains a few ingrowing endothelial capillaries that regress and are rapidly replaced by vascular buds from the host.[48,49] Therefore one cannot consider this state of tolerance to be driven by hemopoietic cells, as is the case in the classical experiments of Billingham et al.[46]

Regrafting experiments using accepted skin grafts of wing donor haplotype placed into naive, host-type recipients confirm that the adult skin grafts can be tolerated for up to 8 months without losing their antigenicity in any way.

It is interesting to note that a peripherally-induced type of tolerance has already been found in amphibians by Flajnik et al.[50] who grafted allogeneic eye anlagen into embryonic frogs and observed long-term survival of an isogenic adult skin transplant in adults.

These experiments raise the question as to whether the foreign antigens of the grafted wing may be transported to the thymus and there be presented for tolerance induction; the uptake of injected antigens by thymic dendritic cells has been demonstrated.[51] Moreover, Mc Donald et al.[52] showed that high doses of Mls-disparate spleen cells administered neonatally resulted in the deletion of $V\beta6$ receptor- expressing cells in the thymus. Since Mls recognition is thought to involve $V\beta6$ TCR[53] these data suggest that splenic antigen presenting cells indeed enter the thymus and induce tolerance by deletion.

However, in the experiments described here the main mechanism responsible for tolerance is probably not clonal deletion, since MLR experiments have revealed that lymphocytes able to proliferate in coculture with cells expressing graft haplotype antigens are still present in the host, with a proliferative response comparable in chimeras and in control birds. Similar observations of split tolerance have already been described in other systems based on neonatally induced tolerance by hematopoietic cell injection in rats and mice.[54,55] All these results point to the existence of peripheral mechanisms for tolerance induction.

Following the intrathymic phase of T lymphocyte differentiation, the population of post-thymic T-cells is not completely devoid of potentially autoreactive clones. Their activation remains therefore a vital challenge for the immunological regulatory mechanisms. A "fail-safe" mechanism based on the production of suppressor cells has been proposed to account for such phenomena.[56] Coutinho and Bandeira[57] propose that, besides the passive mechanism based upon clonal deletion, tolerance induction involves active immunological regulatory processes. In the classical system in which transplantation tolerance is induced by neonatal injection of semi-allogeneic hemopoietic cells, they observe that tolerance always correlates with high levels of T and B lymphocyte activity and then stress the fact that tolerance is "far from being a suppressed state".[58] The establishment of a regulatory functional immune network including host and donor specificities would, according to this model, be the basis for the equilibrium that characterizes the tolerant state. In any event, tolerance to self, far from being the result of a unique mechanism occurring during the intrathymic phase of T cell differentiation, appears rather as a network of regulatory processes involved in the immune function.

REFERENCES

1. B. Glick, T. S. Chang, and R. G. Japp, The bursa of Fabricius and antibody production, *Poultry Sci.* 35:224 (1956).
2. N. L. Warner, A. Szenberg, and F. M. Burnet, The immunological role of different lymphoid organs in the chicken I. Dissociation of immunological responsiveness, *Austr. J. Exper. Biol. Med. Sci.* 40:373 (1962).
3. M. D. Cooper, R. D. A. Peterson, and R. A. Good, Delineation of the thymic and bursal lymphoid systems in the chicken, *Nature* 205:143 (1965).
4. M. D. Cooper, R. D. A. Peterson, M. A. South, and R. A. Good, The functions of the thymus system and the bursa system in the chicken, *J. Exp. Med.* 123:75 (1966).
5. M. D. Cooper, M. M. Schwartz, and R. A. Good, Restoration of gammaglobulin production in agammaglobulinemic chickens, *Science* 151:71 (1966).
6. M. A. S. Moore and J. J. T. Owen, Chromosome marker studies on the development of the haemopoietic system in the chick embryo, *Nature* 208:958 (1965).
7. D. Metcalf and M. A. S. Moore, "Haemopoietic Cells," North Holland Publishing Company, Amsterdam (1971).
8. N. M. Le Douarin and F. V. Jotereau, Origin and renewal of lymphocytes in avian embryo thymuses studied in interspecific combinations, *Nature New Biol.* 246:25 (1973).
9. N. M. Le Douarin and F. V. Jotereau, Tracing of cells of the avian thymus through embryonic life in interspecific chimaeras, *J. Exp. Med.* 142:17 (1975).
10. N. M. Le Douarin, F. Dieterlen-Lièvre, and P. D. Oliver, Ontogeny of primary lymphoid organs and lymphoid stem cells, *Am. J. Anat.* 170:261 (1984).
11. N. M. Le Douarin, Particularités du noyau interphasique chez la Caille japonaise (*Coturnix coturnix japonica*). Utilisation de ses particularités comme "marquage biologique" dans des recherches sur les interactions tissulaires et les migrations cellulaires au cours de l'ontogenèse, *Bull. Biol. Fr. Belg.* 103:435 (1969).
12. N. M. Le Douarin, A biological cell labeling technique and its use in experimental embryology, *Dev. Biol.* 30:217 (1973).
13. B. M. Péault, J. P. Thiery, and N. M. Le Douarin, Surface marker for the hemopoietic and endothelial cell lineages in the quail that is defined by a monoclonal antibody, *Proc. Natl. Acad. Sci. USA* 80:2976 (1983).
14. L. Pardanaud, C. Altmann, P. Kitos, F. Dieterlen-Lièvre, and C. A. Buck, Vasculo-genesis in the early quail blastodisc as studied with a monoclonal antibody recognizing endothelial cells, *Development* 100:339 (1987).
15. N. M. Le Douarin, F. Guillemot, P. Oliver, and B. Péault, Distribution and origin of Ia-positive cells in the avian thymus analysed by means of monoclonal antibodies in heterospecific chimeras, *in*: "Progress in Immunology V," Y. Yamamura and T. Tada, eds., Acad. Press, New York (1983).
16. F. P. Guillemot, P. D. Oliver, B. M. Péault, and N. M. Le Douarin, Cells expressing Ia-antigens in the avian thymus, *J. Exp. Med.* 160:1803 (1984).
17. C. L. H. Chen, L. Lanier Ager, L. Gartland, and M. D. Cooper, Identification of a T3/T cell receptor complex in chickens, *J. Exp. Med.* 16:375 (1986).
18. C. L. Chen, J. Cihak, U. Losch, and M. D. Cooper, Differential expression of two T cell receptors, TcR1 and TcR2, on chicken lymphocytes, *Eur. J. Immunol.* 18:539 (1988).
19. C. H. Chen, J. T. Sowder, J. M. Lahti, J. Cihak, U. Lösch, and M. D. Cooper, TcR3: a third T cell receptor in the chicken, *Proc. Nat. Acad. Sci.* 86:2351 (1989).
20. R. P. Bucy, C. L. Chen, J. Cihak, U. Lösch, and M. D. Cooper, Avian T cells expressing gamma delta receptors localize in splenic sinusoids and in intestinal epithelium, *J. Immunol.* 14:385 (1988).
21. M. M. Chan, C. L. H. Chen, L. Lanier Ager, and M. D. Cooper, Identification of the avian homologues of mammalian CD4 and CD8 antigens, *J. Immunol.* 140:2133 (1988).

22. J. Cihak, H. W. L. Ziegler-Heitbrock, H. Trainer, I. Schranner, M. Merkenschlager, and U. Lösch, Characterization and functional properties of a novel monoclonal antibody which identifies a T cell receptor in chicken, *Eur. J. Immunol.* 18:533 (1988).

23. J. T. Sowder, C.-L. H. Chen, L. L. Ager, M. M. Chan, and M. D. Cooper, A large subpopulation of avian T cells express a homologue of the mammalian T gamma delta receptor, *J. Exp. Med.* 167:315 (1988).

24. D. Char, C. H. Chen, P. Bucy, and M. D. Cooper, Identification of a third T cell receptor (TCR3) in the chicken with a monoclonal antibody, *Fed. Proc.* (Abstr.) 3:485 (1989).

25. F. V. Jotereau and N. M. Le Douarin, Demonstration of a cyclic renewal of the lymphocyte precursor cells in the quail thymus during embryonic and perinatal life, *J. Immunol.* 129:1869 (1982).

26. M. Coltey, F. V. Jotereau, and N. M. Le Douarin, Evidence for a cyclic renewal of lymphocyte precursor cells in the embryonic chick thymus, *Cell Diff.* 22:71 (1987).

27. J. Salaün, A. Bandeira, I. Khazaal, F. Calman, M. Coltey, A. Coutinho, and N. M. Le Douarin, Thymic epithelium tolerizes for histocompatibility antigens, *Science* 247:1471 (1990).

28. J. Salaün, F. Calman, M. Coltey, and N. M. Le Douarin, Construction of chimeric thymuses in the mouse fetus by *in utero* surgery, *Eur. J. Immunol.* 16:523 (1986).

29. F. Jotereau, F. Heuze, V. Salomon-Vie, and H. Gascan, Cell kinetics in the foetal mouse thymus: precursor cell input, proliferation and emigration, *J. Immunol.* 138:1026 (1987).

30. I. Khazaal, J. Salaün, M. Coltey, F. Calman, and N. M. Le Douarin, Restoration of T-cell function in nude mice by grafting the epitheliomesenchymal thymic rudiment, *Cell Diff.* 26:211 (1989).

31. A. R. Ready, E. J. Jenkinson, R. Kingston, and J. J. T. Owen, Successful transplantation across major histocompatibility barrier of deoxyguanosine-treated embryonic thymus expressing class II antigens, *Nature* 310:231 (1984).

32. H. von Boehmer and K. Schubiger, Thymocytes appear to ignore class I major histocompatibility complex antigens expressed on thymus epithelial cells, *Eur. J. Immunol.* 14:1048 (1988).

33. H. Ohki, C. Martin, C. Corbel, M. Coltey, and N. M. Le Douarin, Tolerance induced by thymic epithelial grafts in birds, *Science* 237:1032 (1987).

34. H. Ohki, C. Martin, M. Coltey, and N. M. Le Douarin, Implants of quail thymic epithelium generate permanent tolerance in embryonically constructed quail/chick chimeras, *Development* 104:619 (1988).

35. M. Belo, C. Corbel, C. Martin, and N. M. Le Douarin, Thymic epithelium tolerizes chickens to embryonic grafts of quail bursa of Fabricius, *Internation. Immunol.* 1:105 (1989).

36. N. M. Le Douarin, Ontogeny of hematopoietic organs studied in avian embryo interspecific chimeras, *in*: "Differentiation of Normal and Neoplastic Hematopoietic Cells," B. Clarkson, P. A. Marks, and J. E. Till, eds., Cold Spring Harbor Laboratory, U. S. A. (1978).

37. F. V. Jotereau, E. Houssaint, and N. M. Le Douarin, Lymphoid stem cell homing to the early thymic primordium of the avian embryo, *Europ. J. Immunol.* 10:620 (1980).

38. S. Ben Slimane, F. Houllier, G. G. Tucker, and J. P. Thiery, *In vitro* migration of avian hemopoietic cells to the thymus: preliminary characterization of a chemotactic mechanism, *Cell Diff* 13:1 (1983).

39. S. Champion, B. Imhof, P. Savagner, and J. P. Thiery, The embryonic thymus produces chemotactic peptides involved in the homing of hemopoietic precursors, *Cell* 44: 781 (1986).

40. P. Savagner, B. A. Imhof, K. M. Yamada, and J. P. Thiery, Homing of hemopoietic precursor cells to the embryonic thymus: characterization of an invasive mechanism induced by chemotactic peptides, *J. Cell Biol.* 103:2715 (1986).

41. B. A. Imhof, M. A. Deugnier, J. M. Girault, S. Champion, C. Damais, T. Itoh, and J. P. Thiery, Thymotoxin: A thymic epithelial peptide chemotactic for T-cell precursors, *Proc. Natl. Acad. Sci. USA* 85:7699 (1988).

42. C. Dargemont, D. Dunan, M. A. Deugnier, M. Denoyelle, J. M. Girault, F. Lederer, Kim Ho Diep Le, F. Godeau, J. P. Thiery, and B. A. Imhof, Thymotoxin, a chemotactic protein, is identical to B2 microglobulin, *Science* 246:803 (1989).

43. M. Kinutani and N. M. Le Douarin, Avian spinal cord chimeras: I. Hatching ability and post hatching survival in homo- and heterospecific chimaeras, *Dev. Biol.* 111:243 (1985).

44. M. Kinutani, M. Coltey, and N. M. Le Douarin, Postnatal development of a demyelinating disease in avian spinal cord chimeras, *Cell* 45:307 (1986).

45. M. Kinutani, K. Tan, J. Desaki, M. Coltey, K. Kitaoka, Y. Nagano, Y. Takashima, and N. M. Le Douarin, Avian spinal cord chimeras. Further studies on the neurological syndrome affecting the chimeras after birth, *Cell Diff.* 26:145 (1989).

46. R. E. Billingham, L. Brent, and P. B. Medawar, Actively acquired tolerance to foreign cells, *Nature* 172:603 (1953).

47. C. Corbel, C. Martin, H. Ohki, M. Coltey, L. Hlozaneck, and N. M. Le Douarin, Evidence for peripheral mechanisms inducing tissue tolerance during ontogeny, *Int. Immunol.* 2:33 (1990).

48. F. V. Jotereau and N. M. Le Douarin, The development relationship between osteocytes and osteoclasts: a study using the quail-chick nuclear marker in endochondral ossification, *Dev. Biol* 63:253 (1978).

49. L. Pardanaud, F. Yassine, and F. Dieterlen-Lièvre, Relationship between vasculogenesis and hemopoiesis during avian ontogeny, *Development* 105:473 (1989).

50. M. F. Flajnick, L. Du Pasquier, and M. Cohen, Immune responses of thymus/lymphocyte embryonic chimeras: studies on tolerance and major histocompatibility complex restriction in *Xenopus*, *Europ. J. Immunol.* 15:540 (1985).

51. B. A. Kyewski, C. G. Fathman, and H. S. Kaplan, Intrathymic presentation of circulating non-major histocompatibility complex antigens, *Nature* 308:196 (1983).

52. H. R. Mac Donald, T. Pedrazzini, R. Schneider, J. A. Louis, R. M. Zinkernagel, and H. Hengartner, Intrathymic elimination of Mls[a]-reactive (Vb6[+]) cells during neonatal tolerance induction to Mls[a]-encoded antigens, *J. Exp. Med.* 167:2005 (1988).

53. H. R. Mac Donald, R. Schneider, R. K. Lees, R. C. Howe, H. Acha-Orbea, H. Festenstein, R. M. Zinkernagel, and H. Hengartner, T-cell receptor $V\beta$ use predicts reactivity and tolerance to Mls[a]-encoded antigens, *Nature* 332:40 (1988).

54. R. N. Smith and J. C. Howard, Heterogeneity of the tolerant state in rats with long established skin grafts, *J. Immunol.* 125:2289 (1980).

55. K. M. Mohler and J. W. Streilein, Tolerance to class II major histocompatibility complex molecules is maintained in the presence of endogenous, interleukin 2-producing, tolerogen-specific T lymphocytes, *J. Immunol.* 139:2211 (1987).

56. R. Zamoyska, H. Waldman, and P. Matzinger, Peripheral tolerance mechanisms prevent the development of autoreactive T cells in chimeras grafted with two minor incompatible thymuses, *Europ. J. Immunol.* 19:111 (1989).

57. A. Coutinho and A. Bandeira, Tolerize one, tolerize them all: tolerance is self-assertion, *Immunol. Today* 10:264 (1989).

58. A. Bandeira, A. Coutinho, C. Carnaud, F. Jacquemart, and L. Forni, Transplantation tolerance correlates with high levels of T- and B-lymphocyte activity, *Proc. Natl. Acad. Sci.* 86:272 (1989).

KINETICS OF NEGATIVE AND POSITIVE SELECTION IN THE THYMUS

Pawel Kisielow*# and Harald von Boehmer*

*Basel Institute for Immunology
 CH-4058 Basel, Switzerland
#Institute of Immunology and Experimental Therapy
 Polish Academy of Sciences
 53-114 Wroclaw, Poland

ABSTRACT

Recent experiments show that $CD4^+8^+$ thymocytes represent the critical stage in T cell development at which the specificity of randomly generated $\alpha\beta$ T cell receptors is screened. These cells are deleted when the receptor binds to the MHC molecule plus specific peptide presented by bone marrow derived cells but are rescued from cell death and induced to mature if the receptor binds to the MHC molecule on thymic epithelium in the absence of the specific peptide. Different tolerogens delete $CD4^+8^+$ thymocytes earlier or later during their lifespan and negative selection can occur prior to positive selection. The specificity of the $\alpha\beta$ T cell receptor for either class I or class II thymic MHC molecules determines the $CD4^-8^+$ and $CD4^+8^-$ phenotype of mature T cells.

INTRODUCTION

The recognition of antigen by $\alpha\beta$ T cell receptor ($\alpha\beta$ TCR) requires help from other cell surface molecules: CD4 or CD8 coreceptors on T cells and class I or class II MHC glycoproteins on antigen presenting cells. The role of class I and class II MHC molecules is to bind peptides of processed intracellular and extracellular proteins respectively[1] so that they can be recognized as antigens by $\alpha\beta$ TCR, which exhibits dual specificity: for the antigenic peptide as well as for the peptide binding polymorphic domains of MHC molecules.[2] The CD4 and CD8 coreceptors, which on mature T cells are expressed in a mutually exclusive fashion,[3,4] have specific binding affinity for nonpolymorphic domains of class II and class I MHC molecules respectively.[5,6] As a rule, $CD4^+8^-$ T cells are class II MHC restricted and $CD4^-8^+$ T cells are class I MHC restricted.[7,8] A $CD4^+8^-$ T cell is activated when a class II MHC restricted $\alpha\beta$ TCR recognizing foreign peptide, and a CD4 coreceptor, bind to the same class II MHC molecule. Likewise a $CD4^-8^+$ T cell is activated when a class I MHC restricted TCR and CD8 coreceptor bind to the same class I molecule.[9] As shown by transfection experiments, $CD4^+8^-$ T cells expressing a class I MHC restricted TCR are not activated, even when the stimulator cell expresses both class I and class II MHC molecules.[10] This implies that in order to be activated, T cells must express "fitting" combinations of class I or class II MHC restricted receptors with CD8 and CD4 coreceptors respectively. In the absence of specific peptides, the interaction of $\alpha\beta$ TCRs and their coreceptors with self MHC molecules is without functional consequence which demonstrates the importance of the complete ligand, i.e. MHC plus specific peptide, for the induction of the response in mature T cells.

Mechanisms of Lymphocyte Activation and Immune Regulation III
Edited by S. Gupta *et al.,* Plenum Press, New York, 1991

31

For developmental biologists the perplexing questions are: (A) how, when, and where are TCRs selected so that they are self MHC restricted but self MHC ignoring and in addition coexpressed with "fitting" CD4/CD8 coreceptors? (B) what happens to those T cells whose TCRs, in addition to being self MHC restricted, are also specific for self peptides?

It is believed that the answers to these questions may be obtained by studying the role played by molecules involved in antigen recognition, at the early, immature stages of T cell development. During ontogeny, expression of $\alpha\beta$ TCRs and CD4/CD8 coreceptors begins in the thymus where the immigrant CD4$^-$8$^-$8 TCR$^-$ T cell precursors are induced by thymic microenvironment to differentiate into various subpopulations.[11-14] Genes encoding the β chain of TCR rearrange before genes encoding the α chain[15,16] and the complete $\alpha\beta$ TCR molecules are first expressed at relatively low level on thymocytes bearing both CD4 and CD8 coreceptors.[17] Thymocytes expressing high level of $\alpha\beta$ TCR but only one coreceptor (CD4 or CD8) are first detected two days later.[17] Cells of CD4$^+$8$^+$ $\alpha\beta$ TCR^{low} phenotype constitute the major (about 70%) population of thymocytes.[17,18] Daily, in the murine thymus 40-50 millions of these cells vanish and are replaced by the same number of newly generated cells of this phenotype.[19] The reason for such high turnover of these cells, their developmental potential and the fate of cells disappearing from this compartment, have been the great puzzle of T cell developmental biology.

In recent years as a result of methodological advances allowing tracing of the develop mental pathways of T cells with $\alpha\beta$ TCRs of defined specificity, the CD4$^+$8$^+$$\alpha\beta$ TCR^{low} thymocytes have begun to yield some of their secrets. It appears that they represent the critical stage in T cell development, at which randomly generated $\alpha\beta$ TCRs are screened for self reactivity by intrathymic MHC molecules presenting various self peptides.

In this article we will discuss what we have recently learned about the timing, the mechanisms, and the outcome of this screening procedure, the end result of which is selection of CD4$^+$8$^-$ class II MHC restricted and CD4$^-$8$^+$ class I MHC restricted T cell subsets.

EXPERIMENTAL MODELS PROBING THE MECHANISM OF SELECTION

Recently three different approaches have been used to investigate the role of the molecules specifically involved in antigen recognition during T cell development. One approach consisted of prolonged *in vivo* or *in vitro* -in thymic organ cultures -treatment with large doses of monoclonal antibodies directed against TCR,[20,21] D4,[22-24] class I,[25,26] and class II[27,28] MHC molecules. Another approach involved the study of development of class II MHC restricted T cells expressing the products of certain Vβ genes, which without the contribution from the $\alpha\beta$ chains impart specificity towards so called superantigens like Mls[29,30] and staphylococcal enterotoxin B (SEB).[31] Yet another approach involved transgenic mice with tissue directed expression of the transgene encoded class II MHC molecules,[32-34] β TCR chains[35-37] as well as $\alpha\beta$ TCRs.[38-42] The latter expressed class I[38,40] or class II[41,42] MHC restricted receptors for "conventional" antigens, the specificity of which is determined by both the α and the β TCR chains.

We have studied $\alpha\beta$ TCR transgenic mice obtained by coinjecting into fertilized eggs functionally rearranged α and β TCR genes isolated from CD4$^-$8$^+$ T cell clone, specific for male (H-Y) antigen and restricted by the H-2D^b (class I MHC) molecule.[43,44] In transgenic males, T cells expressing the transgenic receptor are exposed to the complete ligand, i.e., D^b molecules presenting H-Y antigen, whereas in transgenic females they encounter D^b molecules but not H-Y antigen. By analyzing the expression of transgenic and endogenous TCR chains on successive developmental stages defined by the CD4/CD8 phenotype, in transgenic males and transgenic females of different H-2 haplotypes, we could get direct insight into the role of the specificity of $\alpha\beta$ TCR on the one hand, and the role of the encountered TCR ligand on the other hand, in determining the fate of immature thymocytes. This system offered therefore a good opportunity to test the validity of both positive and negative selection concepts, put forward some years ago to explain self MHC restriction and self tolerance, and also to determine the developmental stage at which the selection was taking place. The negative selection concept stated that self tolerance is achieved by deletion of self antigen reactive T cells at the immature stage of their development.[45]

According to the positive selection concept, on the other hand, at some immature stage of intrathymic development, only cells expressing $\alpha\beta$ TCRs recognizing self MHC molecules could be selected for further differentiation and maturation.[46,47] To account for the observed correlated expression of CD4 and CD8 coreceptors with class II and class I MHC restricted TCRs, it was suggested that MHC restriction specificity of $\alpha\beta$ TCRs on positively selected cells determines the CD4/CD8 phenotype of mature T cells.[46]

The problem with accommodating both the positive and the negative selection concepts into one coherent selective theory is that self tolerance is also self MHC restricted[48,49] and MHC molecules are themselves antigens. How then do two opposing processes, one selecting for and another selecting against receptors with specificity for the same molecule, find their compromise and do not nullify each other?

Considering the dual specificity of $\alpha\beta$ TCR in the light of the structural model suggested by Davis and Bjorkman[50] which implies the distinct regions for antigenic peptide and for MHC molecule in TCR binding site, the simplest solution of this dilemma would be to propose that the interaction of the $\alpha\beta$ TCR on immature thymocyte with complete ligand, i.e., polymorphic residues on unprocessed MHC molecules plus antigenic peptide (either MHC or nonMHC derived) would lead to deletion, whereas interaction with MHC molecule in the absence of specific peptide (either MHC or nonMHC derived) would result in positive selection. But are there any MHC molecules devoid of peptides? Would self peptides bound to MHC molecules not interfere with the selection of receptors specific for foreign peptides? Clearly, it had to be studied first, whether in the light of the new information on MHC structure,[51] one could obtain solid evidence for positive selection which was suggested by earlier experiments in hemopoietic chimeras.[52-55] By comparing the T cell development in males and females of our $\alpha\beta$ TCR transgenic mice we could much more rigorously test the propositions of positive and negative selection concepts.

In the following sections we will briefly review the results obtained in transgenic mice which contributed to the elucidation of the selection mechanisms operating in the thymus and discuss them in the context of supporting and sometimes apparently conflicting results obtained by two other approaches mentioned above.

CONTROL OF DIFFERENTIATION OF EARLY INTRATHYMIC PRECURSORS BY $\alpha\beta$ TCR GENE REARRANGEMENT

During ontogeny the surface expression of transgenic TCR molecules occurs one to two days earlier than during development of T cells in normal mice and can be detected on more than 50% of CD4⁻8⁻ intrathymic precursors already at day 15 of gestation.[56] We observed the following consequences of such early expression of transgenes: accelerated growth and differentiation into CD4⁺8⁺ thymocytes[56] and lack of detectable numbers of $\gamma\delta$ TCR expressing thymocytes.[57] These observations suggested that the expression of $\alpha\beta$ TCR genes in the $\gamma\delta$ T cell lineage suppresses, by largely unknown mechanism, the expression of $\gamma\delta$ proteins and that expansion of the early intrathymic precursors and expression of CD4/CD8 coreceptors remains under the control of genes encoding $\alpha\beta$ TCR. The results of experiments in $\alpha\beta$ TCR transgenic mice with severe combined immunodeficiency (SCID) seem to support this last possibility. Whereas in nontransgenic SCID mice, which have very few thymocytes, none expressed CD4 or CD8 coreceptors, in $\alpha\beta$ TCR transgenic SCID mice there were normal numbers of CD4⁺8⁺ thymocytes.[58] However, because lymphoid cells in SCID mice are extremely fragile due to a defect in DNA repair mechanisms, it cannot be excluded that the only effect of the introduced $\alpha\beta$ TCR transgenes is to stabilize CD4⁻8⁻ precursors of CD4⁺8⁺ thymocytes, for instance by reducing the toxic effect of a recombinase.

There is no indication that specific interaction of $\alpha\beta$ TCRs with their ligands is involved in regulating these early events since appearance and accumulation of CD4⁺8⁺ cells is not influenced by $\alpha\beta$ TCR ligands expressed in thymic microenvironment[59,60] and, as recently shown by Krimpenfort et al.[61] does not require expression of functional $\alpha\beta$ TCR on the cell surface.

The first direct evidence that the interaction of CD4$^+$8$^+$ thymocytes with the ligand for αβ TCR plays the regulatory role in their development was obtained in TCR transgenic mice, in which massive deletion of these cells occurred as a result of contact with the complete specific ligand, i.e., the D^b molecule plus H-Y antigen in males of our transgenic mice[38] or L^d in transgenic mice expressing an allospecific (anti L^d) αβ TCR.[60] The 80-90% deficit in the total number of thymocytes in these mice constituted the most direct demonstration of specific deletion of autoreactive lymphocytes at the immature stage of their development, thus providing conclusive experimental evidence for Lederberg's explanation[45] of one mechanism responsible for self tolerance. Other experiments in normal and TCR transgenic mice, tracing the development of T cells expressing superantigen-reactive β TCR chains, after initial failure,[62] have subsequently confirmed that CD4$^+$8$^+$ thymocytes may be the target of negative selection[22,23,31,36] although the deletion appeared much less dramatic, affecting no more than 50% of CD4$^+$8$^+$ thymocytes expressing the appropriate Vβ chain.[31] In some cases the deletion may affect stages just before that of mature T cells, i.e., cells just in the process of expressing one coreceptor at high and the other at low density.[31] It was suggested[36,63] that the large scale deletion of CD4$^+$8$^+$ αβ TCR$^+$ thymocytes by self antigens observed in transgenic mice was the result of the unusually early expression of transgenic TCR chains and that in normal mice a substantial fraction of CD4$^+$8$^+$ αβ TCR$^+$ thymocytes is resistant to deletion.[63] Since in the latter studies different TCR ligands were used (superantigens or anti-TCR antibodies in contrast to class I MHC presented antigens in αβ transgenic mice) it was not clear whether the different extent of deletion reflected the use of different antigens or was due to an artificial situation created in transgenic mice by premature expression of TCRs. In order to address this issue we have explored the fact that H-Y specific, D^b restricted TCR in our αβ transgenic mice contains the β chain, which is specific for the SEB superantigen. This allowed us to compare directly the extent of deletion caused by class I MHC presented antigen on the one hand and by SEB on the other hand. The results showed that in αβ TCR transgenic female mice, SEB deleted CD4$^+$8$^+$ thymocytes to the same extent as in normal mice, thus providing strong evidence that it is the nature of the tolerogenic ligand and not the nature of animals, i.e. normal vs transgenic, which determines the extent of deletion among CD4$^+$8$^+$ thymocytes.[64] Similar results were obtained in αβ TCR transgenic mice expressing D^b restricted receptor specific for LCM virus and in addition recognizing Mls superantigen.[65] Class I MHC presented antigens may also delete similar proportions of CD4$^+$8$^+$ thymocytes in normal and αβ TCR transgenic mice but because there is no receptor marker for class I MHC restricted T cells in normal mice, this could not be tested so far. From the above studies we conclude that in the various experimental models, different tolerogens like class I MHC presented antigens[38,60] or soluble[31] and cell bound superantigens[29,30] can delete CD4$^+$8$^+$ αβ TCR$^+$ thymocytes either early in the outer cortex[66] or late in the inner cortex[33] depending on the way of presentation of the tolerogenic intrathymic ligand to developing T cells. The results of Berg et al.[34] who observed that the same superantigen deleted almost completely CD4$^+$8$^+$ thymocytes in αβ but not in β transgenic mice may not be due, as suggested by the authors, to premature expression of αβ TCR but to the fact that the introduction of their α transgene by breeding might have equipped resulting αβ receptor with additional specificity, might have introduced additional antigens, or both.

Independent of the extent of deletion of CD4$^+$8$^+$ thymocytes, the effect of various antigens on mature cells appeared to be similar in that no cells reactive to the tolerogen were found in the periphery. In the case of our male transgenic mice 90% of peripheral T cells expressed TCRs composed of both α and β transgenic chains but were either CD4$^-$8$^-$ or expressed low levels of CD8 molecule. The number of CD4$^+$8$^-$ T cells, however, was also low despite the fact that their αβ TCRs were not male specific and were composed of transgenic β chains paired with endogenous α chains.[36,67] These results indicated that male specific CD4$^+$8$^+$ cells were the precursors of both male specific CD4$^-$8$^+$ and non-male specific CD4$^+$8$^-$ progeny. This indication has in the meantime been strengthened by a number of independent indirect and direct lines of evidence,[22,23,68,69] which showed that $CD4^+8^+$ thymocytes contain precursors of mature T cells.

The experiments in transgenic mice constituted also the first direct evidence that the interaction of $\alpha\beta$ TCR with intrathymic ligands is involved in positive selection of T cells. By following the fate of $CD4^+8^+$ thymocytes expressing $\alpha\beta$ TCRs with defined restriction specificity, in thymic environments expressing different MHC haplotypes, it was found, that the binding of the TCR to MHC molecules in the absence of nominal antigen was required for generation of mature T cells.[58-60,70] In our $\alpha\beta$ transgenic female mice, expressing the H-Y specific, D^b restricted TCR on immature thymocytes, mature T cells expressing normal levels of CD8 coreceptors and both transgenic TCR chains could develop only in thymuses expressing D^b molecules.[70] These experiments established that maturation of immature $\alpha\beta$ TCR^+ thymocytes depends on the intrathymic expression of the appropriate MHC restricting element but not on the presence of the nominal antigen recognized by $\alpha\beta$ TCR. Altogether these experiments showed that the nature of the ligand, i.e. MHC molecule with or without appropriate peptide, plays a pivotal role in determining whether the developing immature T cell will be positively selected or deleted. They also demonstrated that in contrast to mature T cells, immature thymocytes can receive positive signals through interaction of $\alpha\beta$ TCRs with MHC molecule in the absence of the specific peptide.

THE MHC RESTRICTION SPECIFICITY OF THE $\alpha\beta$ TCR ON DEVELOPING THYMOCYTES DETERMINES THE CD4/CD8 PHENOTYPE OF T CELLS

In transgenic mice expressing the H-Y/D^b specific $\alpha\beta$ TCR, the β transgene suppressed the rearrangement and expression of endogenous $V\beta$ (β_E) genes[35,39] whereas the α transgene (α_T) did not prevent rearrangements of endogenous $V\alpha$ (α_E) genes,[39,64] thus allowing for expression of two different kinds of TCRs: $\beta_T\alpha_T$ (class I MHC restricted, H-Y specific) and $\beta_T\alpha_E$ with the potential to recognize a variety of different ligands including class II MHC presented peptides.

Analysis of the expression of α and β TCR chains on CD4/CD8 subsets of thymocytes and T cells in $H-2^b$ transgenic mice by surface staining with monoclonal antibodies specific for transgenic α (mab T370) and transgenic β (mab F23.1) chains have shown that the great majority of $CD4^+8^+$ thymocytes expressed low levels of both transgenic chains, that $CD4^-8^+$ cells expressed high levels of both transgenic chains whereas $CD4^+8^-$ cells expressed high levels of β_T but no or only low levels of the α_T chain. In non-D^b transgenic mice the majority of $CD4^+8^+$ thymocytes expressed $\beta_T\alpha_T$ TCRs but both $CD4^+8^-$ and $CD4^-8^+$ subsets expressed the $\beta_T\alpha_E$ TCRs.[59,70] These results demonstrated that the interaction of the transgenic $\alpha\beta$ TCR with class I MHC molecules is required for selection of $CD4^-8^+$ T cells and that the selection of $CD4^+8^-$ T cells requires receptors with different restriction specificities. This observation is fully consistent with the proposal[46] that the restriction specificity of self reacting TCRs determines the CD4/CD8 phenotype of positively selected thymocytes. This conclusion has by now been confirmed and extended by studies in several other TCR transgenic mice expressing not only class I MHC restricted[59,60,70] but also class II MHC restricted TCRs.[41,42] In the latter type of transgenic mice it was shown directly that development of $CD4^+8^-$ T cells requires the interaction of $\alpha\beta$ TCR with class II MHC molecules in the thymus.

Experiments of Kruisbeek and colleagues,[25-27] showing that injections of anti class II MHC antibodies inhibited the generation of the $CD4^+8^-$ T cell subset, whereas injections of anti class I MHC antibodies inhibited the generation of the $CD4^-8^+$ T cell subset, are consistent with these conclusions.

It is worthwhile to point out here that receptors specific for superantigens like Mls do not follow these rules,[71,72] i.e. even though the activation by Mls occurs only with $CD4^+8^-$ cells and appears restricted by class II MHC molecules, the receptors are expressed on both $CD4^+8^-$ and $CD4^-8^+$ T cells. This shows that the rules governing the selection of receptors for conventional antigens cannot easily be deduced from experiments with superantigens.

NEGATIVE AND POSITIVE SELECTION: WHICH HAPPENS FIRST?

What is the developmental sequence of positive and negative selection events? Can one distinguish separate developmental stages with different susceptibility to positive and negative selection or can both occur at the same stage, the outcome being determined by the nature of the selecting ligand and possibly the ligand presenting cells? While the deletion of $\alpha\beta$ TCR^{low} CD4$^+$8$^+$ thymocytes constitutes direct evidence that the signals leading to negative selection can be received at this developmental stage, similarly convincing evidence that signals leading to positive selection are also received at this stage is lacking. On the contrary, most of the existing evidence suggests that positive selection, the result of which manifests itself only at the mature stage, expressing one type of coreceptor (CD4 or CD8), is a rather late event in the lifetime of CD4$^+$8$^+$ thymocytes.

In the course of our studies of $\alpha\beta$ TCR transgenic mice we have made several observations indicating that the negative selection can precede positive selection.

As already mentioned, the deletion of immature, male-specific and class I MHC restricted CD4$^+$8$^-$ $\beta_T\alpha_E$ T cells selectable by class II MHC molecules[38,59,67] indicating that the former cells contained precursors of the latter cells. If this was the case, then in CD4$^+$8$^-$ $\beta_T\alpha_E$ T cells one could expect to find RNA transcripts for both endogenous and transgenic α TCR loci. By using appropriate probes distinguishing the transcripts,[64] it was shown that the majority of CD4$^+$8$^-$ T cells expressed both types of transcripts while about fifty percent of the CD4$^-$8$^+$ T cells expressed full size endogenous α TCR transcripts in addition to transgenic α TCR transcripts. Thus, in many CD4$^+$8$^-$ T cells the endogenous α TCR chain was preferred by the transgenic β TCR chain such that they expressed $\beta_T\alpha_E$ TCRs on the surface. Some of these cells expressed even both types of α TCR chains.[64] The possibility that the deletion can affect $\beta_T\alpha_T$ precursors of $\beta_T\alpha_E$ cells undergoing positive selection in the absence of H-Y antigen was further supported by analysis of transgenic offspring of another founder mouse, produced independently by injecting the same gene constructs. The analysis of these mice showed that in female mice not only the CD4$^+$8$^-$ peripheral T cells but also a high proportion of CD4$^-$8$^+$ T cells as well as CD4$^+$8$^+$ thymocytes expressed endogenous α TCR chains. In spite of this, most T cells in the periphery and in the thymus of the male mice expressed transgenic α TCR chains.[64] Thus, the results from these mice show that the earliest CD4$^+$8$^+$ thymocytes are the target of deletion so that there is not time for sufficient α TCR gene rearrangement required for the generation of CD4$^+$8$^+$ cells with endogenous α TCR chains, which then can be positively selected by thymic class I and class II MHC molecules.

The above experiments clearly showed that negative selection can eliminate CD4$^+$8$^+$ precursor cells which otherwise would differentiate further and later become subject to positive selection. Together with the results discussed earlier, concerning the effect of SEB on transgenic thymocytes, they provide little evidence for the existence of subpopulation of early CD4$^+$8$^+$ $\alpha\beta$ TCR^{low} thymocytes inherently resistant to negative selection. The results of Finkel et al.[63] suggesting the existence of such subpopulation could be explained by the inability of antigenic surrogates (superantigens and anti TCR antibodies) used in their studies to crosslink TCR with CD4/CD8 coreceptors which may be critically required for the reception of the signal.

Our experiments in transgenic mice demonstrate in addition that more than one successful rearrangement of α loci is possible in the single clone of developing thymocytes. This would enable CD4$^+$8$^+$ thymocytes to express and to "test" for self reactivity more than one TCR before reaching maturity. If this also happens in normal mice, then this observation suggests very strongly that during the lifetime of CD4$^+$8$^+$ thymocytes, stages susceptible to positive and negative selection do overlap.

INVOLVEMENT OF CD4/CD8 CORECEPTORS IN POSITIVE AND NEGATIVE SELECTION

The involvement of CD4/CD8 coreceptors in the process of intrathymic selection of T cells has been demonstrated by several independent observations in various experiments. The observation made in males of our $\alpha\beta$ TCR transgenic mice, that cells expressing high levels of transgenic, H-Y specific receptor but low levels of CD8 molecules were not deleted, provided the first evidence for the important role of coreceptors during the negative selection process.[38] In addition we found that the thymuses of male mice contained normal numbers of CD4$^-$8$^-$ thymocytes expressing the transgenic TCR.[38,67] Evidence for the critical involvement of CD4 molecules in negative selection was obtained by showing that anti CD4 antibodies could inhibit the deletion of precursors of CD4$^-$8$^+$ cells, expressing receptors specific for superantigen.[22,23] The results of other experiments, in which it was shown that infusion of anti CD4 antibodies selectively blocked the development of CD4$^+$8$^-$ T cells, indicated that CD4 coreceptors also play an important role during positive selection of immature thymocytes.[24] It is not known precisely how coreceptor molecules function during positive and negative selection, but biochemical evidence for their ability to transmit intracellular signals to CD4$^+$8$^+$ thymocytes was recently obtained.[73]

The important general question with regard to the function of CD4/CD8 coreceptors during positive selection is whether the determination of the CD4/CD8 phenotype of mature T cells occurs through instruction or selection. If an instructive mechanism is at work, then the signal received by CD4$^+$8$^+$ thymocyte through $\alpha\beta$ TCR engaged by class I or class II MHC molecules, possibly involving concomitant crosslinking with CD8 or CD4 coreceptors, respectively, would decide which coreceptor gene will be expressed. If, on the other hand, a selective mechanism is responsible for the determination of CD4/CD8 phenotype, this would mean that CD4$^+$8$^+$ thymocytes are switching off the expression of one or the other coreceptor randomly, irrespective of the restriction specificity of their $\alpha\beta$ TCRs. Such cells would die immediately unless the preserved coreceptor is crosslinked with the $\alpha\beta$ TCR by the appropriate class of MHC molecules which rescues the cells from death. In the case of instruction, signaling of CD4$^+$8$^+$ thymocytes through $\alpha\beta$ TCRs interacting with MHC molecules would necessarily be required. In a selection model the signaling could occur after one or the other coreceptor gene had been switched off. At present both possibilities are open to experimental verification.

ROLE OF THE INTRATHYMIC ANTIGEN PRESENTING CELLS IN POSITIVE AND NEGATIVE SELECTION

As already discussed, the nature of the ligand plays a decisive role in determining the fate of immature thymocytes but many earlier experiments indicated that it is also important by which type of stromal cells the ligand is presented. The role of intrathymic antigen presenting cells in positive and negative selection of immature thymocytes, however, was, and partially still is, a controversial issue.

There are at least two kinds of cells in the thymus which can present antigen and express both classes of MHC molecules: epithelial cells and bone marrow derived cells. In $\alpha\beta$ TCR transgenic mice we obtained clear answers with regard to the role of bone marrow derived cells, whereas with regard to the role of epithelial cells conclusive evidence has been obtained in the case of positive selection but not in the case of negative selection.

Repopulation of x-irradiated female recipients of various H-2 haplotypes, with bone marrow from H-2^b transgenic females, resulted in positive selection of T cells bearing transgenic $\alpha\beta$ TCR only in D^b expressing recipients despite the fact that bone marrow derived stromal cells in recipient thymuses of non-H-2^b haplotypes expressed H-2^b.[70] These results demonstrated that positive selection requires the expression of MHC molecules on epithelial cells and that bone marrow derived cells expressing the relevant MHC molecules cannot positively select. The same conclusion was reached by studies in transgenic mice selectively expressing a transgene encoded class II MHC molecules either on epithelium in the thymic cortex or on cells in the medulla.[33,34] Positive selection was observed only in mice expressing class II MHC molecules almost exclusively on thymic epithelium, whereas expression of the same MHC molecules on bone marrow derived cells had no effect.[33,34]

With regard to the negative selection, experiments in $\alpha\beta$ TCR transgenic mice showed that expression of the tolerogenic ligand on bone marrow derived cells is sufficient to cause deletion of CD4$^+$8$^+$ thymocytes. This was again demonstrated in repopulation experiments in

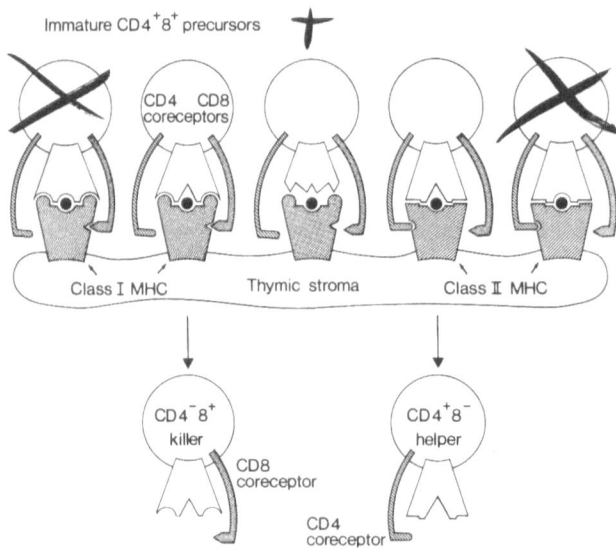

Figure 1. Different immature Cd4$^+$8$^+$ T cells with various
αβ TCRs are shown: if a receptor binds to both self
peptide plus MHC molecule, the cell becomes deleted
(extreme left and extreme right hand side). If a receptor
does not fit precisely the whole ligand but nevertheless
binds to MHC molecule, the cell will not be deleted but
instead induced to mature. Depending on whether the
receptor was engaged by class I or class II restricting
MHC molecule, the mature cell will be of the CD4$^-$8$^+$
or CD4$^+$8$^-$ mature phenotype (second from the left and
second from the right). If the receptor cannot bind to
MHC molecule, the cell will die at an immature stage
(the cell in the middle).

which deletion of CD4$^+$8$^+$ thymocytes expressing H-Y/D^b specific TCRs occurred with
the same efficiency in irradiated H-2^b and non-H-2^b recipients of bone marrow from
H-2^b αβ TCR transgenic males.[64]

The question of whether epithelial cells are able to induce deletion is still controversial because the lack of animals which would express certain antigens exclusively on thymic
epithelium makes it difficult to perform conclusive experiments.

EPILOGUE

Several factors of our experimental system have contributed to the definition of rules
governing positive and negative selection: we had chosen a receptor with specificity for one
antigen in the context of one MHC molecule instead of superantigen specific receptors which
are, as a rule, specific for more than one antigen and "restricted" by more than one MHC
molecule. In addition we could independently monitor β and α TCR chains during the
development in αβ TCR transgenic mice of various H-2 haplotypes in the presence or absence
of specific self antigen. Finally, we could study the selective events in quasi monoclonal
αβ TCR transgenic SCID mice. The rules emerging from these studies, which are confirmed
by studies in five independent lines of αβ TCR transgenic mice and which are also compa-

tible with data obtained in nontransgenic animals, are depicted in the scheme of Figure 1. Different immature $Cd4^+8^+$ T cells with various $\alpha\beta$ TCRs are shown: if a receptor binds to both self peptide plus MHC molecule, the cell becomes deleted (extreme left and extreme right hand side). If a receptor does not fit precisely the whole ligand but nevertheless binds to MHC molecule, the cell will not be deleted but instead induced to mature. Depending on whether the receptor was engaged by class I or class II restricting MHC molecule, the mature cell will be of the $CD4^-8^+$ or $CD4^+8^-$ mature phenotype (second from the left and second from the right). If the receptor cannot bind to MHC molecule, the cell will die at an immature stage (the cell in the middle). What is unknown is the configuration of the MHC ligand inducing positive selection. We do not know whether this ligand can contain any self peptide, a specific "selector" peptide[47] or no peptide at all. Because changes in the residues which are supposed to be in the peptide binding groove of MHC molecules[51] influence positive selection (D. Y. Loh and M. M. Davis, personal communication), it is possible that specific peptides affecting this process are bound by MHC molecules. What is not clear is whether they contribute to, or interfere with, positive selection.

REFERENCES

1. H. M. Grey, A. Sette, and S. Buus, How T cells see antigen, *Sci. Amer.* 261:38 (1989).
2. Z. Dembic, W. Haas, S. Weiss, J. McCubrey, H. Kiefer, H. von Boehmer, and M. Steinmetz, Transfer of specificity by murine α and β T cell receptor genes, *Nature* 320:232 (1986).
3. P. Kisielow, J. A. Hirst, H. Shiku, P. Beverley, M. K. Hoffman, E. A. Boyse, H. F. Ottegen, Ly antigens as markers for functionally distinct subpopulations of thymus derived lymphocytes of the mouse, *Nature* 253:219 (1975).
4. D. P. Dialynas, Z. S. Quan, K. A. Wall, A. Pierres, J. Quintans, M. R. Loken, M. Pierres, and F. Fitch, Characterization of the murine antigenic determinent, designated L3T4a, recognized by monoclonal antibody GD1.5: similarity of L3T4 to the human Leu/T4 molecule, *J. Immunol.* 131:2445 (1983).
5. C. Doyle, J. L. Strominger, Interaction between CD4 and class II MHC molecules mediates cell adhesion, *Nature* 330:256 (1987).
6. A. Norment, R. Salter, P. Parham, V. Engelhardt, and D. Littman, Cell-cell adhesion mediated by CD8 and MHC class I molecules, *Nature* 336:79 (1988).
7. S. Swain, T cell subsets and the recognition of MHC class, *Immunol. Rev.* 74:129 (1983).
8. E. L. Reinherz, S. F. Schlossman, The differentiation and functions of human T cells, *Cell* 19:821 (1980).
9. F. Emmerich, V. Strittmatter, and K. Eichmann, Synergism in the activation of human CD8 T cells by cross-linking the T cell receptor complex with the CD8 differentiation antigen, *Proc. Natl. Acad. Sci. USA* 83:8298 (1986).
10. J. Gabert, C. Langlet, R. Zamoyska, J. R. Parnes, A. Schmitt-Verhulst, and B. Malissen, Reconstitution of MHC class I specificity by T cell receptor and by Lut-2 gene transfer, *Cell* 50:545 (1987).
11. P. Kisielow, W. Leiserson, and H. von Boehmer, Differentiation of thymocytes in fetal organ culture: analysis of phenotypic changes accompanying the appearance of cytolytic and interleukin 2 producing cells, *J. Immunol.* 133: 1117-1123 (1984).
12. R. Kingston, E. J. Jenkinson, and J. J. T. Owen, A single stem cell can recolonize an embryonic thymus producing phenotypically distinct T cell populations, *Nature* 317:811 (1985).
13. H. R. Snodgrass, P. Kisielow, M. Kiefer, M. Steinmetz, and H. von Boehmer, Ontogeny of the T cell antigen receptor within the thymus, *Nature* 313:592 (1985).
14. A. Crisanti, A. colontani, H. R. Snodgrass, and H. von Boehmer, Expression of T cell receptors by thymocytes: *in situ* staining and biochemical analysis, *EMBO J.* 5:2837 (1986).
15. D. H. Raulet, R. D. Garman, H. Saito, and S. Tonegawa, Developmental regulation of T cell receptor gene expression, *Nature* 314:103 (1985).
16. H. R. Snodgrass, Z. Dembic, M. Steinmetz, and H. von Boehmer, Expression of T cell antigen receptor genes during fetal development in the thymus, *Nature* 315:232 (1985).

17. N. Roehm, L. Herron, J. Cambier, D. DiGusto, K. Haskins, J. W. Kappler, and P. Marrack, The major histocompatibility complex restricted antigen receptor on T cells. Distribution on thymus and peripheral T cells. *Cell* 38:577 (1984).

18. R. T. Kubo, W. Born, J. W. Kappler, P. Marrack, and M. Pigeon, Characterization of a monoclonal antibody which detects all murine $\alpha\beta$ T cell receptors, *J. Immunol.* 142: 2736-2742 (1989).

19. D. McPhee, J. Pye, and K. Shortman, The differentiation of T lymphocytes. V. Evidence for intrathymic death of most thymocytes, *Thymus* 1:151 (1979).

20. W. Born, M. McDuffie, N. Roehm, E. Kushnir, J. White, D. Thorpe, J. P. Stefano, J. W. Kappler, and P. Marrack, Expression and role of the T cell receptor in early thymocyte differentiation *in vitro*. *J. Immunol* 138:999 (1987).

21. M. McDuffie, W. Born, P. Marrack, and J. W. Kappler, The role of the T cell receptor in thymocyte maturation: effects *in vivo* of antireceptor antibody. *Proc. Natl. Acad. Sci. USA* 83:8720 (1986).

22. B. J. Fowlkes, R. H. Schwartz, and D. M. Pardoll, Deletion of self-reactive thymocytes occurs at a $CD4^+8^+$ precursor stage, *Nature* 334:620 (1988).

23. H. R. MacDonald, H. Hengarner, and M. Pedrazzini, Intrathymic deletion of self-reactive cells prevented by neonatal anti-CD4 antibody treatment, *Nature* 335:174 (1988).

24. J. C. Zuniga-Pflucker, S. A. McCarthy, M. Weston, D. L. Longo, A. Singer, and A. M. Kruisbeek, Role of CD4 in thymocyte selection and maturation, *J. Exp. Med.* 169: 2085 (1989).

25. J. C. Zuniga-Pflucker, D. L. Long, and A. M. Kruisbeek, Positive selection of $CD4^-/CD8^+$ T cells in the thymus of normal mice, *Nature* 338:76 (1988).

26. S. Marusic-Galesic, D. L. Longo, and A. M. Kruisbeek, Preferential differentiation of T cell receptor specificities based on the MHC glycoproteins encountered during development: evidence for positive selection, *J. Exp. Med.* 169:1619-1628 (1989).

27. A. M. Kruisbeek, J. J. Mond, B. J. Fowlkes, J. A. Carmen, S. Bridges, and D. L. Longo, Absence of the $Lyt2^-$, $L3T4^+$ lineage of T cells in mice treated neonatally with anti-I-A correlates with the absence of intrathymic I-A bearing antigen-presenting cell function, *J. Exp. Med.* 161:1029 (1985).

28. P. Marrack, E. Kushnir, W. Born, M. McDuffie, and J. W. Kappler, The development of helper T cell precursors in mouse thymus, *J. Immunol.* 140:2508 (1988).

29. J. W. Kappler, U. Staerz, J. White, and P. Marrack, Self-tolerance eliminates T cells specific for Mls-modified products of the major histocompatibility complex, *Nature* 332:35 (1988).

30. H. R. MacDonald, R. Schneider, K. Lees, R. C. Howe, H. Acha-Orbea, H. Festenstein, R. M. Zinkernagel, and H. Hengartner, T cell receptor V-beta use predicts reactivity and tolerance to Mls(a)-encoded antigens, *Nature* 332:40 (1988).

31. J. White, A. Herman, A. M. Pullen, R. T. Kubo, J. W. Kappler, and P. Marrack, The $V\beta$ specific superantigen Staphylococcal enterotoxin B: stimulation of mature T cells and clonal deletion in neonatal mice, *Cell* 56:27 (1989).

32. W. vanEwijk, Y. Ron, J. Monaco, J. W. Kappler, P. Marrack, M. LeMeur, P. Gerlinger, B. Durand, C. Benoist, and D. Mathis, Compartmentalization of MHC class II gene expression in transgenic mice, *Cell* 53:357 (1988).

33. C. Benoist and D. Mathis, Positive selection of the T cell repertoire: Where and when does it occur? *Cell* 58:1027 (1989).

34. J. Bill and E. Palmer, Positive selection of $CD4^+$ T cells mediated by MHC class II-bearing stromal cell in the thymus, *Nature* 341:649 (1989).

35. Y. Uematsu, S. Ryser, Z. Dembic, P. Borgulya, P. Krimpenfort, A. Berns, H. von Boehmer, and M. Steinmetz, In transgenic mice the introduced functional T cell receptor β gene prevents expression of endogenous β genes, *Cell* 52:831 (1988).

36. L. J. Berg, B. Fazekas de St. Groth, A. M. Pullen, and M. M. Davis, Phenotypic differences between $\alpha\beta$ and β T cell receptor transgenic mice undergoing negative selection, *Nature* 340:559 (1989).

37. H. Pircher, T. K. Mak, R. Lang, W. Ballhausen, E. Ruedi, H. Hengartner, R. M. Zinkernagel, and K. Burki, T cell tolerance to Mls^a encoded antigens in T cell receptor $V\beta8.1$ chain transgenic mice, *EMBO J.* 8:719 (1989).

38. P. Kisielow, H. Bluthmann, U. Staerz, M. Steinmetz, and H. von Boehmer, Tolerance in T-cell receptor transgenic mice involves deletion of nonmature CD4$^+$8$^+$ thymocytes, *Nature* 333:742 (1988).

39. H. Bluthman, P. Kisielow, Y. Uematsu, M. Mallissen, P. Krimpenfort, A. Berns, H. von Boehmer, and M. Steinmetz, T cell specific deletion of T cell receptor transgenes allows functional rearrangement of endogenous α and β genes, *Nature* 334:156 (1988).

40. W. C. Sha, C. A. Nelson, R. D. Newberry, D. M. Kranz, J. H. Russel, and D. Y. Loh, Selective expression of an antigen receptor on CD8-bearing T lymphocytes in transgenic mice, *Nature* 335:271 (1988).

41. L. J. Berg, A. M. Pullen, B. Fazekas de St. Groth, D. Mathis, C. Benoist, and M. M. Davis, Antigen/MHC-specific T cells are preferentially exported from the thymus in the presence of their MHC ligand, *Cell* 58:1035 (1989).

42. J. Kaye, M. L. Hsu, M. E. Sauron, J. C. Jameson, R. J. Gascoigne, and S. M. Hedrick, Selective development of CD4 T cell in transgenic mice expressing a class II MHC-restricted antigen receptor, *Nature* 341:746 (1989).

43. H. von Boehmer, H. S. Teh, and P. Kisielow, Thymus selects the useful, neglects the useless and destroys the harmful, *Immunol. Today* 10:57 (1989).

44. H. von Boehmer, Developmental biology of T cells in T-cell receptor transgenic mice, *Ann. Rev. Immunol.* (1990, in press).

45. J. Lederberg, Genes and antibodies, *Science* 129:1649 (1959).

46. H. von Boehmer, The selection of the $\alpha\beta$ heterodimeric T cell receptor for antigen, *Immunol. Today* 7:333 (1986).

47. P. Marrack and J. W. Kappler, The T cell receptor, *Science* 238:1073 (1987).

48. P. Matzinger, R. Zamoyska, and H. Waldmann, Self-tolerance is H-2 restricted, *Nature* 308:738 (1984).

49. H. G. Rammensee and M. Bevan, Evidence from *in vitro* studies that tolerance to self is MHC restricted, *Nature* 308:741 (1984).

50. M. M. Davis and P. Bjorkman, T cell antigen receptor genes and T cell recognition, *Nature* 334:395 (1988).

51. P. Bjorkman, M. A. Saper, B. Samaoui, W. S. Bennett, J. L. Strominger, and D. C. Wiley, Structure of the human class I histocompatibility antigen, HLA-A2, *Nature* 329:506 (1987).

52. M. Bevan, In a radiation chimaera, host H-2 antigens determine immune responsiveness of donor cytotoxic cells, *Nature* 269:417 (1977).

53. H. von Boehmer, W. Haas, and N. K. Jerne, Major histocompatibility complex-linked immune-responsiveness is acquired by lymphocytes of low responder mice differentiating in thymus of high-responder mice, *Proc. Natl. Acad. Sci. USA* 75:2439 (1978).

54. R. M. Zinkernagel, G. Callahan, A. Althage, S. Cooper, P. Klein, and J. Klein, On the thymus in the differentiation of "H-2 self-recognition" by T cells: Evidence for dual recognition? *J. Exp. Med.* 147:882 (1978).

55. D. Loh and J. Sprent, Identity of cells that imprint H-2 restricted T cell specificity in the thymus, *Nature* 319:672 (1986).

56. H. S. Teh, H. Kishi, B. Scott, P. Borgulya, H. von Boehmer, and P. Kisielow, Early deletion and late positive selection of T cells expressing a male-specific receptor in T cell receptor transgenic mice, *Developmental Immunol.* (1990, in press).

57. H. von Boehmer, M. Bonneville, I. Ishida, S. Ryser, G. Lincoln, R. T. Smith, H. Kishi, B. Scott, P. Kisielow, and S. Tonegawa, Early expression of a T cell receptor β chain transgene suppresses rearrangement of the Vγ4 gene segment, *Proc. Natl. Acad. Sci. USA* 85:9729 (1988).

58. B. Scott, H. Bluthmann, H. S. Teh, and H. von Boehmer, The generation of mature T cells requires interaction of the $\alpha\beta$ T cell receptor with major histocompatibility complex antigens and the $\alpha\beta$ T cell receptor determine the CD4/CD8 phenotype of T cells, *Nature* 335:229 (1988).

59. H. S. Teh, P. Kisielow, B. Scott, H. Kishi, Y. Uematsu, H. Bluthmann, and H. von Boehmer, Thymic major histocompatibility complex antigens and the $\alpha\beta$ T cell receptor determine the CD4/CD8 phenotype of T cells, *Nature* 335:229 (1988).

60. W. C. Sha, C. A. Nelson, R. D. Newberry, D. M. Kranz, J. H. Russell, and D. Y. Loh, Positive and negative selection of an antigen receptor on T cells in transgenic mice, *Nature* 336:73 (1988).

61. P. Krimpenfort, B. J. Ossendorp, C. Melief, and A. Berns, T cell depletion in transgenic mice carrying a mutant gene for TCR-β, *Nature* 341:742 (1989).

62. J. W. Kappler, N. Roehm, and P. Marrack, T cell tolerance by clonal elimination in the thymus, *Cell* 49:273 (1987).

63. T. H. Finkel, J. C. Cambier, R. T. Kubo, W. K. Born, P. Marrack, and J. W. Kappler, The thymus has two functionally distinct populations of immature $\alpha\beta^+$ cells: one population is deleted by ligation of $\alpha\beta$ TCR, *Cell* 58:1047 (1989).

64. P. Borgulya, H. Kishi, B. Scott, Y. Uematsu, A. Berns, and H. von Boehmer, Positive and negative selection of T cells: which comes first? (Submitted)

65. D. A. Ferrick, P. S. Ohashi, V. Wallace, M. Schilham, and T. W. Mak, Thymic ontogeny and selection of α, β, and $\gamma\delta$ T cells, , *Immunol. Today* 10:403 (1989).

66. W. van Ewijk, P. Kisielow, and H. von Boehmer, Immunohistology of T cell differentiation in the thymus of HY specific $\alpha\beta$ T cell receptor transgenic mice, *Eur. J. Immunol.* 20:129 (1990).

67. H. S. Teh, H. Kishi, B. Scott, and H. von Boehmer, Deletion of autospecific T cells in T cell receptor transgenic mice spares cells with normal TCR levels and low levels of CD8 molecules, *J. Exp. Med.* 169:795 (1989).

68. C. Penit and F. Vasseur, Sequential events in thymocyte differentiation and thymus regeneration revealed by a combination of bromodeoxyuridine DNA labelling and antimitotic drug treatments, *J. Immunol.* 140:3315 (1989).

69. J. Nicolic-Zugic and M. Bevan, Thymocytes expressing CD8 differentiate into CD4$^+$ cells following intrathymic injection, *Proc. Natl. Acad. Sci. USA* 85:8663 (1988).

70. P. Kisielow, H. S. Teh, H. Bluthmann, and H. von Boehmer, Positive selection of antigen-specific T cells in thymus by restricting MHC molecules, *Nature* 335:730 (1988).

71. M. Blackman, P. Marrack, and J. Kappler, Influence of the MHC on positive thymic selection of Vβ17$^+$ T cells, *Science* 244:214 (1989).

72. H. R. MacDonald, R. K. Lees, R. Schneider, R. Zinkernagel, and H. Hengartner, Positive selection of CD4$^+$ thymocytes controlled by MHC class II gene products, *Nature* 336:471 (1988).

73. A. Villette, J. C. Zuniga-Pflucker, J. B. Bolen, and A. M. Kruisbeek, Engagement of CD4 and CD8 expressed on immature thymocytes induces activation of intracellular tyrosine phosphorylation pathways, *J. Exp. Med.* 170:1671 (1989).

LIFE AND DEATH OF A T CELL

Dennis Y. Loh

Howard Hughes Medical Institute
Department of Medicine, Genetics, Molecular Microbiology
Washington University School of Medicine
St. Louis, Missouri

INTRODUCTION

Thymus-derived lymphocytes or T cells recognize nominal antigens in the context of the products of the self major histocompatibility locus (MHC)--a phenomenon called MHC restriction. In addition, T cells must be self-tolerant to antigens that are endogenous to the organism in order to avoid an autoimmune condition. How the developing T cells acquire these two characteristics is the main theme of my laboratory.

T cells recognize the universe of antigens using a cell surface receptor called the T cell receptor (TCR). The TCR is composed of a heterodimeric protein structure and the genes encoding each polypeptide chain undergo tissue- and stage-specific DNA rearrangement in the thymus. The control of these events is thought to be similar to those of the immunoglobulin gene rearrangements although very little definitive data is available at this time. Once the TCR genes have undergone successful DNA rearrangements, it is thought that the developing T cells undergo cellular selection whose outcome is primarily determined by the nature of the TCR on the cell surface. In particular, the thymocytes undergo positive selection to "learn" self MHC recognition leading to MHC restriction and negative selection to eliminate self-reactive thymocytes. Thus, through these selection steps, the peripheral repertoire is insured to be self-tolerant and MHC restricted.

TCR TRANSGENIC MOUSE APPROACH TO THYMOCYTE DEVELOPMENT

One of the major problems in trying to understand thymocyte development has been the lack of a reliable *in vitro* model that can recapitulate the entire normal thymocyte develop ment. Particularly, lack of fully differentiating normal clonal cell lines or tumor cell model has made it difficult to study differentiation at the individual cellular level. In addition, because a phenomenon such as self-tolerance can be best studied in an intact animal, the usefulness of an *in vitro* model may have inherent limitations. To overcome these difficulties, we decided to create a TCR transgenic mouse whose receptor specificity is known. In addition, the availability of the anti-clonotypic monoclonal antibody (mAb) would allow us to follow the fate of the transgenic TCR bearing cells using microfluorocyto-metric assays (MFC).

Alloreactive cytotoxic T-cell, 2C, originated from a BALB.B (H-2b) mouse, and it has a specificity for the H-2Ld antigen. The 2C TCR uses a Vβ8.2 which can be followed by the mAb F23.1 while mAb 1B2 is anti-clonotypic. Initially, rearranged TCR genes were cloned

Mechanisms of Lymphocyte Activation and Immune Regulation III
Edited by S. Gupta *et al.*, Plenum Press, New York, 1991

43

from various recombinant phage libraries and put together in expressible forms using cosmid vectors. Earlier failures to obtain sufficient expression in transgenic mice forced us to use very large DNA constructs containing the TCR transcriptional enhancers. The completed cosmid constructs were microinjected into one-celled mouse embryos and transgenic mice were obtained. After the establishment of the founder animals, appropriate matings were performed to test for thymic and peripheral (splenic) expression patterns of the transgenic TCR bearing cells.[1]

SELF-TOLERANCE IS ACHIEVED BY CLONAL DELETION

2C is alloreactive against H-2Ld. Thus we could test how self-tolerance may be achieved by backcrossing the transgenic mouse to one that bears H-2d. When this was done in both H-2bXd and H-2d, T cells bearing the 2C TCR in conjunction with CD8 molecules were deleted (the presence of CD8 molecule is necessary for proper recognition for this cytotoxic T cell.). In contrast, the 2C TCR bearing T cells emerged as single CD8$^+$ but not CD4$^+$ cells in H-2b animals. In fact, when thymocytes in the negatively selecting setting were studied, there was complete deletion of the transgenic TCR bearing cells even among the CD4$^+$CD8$^+$ population. Thus it appears that deletion occurs at or prior to the double positive stage of thymocyte development. Our results obtained here, when generalized, would imply that self-tolerance against antigens known to be present in the thymus is achieved by the actual clonal deletion of those self-reactive cells in the thymus.[2]

ORIGIN OF SELF-MHC RESTRICTION

While the clonal deletion of the transgenic TCR bearing cells was somewhat expected, we were surprised when we backcrossed transgenic mice into H-2s mice. 2C does not recognize H-2s cells as an alloreactive target and yet very few clonotype positive CD8$^+$ cells were found in its periphery. To determine the genetic basis of this finding, mice of the H-2bXs were examined. These mice were found to have the phenotype identical to H-2b mice. Thus, rather than postulating that the presence of H-2s was suppressing or deleting the 2C TCR cells, we proposed that the presence of H-2b in the thymus led to the normal development of 2C TCR bearing cells in the periphery as functional CD8$^+$ T cells. We call this the positive selection model of T cell development.[2] When generalized, this model explains how self-MHC restriction may arise. Every T cell developing in the thymus must be able to interact with a self-MHC element in order to become a mature single-positive T cell. Because this step is a mandatory step during development, it necessarily skews the peripheral repertoire towards self-MHC recognition, and hence self-MHC restriction. However, we assume that those developing thymocytes whose TCR has even higher interaction with self-MHC will be eliminated. Unfortunately, we currently cannot measure the "affinity" of the TCR/MHC interaction and thus are unable to test if such an affinity model is correct.

STRUCTURE-FUNCTION STUDIES TO ELUCIDATE THE REQUIREMENTS FOR SELF-RECOGNITION IN THE THYMUS

Having established that an element in H-2b is an absolute requirement for positive selection to occur, we sought to identify the specific gene product(s) responsible for this process. Once again, we backcrossed the transgenic mice to various intra-H-2 recombinant mice and identified that the presence of H-2Kb was necessary and sufficient for positive selection to occur. We detected no significant differences in the ability of H-2Kb to positively select as a function of any other background genes.

To characterize the structural requirements of the H-2Kb molecule responsible for positive selection, we utilized the existence of the bm mutant mice. The bm mutants were originally detected by skin graft rejection and were identified to be different from the parental mice at the H-2Kb locus. The amino acid sequences of the mutation have been deciphered and the impact of these changes on the H-2 molecule can be predicted based on the crystallographic data of the MHC Class I molecule, H-LA A2. By ascertaining the ability or inability of the various H-2Kb mutants to positively select, we can determine the structural requirements for molecular recognition that lead to positive selection. In all, we analyzed the effect of bm1, bm3, bm7, bm8, bm10, and bm11 mutations on positive selection. Of these, only bm7 behaved similarly to the wild-type H-2Kb leading to positive selection. Bm1 and bm10 no longer could be recognized for positive selection and behaved similarly to previously characterized H-2s mice. Similar data were obtained with bm8 mice though the effect was not as dramatic. Surprisingly, bm3 or bm11 resulted in negative selection of the 2C TCR bearing cells in the thymus. In other words, the presence of the bm3 or bm11 molecules in the thymus led to the clonal deletion of the 2C bearing T cells.[3]

Thus, the bm mutants studied here display the entire spectrum of possible interaction between the 2C TCR bearing thymocyte and the products of the H-2K locus. The data imply several important characteristics of the positive selection process. Because changes in the α-1 and α-2 helixes as well as the floor of the peptide-binding groove affect selection, it implies that the TCR recognizes the entire H-2 molecule including the putative bound peptide during selection. In addition, very subtle changes in the H-2 molecule can lead to dramatic results, transforming a positively selecting element into one that leads to clonal deletion (as observed in bm3 and bm11). This suggests that positive and negative selection may be a different manifestation of a fundamentally similar process as far as elements of molecular recognition are concerned. Nonetheless, the outcomes of positive and negative selections are quite different. In one case, the cells emerge as MHC-restricted mature T cells whereas in the latter case, the emerging T cells are clonally deleted from the organism in the thymus. It is the elucidation of the exact mechanism that leads to this selective life and death that will be the central theme for many investigators, including ourselves, for many years to come.

ACKNOWLEDGEMENT

Work reported here was funded by Howard Hughes Medical Institute. D. Y. L. is an associate investigator of HHMI.

REFERENCES

1. W. C. Sha, C. A. Nelson, R. D. Newberry, D. M. Kranz, J. H. Russell, and D. Y. Loh, Selective expression of an antigen receptor on CD8-bearing T lymphocytes in transgenic mice, *Nature* 335:271 (1988).
2. W. C. Sha, C. A. Nelson, R. D. Newberry, D. M. Kranz, J. H. Russell, and D. Y. Loh, Positive and negative selection of an antigen receptor on T cells in transgenic mice, *Nature* 336:73 (1988).
3. W. C. Sha, C. A. Nelson, R. D. Newberry, J. K. Pullen, L. R. Pease, J. H. Russell, and D. Y. Loh, Positive selection of transgenic receptor-bearing thymocytes by K[b] mutations that involve peptide binding, *Proc. Natl. Aca. Sci. USA* (in press, 1990).

FIRST WAVE FETAL THYMOCYTES EXPRESSING V3JγlCγl-Vδ1Jδ2Cδ T CELL RECEPTORS ARE NOT REQUIRED FOR αβ T CELL RECEPTOR REARRANGEMENT AND EXPRESSION

David A. Ferrick, Donald Gajewski and Tak W. Mak

Ontario Cancer Institute
Departments of Medical Biophysics and Immunology
University of Toronto
Toronto, Ontario, Canada

SUMMARY

We have determined the fetal expression of Vγ3, δ and β TcRs in mice transgenic for the Vγ1.1J4Cγ4 TcR chain. The first wave thymocytes appearing at day 14 and disappearing by day 17 in normal mice was absent from the transgenic mice. However, both mice had an almost identical number of γδ-bearing thymocytes throughout gestation. Therefore, it is most likely that the Vγ3JγlCγl chain was replaced in the transgenic mice by the Vγ1.1Jγ4Cγ4 transgene. The appearance, although slightly earlier for the transgenic mice, of αβ-bearing thymocytes was also very similar between transgenic and control mice during gestation. These data suggest that whatever the role of the first wave thymocytes expressing V3-Vδ1 TcRs is, it most likely is not required for the rearrangement, expression and maturation of the αβ TcR repertoire. We are currently analyzing a series of γδ transgenic mice to determine whether other restricted populations of γδ-bearing T cells are involved in specific aspects of immune development or function.

INTRODUCTION

For some time now the existence of a second T cell receptor (TcR) composed of δ and γ chains has been known.[1-6] However, the nature of the ligands recognized by these γδ TcRs remains enigmatic. Two of the most salient features of the γ and δ chain genes are their early expression in fetal ontogeny[7,8] (before the appearance of αβ receptors) and their temporal and spatial skewing towards a restricted combination of variable (V), diversity (D), joining (J) and constant (C) gene segments.[9-16]

Recent studies aimed at characterizing the newest TcR chain genes, γ and δ, have shown a pattern that is not convergent with respect to the well-understood α and β loci. Whereas the rearrangement and expression of αβ heterodimeric receptors are random until they are exposed to the selective pressures of the thymus and periphery,[17] γδ TcR rearrangement and expression appear to be much more regulated and restricted, especially early in fetal ontogeny.[18] There are also many fewer gene segments available to the γ and δ loci compared with α and β.[18] In addition, γδ TcR expression precedes αβ in T cell ontogenic which has led to the speculation that one of their functions may be involved with the appearance and maturation of the αβ TcR repertoire.[18] Finally, with respect to classical recognition of foreign antigen in the context of either Class I or Class II MHC molecules, γδ-bearing cells appear to predominantly recognize uncommon MHC-like molecules (e.g., Qa and TL in mouse, CD1 in humans)[19-21] complexed with as yet undefined peptides.

Mechanisms of Lymphocyte Activation and Immune Regulation III
Edited by S. Gupta *et al.*, Plenum Press, New York, 1991

47

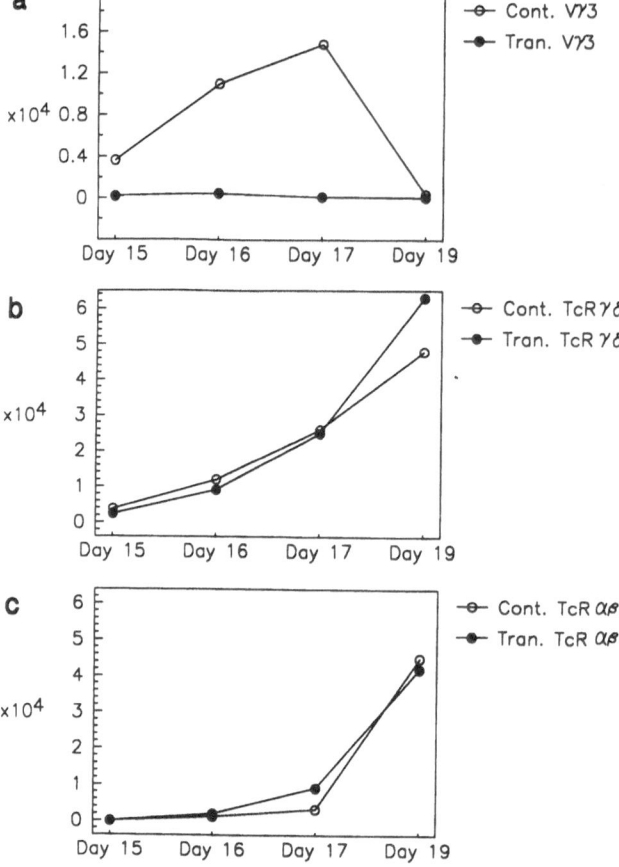

Figure 1. TcR expression during fetal ontogeny on thymocytes from TcR Cγ4 transgenic and control mice. The total number of Vγ3$^+$ (a), TcR γδ$^+$ (b), and TcR αβ$^+$ (c) thymocytes per embryo for control (open circles) and transgenic (closed circles) mice.

One of the first examples of restricted γδ expression was described by W. Havran and J. P. Allison.[9] They were able to show that the first wave of thymocytes to enter the thymic rudiment in mouse at day 14 of fetal gestation expressed almost exclusively one receptor pair, Vγ3Jγ1Cγ1-Vδ1Jδ2Cδ. In fact, there was almost no N-region diversity observed for this δ heterodimer and, hence, suggested that almost all of the lymphocytes present during this period of fetal development expressed one predominant TcR. The Vγ3-Vδ1 receptor expressed on first wave thymus immigrants was the major receptor until day 16 when its expression decreased to 50%.[9] From then on other δ receptors predominated and at around day 17 αβ-bearing thymocytes began to appear in the thymus.[9] The expansion of αβ-bearing thymocytes continued until birth, at which time they represented a majority of the TcRs present and, thereafter, γδ-bearing thymocytes only accounted for a minor fraction of the total TcRs present (1-5%) in mice.[18]

The appearace and disappearance of this rather homogenous wave of Vγ3-Vδ1 thymocytes prior to the emergence of αβ-expressing cells has led to speculation that this restricted γδ population may be associated with the rearrangement of the α locus and, hence, the appearance of αβ-expressing thymocytes. This possibility is supported indirectly by the recent finding of A. Winoto and D. Baltimore (personal communication) describing the existence of α locus silencers that prevent expression of α genes in γδ cells. Presumably something must occur during early fetal gestation to release developing thymocytes from the inhibitory effects of these α silencers so that α locus rearrangement followed by $\alpha\beta$ TcR surface expression can be initiated. The inverse relationship between the expression of Vγ3-Vδ1 receptors and the appearance of $\alpha\beta$ thymocytes (Figure 1a,b open circles) may suggest a link between first wave $\gamma\delta$ thymocytes and $\alpha\beta$ expression.

In the present study, mice transgenic for the Vγ1.1Jγ4Cγ4 TcR chain were devoid of thymocytes bearing Vγ3-Vδ1 receptors during fetal development as measured by an antibody specific for this receptor. Therefore, using these transgenic mice we were able to assess the effect of the absence of this first wave population of $\gamma\delta$ thymocytes on the appearance of $\alpha\beta$ cells.

Materials and Methods

Mice. All mice were bred and maintained in our mouse colony. The generation of the Vγ1.1Jγ4Cγ4 transgenic mice are described elsewhere.[22] CD1 mice (H-2^q) were purchased from Charles River.

Antibodies. FITC conjugated to goat anti-hamster Ig antibody was purchased from Miles Scientific. Strepavidin conjugated to phycoerythrin or FITC was purchased from Becton Dickinson. The anti-Vγ3 monoclonal antibody (536) was a generous gift from J. Allison. The monoclonal antibody from the hybridoma H57-595, pan reactive for the mouse $\alpha\beta$ TcR, was a gift from R. Kubo. The antibody, UC7-13D5, pan reactive for the $\gamma\delta$ TCR, was kindly supplied by J. Bluestone.

Antibody Staining and Flow Cytometry. Isolated fetal thymocyte cell suspensions were enriched for viable lymphocytes by centrifugation on a density gradient (Lympholyte M, Cedarlane Laboratories), washed in PBS and stained with saturating concentrations of antibody. The cells were washed and analyzed by flow cytometry on a FACScan cell analyzer with an Argon-Ion laser (488 nm) (Becton Dickinson Immunocytometry Systems). The data were analyzed using the Consort 30 software (Becton Dickinson).

RESULTS AND DISCUSSION

In this study we analyzed first wave thymocytes expressing Vγ3Jγ1Cγ1- Vδ1Jδ2Cδ receptors[9] in mice transgenic for the TcR chain Vγ1.1Jγ4Cγ4. As can be seen in Figure 1, until day 17 of gestation Vγ3-expressing thymocytes are the major population in normal mice (Figure 1a,b open circles). In contrast, transgenic mice do not express this receptor in fetal thymus at any of the time points tested (Figure 1a, b closed circles), even though they have an equal number of $\gamma\delta^+$ cells when compared to controls (Figure 1b). It can be inferred from the previous statement that the majority of δ cells present in the transgenic thymus from Day 15-17 most likely express the transgene receptor since no other $\gamma\delta$ receptors are present in normal fetal thymus this early in development.

It has been suggested that the Vγ3-Vδ1 population found in fetal thymus seeds the epidermis since, in adult mice, lymphocytes residing in the skin (referred to as dendritic epidermal cells) predominantly express Vγ3-Vδ1 receptors that closely resemble those found in fetal thymus.[14] In the future it will be interesting to determine what receptor is expressed in the skin of the Cγ4 transgenic mice and what effects this may have on the function of these cells in the epidermis.

Figure 1c shows the appearance of thymocytes expressing $\alpha\beta$ TcR chains. As can be seen, both the control and transgenic mice have a very similar pattern. However, the transgenic mice do show a slightly earlier appearance of significant numbers of $\alpha\beta$ T cells compared to controls beginning at day 16. Based on northern blot analysis and staining with various Vβ specific antibodies (data not shown) both the transgenic and control are indistinguishable from each other. Therefore, it appears that the lack of this first wave of Vγ3-Vδ1 thymocytes does not hinder the appearance and selection of $\alpha\beta$ T cell receptors in the thymus or their transit to the periphery.

Based on the findings presented in this study there doesn't appear to be a requirement for the presence of first wave thymocytes expressing Vγ3 and Vδ1 TcR chains prior to the appearance of $\alpha\beta$ thymocytes. Since cells expressing these receptors are only seen in the epidermis of adult mice it suggests that they may perform a specific function in the skin by virtue of the unique $\gamma\delta$ receptor that they express. If this is true, then perhaps other $\gamma\delta$ heterodimers that have been shown to be regulated in a spatial and temporal manner[18] will also have specifically defined functions based on when and where they are expressed. This is in contrast to $\alpha\beta$-bearing T cells which circulate freely throughout the body, constantly probing cells for the presence of their unique ligand consisting of a foreign peptide and MHC. In the future it will be interesting to see whether mice transgenic for other $\alpha\beta$ receptors will show a predominant phenotype associated with the biological function of lymphocytes that express a specific $\gamma\delta$ receptor heterodimer.

ACKNOWLEDGMENTS

We thank Drs. J. Bluestone, J. Allison, and F. Kubo for their generous gift of monoclonal antibodies and Irene Ng for technical assistance.

This work was supported in part by grants from the National Cancer Institute of Canada, the Medical Research Council of Canada (MT-4519, MT-3017) and a special grant from the University of Toronto. D. A. Ferrick is a recipient of a Cancer Research Institute fellowship (New York, N. Y.).

REFERENCES

1. M. B. Brenner, J. MacLean, D. P. Dialynas, J. L. Strominger, J. A. Smith, F. L. Owen, J. G. Seidman, D. Ip, F. Rosen, and M. S. Krangel, Identification of a putative second T-cell receptor, *Nature* 322:145 (1986).
2. H. Saito, D. M. Krantz, Y. Takagaki, A. Hayday, H. Eisen, and S. Tonegawa, A third rearranged and expressed gene in a clone of cytotoxic T lymphocytes, *Nature* 312:36 (1984).
3. Y.-H. Chien, M. Iwashima, K. B. Kaplan, J. F. Elliot, and M. M. Davis, A new T-cell receptor gene located in the alpha locus and expressed early in T cell differentiation, *Nature* 327:677 (1987).
4. Y.-H. Chien, M. Iwashima, D. A. Wittstein, K. B. Kaplan, J. F. Elliot, W. Born, and M. M. Davis, T-cell receptor δ gene rearrangements in early thymocytes, *Nature* 330:722 (1987).
5. I. Iwamoto, P. Ohashi, C. Walker, F. Rupp, H. Yoho, H. Hengartner, and T. W. Mak, The murine γ chain genes in B10 mice: Sequence and expression of new constant and variable genes, *J. Exp. Med.* 163:1203 (1986).
6. R. Cron, R. Koning, W. L. Maloy, D. Pardoll, J. E. Coligan, and J. A. Bluestone, Peripheral murine CD3$^+$, CD4$^-$, CD$^-$ T lymphocytes express novel T cell receptor δ structures, *J. Immunol.* 141:1074 (1988).
7. D. M. Pardoll, B. J. Fowlkes, J. A. Bluestone, A. Kruisbeek, W. L. Maloy, J. E. Coligan, and R. H. Schwartz, Differential expression of two distinct T cell receptors during thymocyte development, *Nature* 326:79 (1987).
8. W. L. Havran and J. P. Allison, Developmentally ordered appearance of thymocytes expressing different T-cell antigen receptors, *Nature* 335:443 (1988).
9. R. L. O'Brien, M. P. Happ, A. Dallas, E. Palmer, R. Kubo, and W. K. Born, Stimulation of a major subset of lymphocytes expressing T cell receptor $\gamma\delta$ by an antigen derived from *Mycobacterium* tuberculosis, *Cell* 57:667 (1989).
10. M. B. Brenner, J. MacLean, H. Scheft, J. Riberdy, S. Ang, J. Siedman, P. Devlin, and M. S. Krangel, Two forms of the T cell receptor γ protein found on peripheral blood cytotoxic T lymphocytes, *Nature* 325:145 (1987).
11. D. M. Asarnow, W. A. Kuziel, M. Bonyhadi, R. E. Tigelaar, P. W. Tucker, and J. P. Allison, Limited diversity of $\gamma\delta$ antigen receptor genes of Thy-1$^+$ dendritic epidermal cells, *Cell* 55:837 (1988).
12. G. Steiner, F. Koning, A. Elbe, E. Tschachler, W. M. Yokoyama, E. M. Shevach, G. Stingl, and J. E. Coligan, Characterization of T cell receptors on resident murine dendritic epidermal T cells, *Eur. J. Immunol.* 18:1323 (1988).

13. T. Goodman and L. Lefrancois, Expression of the γ-δ T cell receptor on intestinal CD8$^+$ intraepithelial lymphocytes, *Nature* 333:855 (1988).

14. D. M. Asarnow, W. A. Kuziel, M. Bonyhadi, R. E. Tigelaar, P. W. Tucker, and J. P. Allison, Limited diversity of $\gamma\delta$ antigen receptor genes of thy-1$^+$ dendritic epidermal cells, *Cell* 55:837 (1988).

15. J. F. Elliott, E. P. Rock, P. A. Patten, M. M. Davis, and Y.-H. Chien, The adult T-cell receptor δ-chain is diverse and distinct from that of fetal thymocytes, *Nature* 331:627 (1988).

16. R. D. Garman, P. J. Doherty, and D. H. Raulet, Diversity, rearrangement and expression of murine T cell gamma genes, *Cell* 45:733 (1986).

17. R. H. Schwartz, Acquisition of immunologic self-tolerance, *Cell* 57:1073 (1989).

18. D. A. Ferrick, P. S. Ohashi, V. Wallace, M. Schilham, and T. W. Mak, Thymic ontogeny and selection of $\alpha\beta$ and $\gamma\delta$ T cells, *Immunology Today* (in press, 1990).

19. M. Bonneville, K. Ito, E. G. Krecko, S. Itohara, D. Kappes, I. Ishida, O. Kanagawa, C. A. Janeway, Jr., and D. B. Murphy, Recognition of a self major histocompatibility complex TL region product by $\gamma\delta$ T-cell receptors, *Proc. Natl. Acad. Sci. USA* 86:5928 (1989).

20. D. Vidovic, M. Roglic, K. McKune, S. Guerdner, C. MacKay, and Z. Dembic, Qa-1 restricted recognition of foreign antigen by a $\gamma\delta$ T-cell hybridoma, *Nature* 340:646 (1989).

21. S. Porcelli, M. B. Brenner, J. L. Greenstein, S. P. Balk, C. Terhorst, and P. A. Bleicher, *Nature* (in press, 1989).

22. D. A. Ferrick, S. R. Sambhara, W. Ballhausen, A. Iwamoto, H. Pircher, C. L. Walker, W. M. Yokoyama, R. G. Miller, and T. W. Mak, T cell function and expression are dramatically altered in T cell receptor Vγ1.1Jγ4Cγ4 transgenic mice, *Cell* 57:483 (1989).

DIVERSITY, DEVELOPMENT, LIGANDS, AND PROBABLE FUNCTIONS

OF γδ T CELLS

Susumu Tonegawa, Anton Berns*, Marc Bonneville, Andrew G. Farr#, Isao Ishida, Kouich Ito, Shigeyoshi Itohara, Charles A. Janeway, Jr.†, Osami Kanagawa§, Ralph Kubo**, Juan J. Lafaille, Donal B. Murphy^, Nobuki Nakanishi, Yohtaro Takagaki, and Sjek Veebeek*

Howard Hughes Medical Institute
Center for Research, Department of Biology
Massachusetts Institute of Technology, Cambridge, Massachusetts

*The Netherlands Cancer Institute
Division of Molecular Genetics and Department of Chemistry
University of Amsterdam, The Netherlands

#Department of Biological Structure
University of Washington, Seattle, Washington

†Howard Hughes Medical Institute
Yale University Medical School, New Haven, Connecticut

§Eli Lilly Research Laboratories
La Jolla, California

**National Jewish Center of Immunology and Respiratory Medicine
Denver, Colorado

^The Wadsworth Center
New York State Department of Health, Albany, New York

INTRODUCTION

The most critical step in the vertebrate immune response is the recognition of antigens by lymphocytes. This task is accomplished by two sets of glycoproteins, immunoglobulins and T cell antigen-receptors (TCRs). The most extraordinary feature of these proteins is their structural variability, much of which originates from the ability of the encoding gene segments to undergo somatic rearrangement.[1] All TCRs were initially thought to be composed of a heterodimeric protein composed of α and β subunits. However, the search for the genes encoding these polypeptides led to the identification of a third rearranging gene[2,3] which was later shown to code for one of the two subunits of another heterodimeric, TCR γδ.[4-6]

Despite the striking similarities in the overall structure of their genes and polypeptide chains, TCR γδ and the T cells which express it are significantly different from their αβ counterparts. For instance, γδ T cells are detected in both the thymus[5,7,8] and peripheral lymphoid organs[4,9] in relatively low numbers (< 5% of T cells), but predominate (50-100%) within epithelia, such as epidermis[10,11] and small intestine.[12,13] In contrast to most αβ T

Mechanisms of Lymphocyte Activation and Immune Regulation III
Edited by S. Gupta *et al.*, Plenum Press, New York, 1991

53

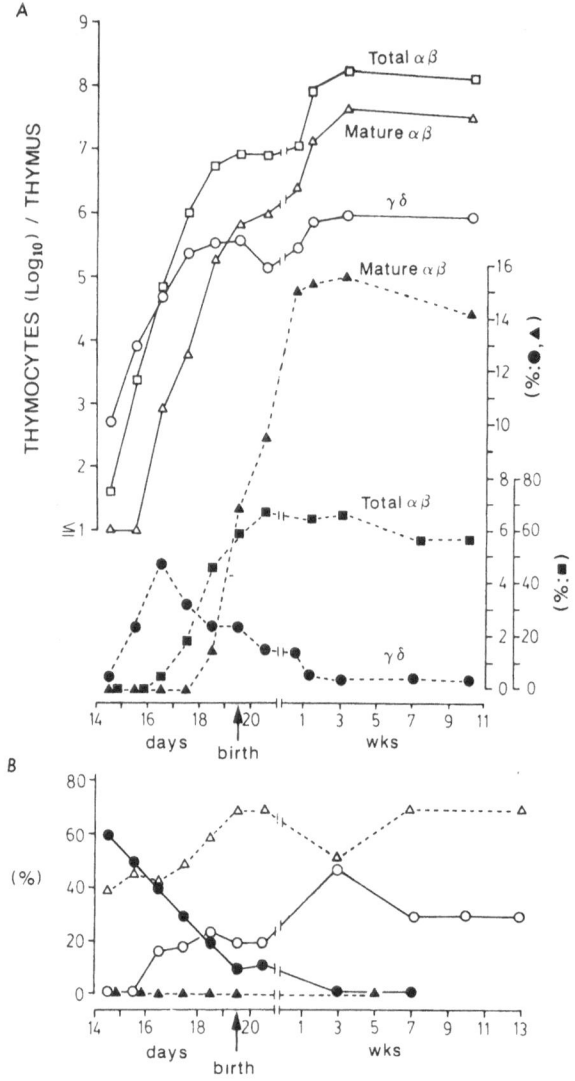

Figure 1. Ontogeny of TCR $\gamma\delta$, $\alpha\beta$ thymocytes in C57BL/6 mice. Thymocytes of mice at different ages were stained with anti $\gamma\delta$ (3A10) or anti $\alpha\beta$ (H57-597) biotin conjugates followed by a streptavidin PE conjugate together with an anti-CD3 (2C11) FITC conjugate. The samples were analyzed with FACScan (Becton-Dickinson) using FACSCAN software.

cells, the majority of δ T cells in the thymus and spleen do not express either CD4 or CD8.[4,5,7-9] Neither the specificity of $\gamma\delta$ TCR recognition nor $\gamma\delta$ T cell function in immune defense is understood. In this paper we summarize results of our recent studies on the diversity, development, and specificity of $\gamma\delta$ T cells, and discuss possible functions of these cells.

RESULTS

Anti Mouse δ TCR mAbs

To study the nature of the TCR δ and the role of γδ T cells we generated three hamster anti-mouse monoclonal antibodies (mAbs) directed against the native receptor: mAb 3A10 specific for a Cγ constant region determinant, mAb 8D6 specific for $V\gamma_4$- and $V\gamma_5$-encoded γδ TCR, and mAb 5C10 specific for a KN6 TCR idiotope.[14].

Appearance of αβ and γδ T Cells in the Developing and Mature Thymus and in Peripheral Lymphoid Organs

Using appropriate mAbs, we determined the number of thymocytes bearing γδ TCR (abbreviated hereafter as γδ-thymocytes) or αβ TCR (αβ-thymocytes) as a function of age (Figure 1). γδ-thymocytes were detected at the 14.5th day of gestation (E14.5), increased until E17.5, remained at this level until birth, dipped, and then gradually increased to the adult level of one million cells per thymus. In relative terms, only 0.4 to 0.6% of total thymocytes at E14.5 are γδ-thymocytes, increasing rapidly to its highest value (5%) at E16.5, then gradually decreasing through embryonic life and the first postnatal week until it reaches a stationary adult level of 0.3 to 0.5% at about ten days after birth. αβ-thymocytes are rare at E14.5 but outnumber γδ-thymocytes after E16.5. By E18.5 they are the dominant thymocyte population. Two color analysis of thymocytes showed that αβ and γδ TCR double positive cells must be extremely rare, or do not exist at all. Thus a mechanism appears to exist that restricts the surface expression of TCRs to only one of the two types.

We also determined the frequency of γδ T cells in peripheral lymphoid organs (i.e., spleen, lymph node, and blood) as well as their CD4, CD8 phenotype. In adult mice (7-13 wk old), the fraction of γδ T cells among CD3[+] T cells is no more than 3% in any of the three sites studied. The majority of these γδ T cells do not express CD4 nor CD8.

Table 1. Distribution of γδT Cells in Mouse

Organs	Presence of γδT cells in mouse		Contact with epithelial cells	No. of cells per animal	Major Vγ gene segments used	Diversity of γδ TCR
	normal	nude				
Digestive system						
Tongue	+		+		6	−
Esophagus	+/−		+			
Stomach	+		+			
Small intestine	+	+	+	1×10^6	7	+
Large intestine	+	+	+	1×10^5	7, 4	
Liver	−					
Pancreas	−					
Reproductive system						
Ovary	+/−					
Uterus	+	−	+	1×10^5	6	−
Vagina	+	−	+	1×10^5	6	−
Testis	−					
Epididymis	+/−		−			
Seminal vesicle					7, 5	
Urinary system						
Kidney	−					
Bladder	+	−	+			
Others						
Skin	+	−	+	5×10^6	5	−
Lung					7, 4, 6	
Brain	−					
Heart	−					
Lymphoid organs						
Thymus	+			1×10^6	4, 7	+
Spleen	+	−		8×10^5	4, 7, 6	+
Lymph node	+	−		1×10^5	4, 7, 6	+
Blood	+			6×10^4	4, 7, 6	+

Intrathymic Distribution of γδ T Cells Alters Drastically During the First Postnatal Week

We analyzed sections of fetal and adult thymuses using immunohistochemical technique, and observed a striking co-localization of $γδ^+$ thymocytes and medullary epithelial cells during late fetal and neonatal periods of development. In contrast, γδ thymocytes were scattered throughout cortical and medullary areas of the thymus and most concentrated in the subcapsular areas of the thymus in the thymuses of adult mice. As shown below, the pattern of γδ TCR expression in the thymus is limited in diversity and developmentally ordered. Thus these histochemical results indicate that changes in receptor repertoire correlates with the changes in the intrathymic distribution of γδ thymocytes.

γδ T Cells are Present in Many Organs Containing Epithelia

Since γδ T cells are relatively scant in peripheral lymphoid organs, we searched for these cells in sections of various organs and tissues (Table 1). The γδ T cells were most abundant in small intestine (i-IEL), moderately abundant in tongue (t-IEL), stomach, and large intestine, and very scarce in esophagus. Most of these γδ T cells were associated with respective epithelial cells (Table 1). γδ T cells were also found to be associated with the epithelia of the uterus, vagina, and bladder. In contrast, no γδ T cells were found in liver, pancreas, kidney, nor brain. Unlike the human and chicken epidermis[15,16] the mouse epidermis is the site of a major γδ T cell subset (DEC or s-IEL).

For comparison, adjacent sections were stained with the mAb against αβ TCR. These T cells were preferentially localized in the lamina propia of the intestines, in the dermis of the skin, in the connective tissue of the vagina, and in the endometrium and myometrium of the uterus. Clearly, the majority of $CD3^+$ intraepithelial lymphocytes are γδ T cells.

We extended our immunohistological analyses to 8 week old athymic nude mice (Table 1), but found no γδ T cells except in the small intestines, suggesting a thymus dependency of most of these T cells.

γδ T Cells of Different Peripheral Sites Utilize Different Vγ Gene Segments to Encode Their TCR

Previous studies have demonstrated that s-IEL (DEC) and i-IEL use distinct Vγ gene segments, V_5[17] and V_7,[18] respectively, to encode the γ subunits of their TCR. To determine whether there was a similar preferential usage of specific Vγ gene segments by γδ T cells present in other sites, we analyzed the DNA extracted from crude lymphocyte preparations from these sites by the PCR (polymerase chain reaction) technique[19] followed by the Southern blot method (Table 1). Among the results obtained, the most conspicuous were the vagina, uterus, and tongue in which $Vγ_6$-$Vγ_1$ rearrangement was abundant. Cloning and sequencing of the PCR products indicated that most (12/14) $Vγ_6$-$Jγ_1$ clones isolated from vaginae and uteri contained in-frame junctions with an identical nucleotide sequence.[20] These results strongly suggest that most γδ T cells in female reproductive organs use the single $V_6 J_1 C_1$ γ gene to encode the γ subunits of their TCR. Interestingly, γδ T cells associated with the tongue (t-IEL) also use the same γ gene.[20]

γδ T Cells Associated With Some Epithelial Organs are Primarily Derived From Fetal Thymocytes and Carry an Entirely Homogeneous γδ TCR

It was previously shown that the utilization of γ and δ V gene segments is developmentally ordered in thymocytes.[14,21,22] γδ thymocytes from early fetuses (e.g., E15) preferentially express TCR encoded by $V_5 J_1 C_1$ γ and $V_1 D_2 J_2 C$ δ genes.[21,22] We found that these TCR are entirely homogeneous[23] and that they are identical to the γδ TCR expressed on s-IEL.[17] Thus s-IEL probably originate from this first wave of fetal γδ thymocytes. At late fetal and newborn stages most γδ thymocytes bear TCR encoded by another pair of γ and δ genes, $V_6 J_1 C_1$ γ and $V_1 D_2 J_2 C$ δ[22] which are also structurally homogeneous.[23] Since the nucleotide sequence of this $V_6 J_1 C_1$ γ gene is identical to the sequence of the sole in-frame joined $V_6 J_1 C_1$ γ gene present in r-IEL and t-IEL,[20] we suspected that these cells are derived from the second wave of fetal γδ thymocytes. If this were the case the δ chains of these T cell populations also should be identical. Indeed, this turned out to be the case.[20] We conclude from these results that r-IEL are derived from the late fetal wave of thymocytes.

γδ TCR Expressed in an Adult Thymus and Peripheral Lymphoid Organs are Diverse

Unlike the early and late fetal thymocytes or s-IEL, and t-IEL, adult γδ thymocytes and γδ T cells of peripheral lymphoid organs utilize various combinations of Vγ and Vδ gene segments to encode their TCR. Furthermore abundant V-(D)-J junctional diversity occurs among the TCR of these adult γδ T cells.[24-26] Taken together these results indicate that in addition to the difference in the gene segements utilized for surface expression, there is a drastic fetal vs adult shift in the extent of the junctional diversity in the γδ TCR. Since i-IEL preferentially use $V\gamma_7$ gene segment[18] which is rarely utilized by fetal thymocytes but is used by some adult thymocytes, and since the TCR genes assembled in adult thymocytes and i-IEL show similar extensive deletions and insertions in the junctions,[18,25] some adult thymocytes may home to intestinal epithelia.

Recognition of a Self Histocompatibility Complex TL Region Product by γδ T Cell-Receptors

In order to understand the function of γδ T cells it is essential to identify the putative ligand of the TCR. For this purpose we prepared a number of γδ T cell hybridomas from fetal and adult thymocytes and screened them for specificity using a growth inhibition assay based on the observation[27] that crosslinking of the αβ TCR, which usually promotes growth of normal T cells, results in growth inhibition of T cell hybridomas. Assuming a similar effect of γδ TCR crosslinking by a ligand, we cocultivated the γδ T cell hybridomas with a variety of irradiated (6,000 rads) test cells and measured the effect of these cells on the incorporation of ^{3}H-thymidine into the DNA of the hybridomas. We identified one γδ T hybridoma, KN6, whose growth was inhibited by syngeneic (C57BL/6) but not allogeneic (BALB/c, CBA/J and AKR/J) spleen cells, thymocytes, peritoneal and macrophages.[28] Abelson transformed B6 T cell line 2052C, and an embryonal carcinoma cell line, PCC3. PCC3 also inhibited KN6 growth. The growth inhibition was not observed with TCR-negative variants of KN6, indicating that it was mediated by the TCR. This conclusion was further supported by the finding that the KN6 growth inhibition was blocked in a dose-dependent fashion by the anti γδ TCR mAb 3A10 as well as by the anti KN6 TCR clonotypic mAb 5C10.[28]

Table 2. The Ligand Recognized by KN6 is Controlled by MHC Genes Telomeric to the Q Region

Strain[a]	Region[b]				Phenotype[c]	Percent Proliferation[d]
	K	D	Q	TL		
A.BY/SnJ	b	b	b	b	+	−75
C57BL/6J (B6)	b	b	b	b	+	−76±12
C57BL/10SnJ (B10)	b	b	b	b	+	−89±6
A.CA/Sn	f	f	f	f	±?	−37±19
B10.M/Sn	f	f	f	f	±?	−36±29
B10.D2/nSnJ	d	d	d	d	−	+30±17
B10.BR/SgSnJ	k	k	k	k	−	−5±12
P/J	p	p	p	p	−	+31±27
B10.G/Sg	q	q	q	q	−	+8±6
B10.RIII(7INS)/SnJ	r	r	r	r	−	+11±2
B10.S/Sg	s	s	s	s	−	+8±23
A/Boy	a	a	a	a	−	+4±8
B6/Boy	b	b	b	b	+	−85±4
A-Tla^a/Boy	a	a	a	b	+	−66±14
B6-$\overline{Tla}^a$/Boy	b	b	b	a	−	−2±13

^{a}Other + strains include: BALB.B/Li, B10.A(2R)SgSnJ, B10.A(R149)-Tla^b/Mrp, B10.A(R410)-Tla^b/Mrp, B10.P(13R)/Sg, B10.SM(70NS)/Sn, and TBR2. Other − strains include: AKR/J, A/WySnJ, BALB/c (what subline??), B6-H-2^k/Boy, B6.K1/Fla, B6.K2/Fla, B10.A/SgSnJ, B10.PL(73NS)/Sn, B10(R297)-$\overline{Tla}^a$/Mrp, B10(R310)-Tla^a/Mrp, C3H/HeJ and MA.My/J. ^{b}Haplotype origin of region. ^{c}+ = strong inhibition. ±=possible intermediate inhibition. Results variable and not consistent. − = no inhibition. ^{d}Data shown are with a single clone, KN6-7. All experiments done at least twice, except with A.BY, which also tested + with the original KN6 line.

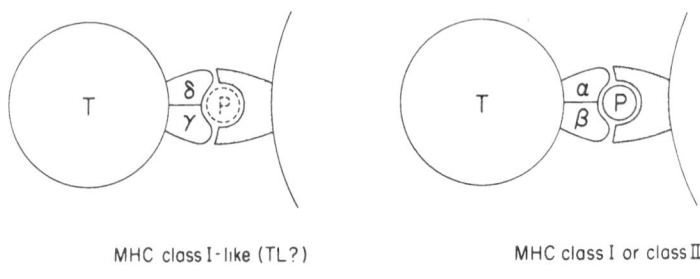

MHC class I-like (TL?)   MHC class I or class II

Figure 2. Comparison of the proposed recognition by a $\gamma\delta$ T cell
with the established recognition by a $\alpha\beta$ T cell.

In order to map the gene, the KN6 hybridoma was screened for reactivity with a panel of spleen cells from various mouse strains (Table 2). Linkage to the MHC was demonstrated by results with congenic strains which differ only at H-2. Furthermore, the analysis of recombinant strains showed that the gene controlling the KN6 ligand is located in or distal to the TL region (Table 2).

Fate of Self-Reactive $\gamma\delta$ Thymocytes in the KN6 Transgenic Mice

Recent studies suggest that T cell tolerance is accomplished, at least in part, by intrathymic deletion of self-reactive $\alpha\beta$ T cells.[29-31] We have begun studies to determine whether or not a similar deletion of self-reactive $\gamma\delta$ T cells occurs. Since the KN6 $\gamma\delta$ TCR recognize syngeneic (TL^b) thymocytes and splenocytes we analyzed thymocytes and splenic T cells from the KN6 $\gamma\delta$ transgenic mice[32] expressing MHC^b or MHC$^{k/d}$ haplotype. Nearly all $\gamma\delta$ T cells of these mice bore TCR encoded by the transgenes regardless of the MHC haplotype of the host (i.e., MHC^b or MHC$^{k/d}$), and the number of $\gamma\delta$ T cells present in either thymus or spleen was not lower in the MHC^b mice than in MHC$^{k/d}$ mice. Thus there was no obvious deletion of self-reactive $\gamma\delta$ T cells. $\gamma\delta$ thymocytes isolated from MHC^b hosts but not MHC$^{k/d}$ hosts propagate rapidly *in vitro* in the presence of "Con A supernatant" or recombinant IL-2. $\gamma\delta$ thymocytes derived from MHC$^{k/d}$ transgenic mice require, in addition to the lymphokine, stimulation by TL^b ligand-bearing cells. These results indicate that the $\gamma\delta$ thymocytes recognize and respond to the putative TL^b ligand *in vitro* in the thymus of the KN6 transgenic mice.

DISCUSSION

Despite the considerable information which has accumulated during the past few years on $\gamma\delta$ TCR and $\gamma\delta$ T cells, the biological functions of these cells remain to be determined. Below, we wish to present some thoughts on this issue.

There is considerable evidence indicating that $\gamma\delta$ T cells recognize molecules similar to but distinct from the MHC class I molecules that are utilized to present antigen-derived peptides to $\alpha\beta$ T cells (this paper).[28,33] While there also are reports that some $\gamma\delta$ T cells recognize allogeneic K- or D-region class I molecules or the I-region class II molecules, these specificities may have resulted from crossreactivity, for strong selective pressure was applied to detect them.[33,34] Indeed, the recent finding (Y. Utsunomiya, S. Itohara, R. Kubo, S. Tonegawa, and O. Kanagawa, unpublished results) that the subunit structure of CD8 molecules expressed on activated i-IEL and splenic $\gamma\delta$ T cells is different from that expressed on $\alpha\beta$ T cells supports the hypothesis that $\gamma\delta$ and $\alpha\beta$ T cells recognize distinct sets of class I (or class I-like) molecules (Figure 2).

If $\gamma\delta$ T cells generally recognize distinct class I (or class I-like) molecules, the next question is whether or not these class I molecules present peptides to $\gamma\delta$ TCR, as do the MHC class I K, D, and L molecules to $\alpha\beta$ TCR. While no direct evidence is available at the moment, the structural similarities both between the $\alpha\beta$ and $\gamma\delta$ TCR and between the class I (or class I-like) molecules recognized by the two types of TCR suggest that the ligands for $\gamma\delta$ TCR generally include antigen-derived peptides (Figure 2).

Are $\gamma\delta$ T cells, then, specific to a variety of antigens as are $\alpha\beta$ T cells, and the repertoire of the antigens recognized by $\gamma\delta$ TCR more or less identical to that of the antigens recognized by $\alpha\beta$ TCR? Here it would be useful to consider separately the two types of $\gamma\delta$ T cell subsets--one with a high degree of TCR diversity such as i-IEL and splenic $\gamma\delta$ T cells, and the other with invariant TCR such as s-IEL and r-IEL. In the former case it is likely that the antigens (hence the peptides derived from them) recognized by $\gamma\delta$ T cells are structurally diverse. However, the antigenic repertoire for $\gamma\delta$ T cells may be different from that for $\alpha\beta$ cells. Perhaps the class I molecules recognized by $\gamma\delta$ TCR (hereafter abbreviated as RE$\gamma\delta$, antigen-restriction elements for $\gamma\delta$ TCR) have evolved to present a special set of antigens to the immune system. The selectivity of RE$\gamma\delta$ for a particular set of antigens might be explained by postulating a special intracellular pathway of peptide loading for RE$\gamma\delta$ and/or common structural features of the presented peptides. A hint to which set of proteins may be presented efficiently by RE$\gamma\delta$ has come from recent experiments that have shown the recognition of mycobacterial heat shock-like proteins by some $\gamma\delta$ T cells.[35-38] It may be that $\gamma\delta$T cells with diverse TCR are primarily directed to a variety of mycobacteria and parasitical protozoa known to constitutively produce structurally related but distinct heat shock-like proteins. The effector function of the $\gamma\delta$ T cells is unknown, but one might speculate that these cells play a role in the initiation or regulation of defense reactions by lymphokine secretion or elimination of undesirable cells such as persistent antigen-presenting cells.

In contrast to a $\gamma\delta$ T cell subset capable of recognizing structurally diverse ligands, a $\gamma\delta$ T cell subset with an invariable TCR must recognize an equally invariable ligand. Despite this important difference in the diversity of the ligands, it is likely that the basic composition (i.e., a peptide-RE complex) of the ligand recognized by the invariant $\gamma\delta$ TCR is the same as that of the ligands recognized by diverse $\gamma\delta$ TCR. Assuming that this is indeed the case, the key issue is again the origin of the peptide. Since it does not make sense for an organism to secure an entire subset of T cells localized in an organ for the protection against just a single foreign antigen, we should consider an alternative role and/or mechanism for the undiversified $\gamma\delta$ T cells. Perhaps these $\gamma\delta$ T cells recognize a tissue-specific host antigen whose synthesis may be induced in the epithelial cells by a variety of unfavorable stimuli such as viral infections, toxic chemicals, radiation, heat shock, and malignancy.[39] One candidate of such a host antigen is a stress protein whose synthesis may be induced in a tissue-specific fashion. Since at least some of these proteins are known to be structurally related to mycobacterial heat shock-like proteins, it is conceivable that the peptides derived from them could also effectively bind to and be presumed by RE$\gamma\delta$. Since the postulated antigen is induced only under specific conditions, autoreactivity of these $\gamma\delta$ T cells may not have any hazardous consequences but rather play an important regulatory and/or effector role. The peculiar specificity of $\gamma\delta$ T cells envisaged in our model is somewhat reminiscent of the specificity of another class of lymphocytes, those called CD5 B cells. These B cells also appear to preferentially recognize common bacterial antigens as well as self antigens.

ACKNOWLEDGEMENTS

We wish to thank Amy DeCloux, Edvins Krecko, Carol Browne, and Sang Hsu for excellent technical assistance and Elly Basel for typing the manuscript. We also thank Werner Haas for useful comments. This work was supported by grants from Howard Hughes Medical Institute, N.I.H., American Cancer Society and Ajinomoto Co., Ltd., to S. T. as well as the NIH CORE P30-CA14051. We also thank FUJI Photo Film Co. for the use of the BA100-Bioimage Analyzer.

REFERENCES

1. S. Tonegawa, Somatic generation of antibody diversity, *Nature* 302:575 (1983).
2. H. Saito, D. M. Kranz, Y. Takagaki, A. C. Hayday, H. N. Eisen, and S. Tonegawa,

Complete primary structure of a heterodimeric T-cell receptor deduced from cDNA sequences, *Nature* 309:757 (1984).

3. A. C. Hayday, H. Saito, S. D. Gillies, D. M. Kranz, G. Tanigawa, H. N. Eisen, and S. Tonegawa, Structure organisation and somatic rearrangement of T cell "gamma genes", *Cell* 40:259 (1985).

4. M. B. Brenner, J. McLean, D. P. Dialynas, J. L. Strominger, J. A. Smith, F. L. Owen, J. G. Seidman, S. Ip, F. Rosen, and M. S. Krangel, Identification of a putative second T-cell receptor, *Nature* 322:145 (1986).

5. I. Bank, R. A. DePinho, M. B. Brenner, J. Cassimeris, F. W. Alt, and L. Chess, A functional T3 molecule associated with a novel heterodimer on the surface of immature human thymocytes, *Nature* 322:179 (1986).

6. A. Weiss, M. Newton, and D. Crommie, Expression of T3 in association with a molecule distinct from the T-cell antigen receptor heterodimer, *Proc. Natl. Acad. Sci. USA* 83:6998 (1986).

7. A. M. Lew, D. M. Pardoll, W. L. Maloy, B. J. Fowlkes, A. Kruisbeek, S.-F. Cheng, R. N. Germain, J. A. Bluestone, R. H. Schwartz, and J. E. Coligan, Characterization of T cell receptor gamma chain expression in a subset of murine thymocytes, *Science* 234:1401 (1986).

8. N. Nakanishi, K. Maeda, K. Ito, M. Heller, and S. Tonegawa, T γ protein is expressed on murine fetal thymocytes as a disulphide-linked heterodimer, *Nature* 325:720 (1987).

9. K. Maeda, N. Nakanishi, B. L. Rogers, W. G. Haser, K. Shitara, H. Yoshida, Y. Takagaki, A. A. Augustin, and S. Tonegawa, Expression of Tγ gene products on the surface of peripheral T cells and T cell blasts generated by allogeneic mixed lymphocyte reaction, *Proc. Natl. Acad. Sci. USA* 84:6536 (1987).

10. G. Stingl, K. C. Gunter, E. Tschachler, H. Yamada, R. I. Lechler, W. M. Yokoyama, G. Steiner, R. N. Germain, and E. M. Shevach, Thy1[+] dendritic epidermal cells belong to the T cell lineage, *Proc. Natl. Acad. Sci. USA* 84:2430 (1987).

11. W. A. Kuziel, A. Takashima, M. Bonhadi, P. R. Bergstresser, J. P. Allison, R. E. Tigelaar, and P. W. Tucker, Regulation of T-cell receptor γ-chain RNA expression in murine Thy-1[+] dendritic epidermal cells, *Nature* 328:263 (1987).

12. M. Bonneville, C. A. Janeway, Jr., K. Ito, W. Haser, I. Ishida, N. Nakanishi, and S. Tonegawa, Intestinal intraepithelial lymphocytes are a distinct set of γδ T cells, *Nature* 336:479 (1988).

13. T. Goodman and L. Lefrancois, Expression of the γδ T-cell receptor on intestinal CD8[+] intraepithelial lymphocytes, *Nature* 333:855 (1988).

14. S. Itohara, N. Nakanishi, O. Kanagawa, R. Kubo, and S. Tonegawa, Monoclonal antibodies specific to native murine T cell receptor γδ: Analysis of γδ T cells in thymic ontogeny and peripheral lymphoid organs, *Proc. Natl. Acad. Sci. USA* 86:5094 (1989).

15. R. P. Bucy, C.-L. H. Chen, J. Cihak, U. Löch, and M. D. Cooper, Avian T cells expressing γδ receptors localize in the splenic sinusoids and the intestinal epithelium, *J. Immunol.* 141:2200 (1988).

16. V. Groh, S. Porcelli, M. Fabbi, L. L. Lanier, L. J. Picker, T. Anderson, R. A. Warnke, A. K. Bhan, J. L. Strominger, and M. B. Brenner, Human lymphocytes bearing T cell receptor γδ are phenotypically diverse and evenly distributed throughout the lymphoid system, *J. Exp. Med.* 169:1277 (1989).

17. D. M. Asarnow, W. A. Kuziel, M. Bonyhadi, R. E. Tigelaar, P. W. Tucker, and J. P. Allison, Limited diversity of γδ antigen receptor genes of Thy-1[+] dendritic epidermal cells, *Cell* 55:837 (1988).

18. Y. Takagaki, A. DeCloux, M. Bonneville, and S. Tonegawa, γδ T cell receptors on murine intestinal intra-epithelial lymphocytes are highly diverse, *Nature* 339:712 (1989).

19. R. K. Saiki, D. H. Gelfand, S. Stoffel, S. J. Scharf, R. Higuchi, G. T. Horn, K. B. Mullis, and H. A. Erlich, Primer-directed enzymatic amplification of DNA with a thermostable DNA polymerase, *Science* 239:487 (1988).

20. S. Itohara, A. G. Farr, J. J. Lafaille, M. Bonneville, Y. Takagaki, and S. Tonegawa, A γδ thymocyte subset with homogenous T cell receptors homes to certain mucosal epithelia, *Nature* 343:754 (1989).

21. W. L. Havran and J. P. Allison, Developmentally ordered appearance of thymocytes expressing different T-cell antigen receptors, *Nature* 335:443 (1988).

22. K. Ito, M. Bonneville, Y. Takagaki, N. Nakanishi, O. Kanagawa, E. Krecko, and S. Tonegawa, Different $\gamma\delta$ T-cell receptors are expressed on thymocytes at different stages of development, *Proc. Natl. Acad. Sci. USA* 86:631 (1989).

23. J. J. Lafaille, A. DeCloux, M. Bonneville, Y. Takagaki, and S. Tonegawa, Junctional sequences of T cell receptor γ and δ genes: Implications for $\gamma\delta$ T cell lineages and novel intermediates of V-(D)-J joinings, *Cell* 59:859 (1989).

24. A. J. Korman, S. M. Galesic, D. Spencer, A. M. Kruisbeek, and D. Raulet, Predominant variable region gene usage by γ/δ T cell receptor-bearing cells in the adult thymus, *J. Exp. Med.* 168:1021 (1988).

25. Y. Takagaki, N. Nakanishi, I. Ishida, O. Kanagawa, and S. Tonegawa, T cell receptor γ and δ genes preferentially utilized by adult thymocytes for the surface expression, *J. Immunol.* 142:2112 (1989).

26. M. J. Lacy, L. K. McNeil, M. E. Roth, and D. M. Kranz, T-cell receptor δ-chain diversity in peripheral lymphocytes, *Proc. Natl. Acad. Sci. USA* 86:1023 (1989).

27. J. D. Ashwell, P. E. Cunningham, P. D. Noguchi, and D. Hernandez, Cell growth cycle block of T cell hybridomas upon activation with antigen, *J. Exp. Med.* 165:173 (1987).

28. M. Bonneville, K. Ito, E. G. Krecko, S. Itohara, D. Kappes, I. Ishida, O. Kanagawa, C. A. Janeway, Jr., D. B. Murphy, and S. Tonegawa, Recognition of a self MHC TL region product by $\gamma\delta$ T cell receptors, *Proc. Natl. Acad. Sci. USA* 86:5928 (1989).

29. J. W. Kappler, N. Roehm, and P. Marrack, T cell tolerance by clonal elimination in the thymus, *Cell* 49:273 (1987).

30. P. Kisielow, H. Bluthmann, U. D. Staerz, M. Steinmetz, and H. von Boehmer, Tolerance in T-cell-receptor transgenic mice involves deletion of nonmature CD4$^+$8$^+$ thymocytes, *Nature* 333:742 (1988).

31. H. R. MacDonald, R. Schneider, R. K. Lees, R. C. Howe, H. Acha-Orbea, H. Festenstein, R. M. Zinkernagel, and H. Hengartner, T cell receptor Vβ use predicts reactivity and tolerance to MLs^a-encoded antigens, *Nature* 323:40 (1988).

32. I. Ishida, S. Verbeek, M. Bonneville, A. Berns, and S. Tonegawa, T cell receptor $\gamma\delta$ transgenic mice suggest a role of a γ gene silencer in the generation of $\alpha\beta$ T cells, *Proc. Natl. Acad. Sci. USA* 87:3067-3071 (1990).

33. J. A. Bluestone, R. Q. Cron, M. Cotteman, B. A. Houlden, and L. A. Matis, Structure and specificity of TCR $\gamma\delta$ on major histocompatibility complex antigen specific CD3$^+$, CD4$^-$, CD8$^-$ T lymphocytes, *J. Exp. Med.* 168:1989 (1988).

34. L. A. Matis, R. Cron, and J. A. Bluestone, Major histocompatibility complex-linked specificity of $\gamma\delta$ receptor-bearing T lymphocytes, *Nature* 33:262 1987.

35. J. Holoshitz, F. Koning, J. E. Coligan, J. deBruyn, and S. Strober, Isolation of CD4$^-$ CD8$^-$ mycobacteria-reactive T lymphocyte clones from rheumatoid arthritis synovial fluid, *Nature* 339:226 (1989).

36. E. M. Janis, S. H. E. Kaufmann, R. H. Schwartz, and D. M. Pardoll, Activation of $\gamma\delta$ T cells in the primary immune response to Mycobacterium tuberculosis, *Science* 244:713 (1989).

37. R. L. Modlin, C. Pirmez, F. M. Hofman, V. Torigian, K. Uyemura, T. H. Rea, B. R. Bloom, and M. B. Brenner, Lymphocytes bearing antigen-specific $\gamma\delta$ T-cell receptors accumulate in human infectious disease lesions, *Nature* 339:544 (1989).

38. R. L. O'Brien, M. P. Happ, A. Dallas, E. Palmer, R. Kubo, and W. Born, Stimulation of a major subset of lymphocytes expressing T cell receptor $\gamma\delta$ by an antigen derived from Mycobacterium tuberculosis, *Cell* 57:667 (1989).

39. C. A. Janeway, Jr., B. Jones, and A. Hayday, Specificity and function of T cells bearing γ/δ receptors, *Immunol. Today* 9:73 (1988).

γδ T CELLS IN MURINE EPITHELIA: ORIGIN, REPERTOIRE, AND FUNCTION

James P. Allison, David M. Asarnow, Mark Bonyhadi, Amy Carbone, Wendy L. Havran, Diphankar Nandi, and Janelle Noble

Division of Immunology, Department of Molecular and Cell Biology, and Cancer Laboratory, University of California, Berkeley, CA

INTRODUCTION

The overwhelming majority of T cells in the thymus and peripheral lymphoid organs express an antigen receptor (TCR) comprised of αβ heterodimers in association with the CD3 complex. The repertoire of αβ T cells is shaped for the recognition of peptide antigens in the context of MHC Class I or Class II antigens by cells bearing CD8 or CD4, respectively. A minor population of T cells in the lymphoid organs express TCR composed of δ heterodimers, and these cells typically do not express CD4, although a small fraction expresses CD8.[1] In certain epithelial tissues of the mouse, however, γδ T cells comprise the bulk, if not the entirety, of the T cells. There is a tight correlation between γ gene usage and epithelial localization as well as differences in the potential TCR repertoire in the different epithelia, two features which raise important questions as to the origin, homing, selection, and function of these cells.

CELLS BEARING γδ TCR APPEAR AS A SERIES OF WAVES DURING ONTOGENY

It has been previously demonstrated that Vγ and Vδ gene segments rearrange and are expressed in an ordered manner during development, with Vγ3, Vγ4, and Vγ1 transcripts dominating the early fetal thymus.[2,3] (Note: the nomenclature used here for the γ gene is that of Garman et al.).[2] Using a panel of monoclonal antibodies, we established that the earliest CD3[+] cells to emerge in the fetal thymus expressed Vγ3, and that this first wave was followed by additional waves of γδ and finally, αβ T cells.[4] We have recently extended and refined this analysis using additional antibodies (prepared by us or generously supplied by Jeffrey Bluestone, Leo LeFrancois, and Ralph Kubo) as well as sequence determination of cDNA amplified by the polymerase chain reaction (PCR). As can be seen in Figure 1, the wave of Vγ3[+] cells is followed successively by Vγ4[+], Vγ2[+], and finally by cells expressing other γδ TCR or αβ TCR. Similar results have been reported by Itohara et al.[5] In addition to establishing that cells bearing different γδ V segments arise in a highly ordered manner during the early stages of fetal development, these findings were of interest because the products of the earliest two waves, Vγ3 and Vγ4, were not detectable in adult lymphoid tissues.[2] This suggested that the function of cells bearing these TCR might be limited to the fetal period, or that the cells emigrated to non-lymphoid sites following their emergence from the thymus.

Mechanisms of Lymphocyte Activation and Immune Regulation III
Edited by S. Gupta *et al.*, Plenum Press, New York, 1991

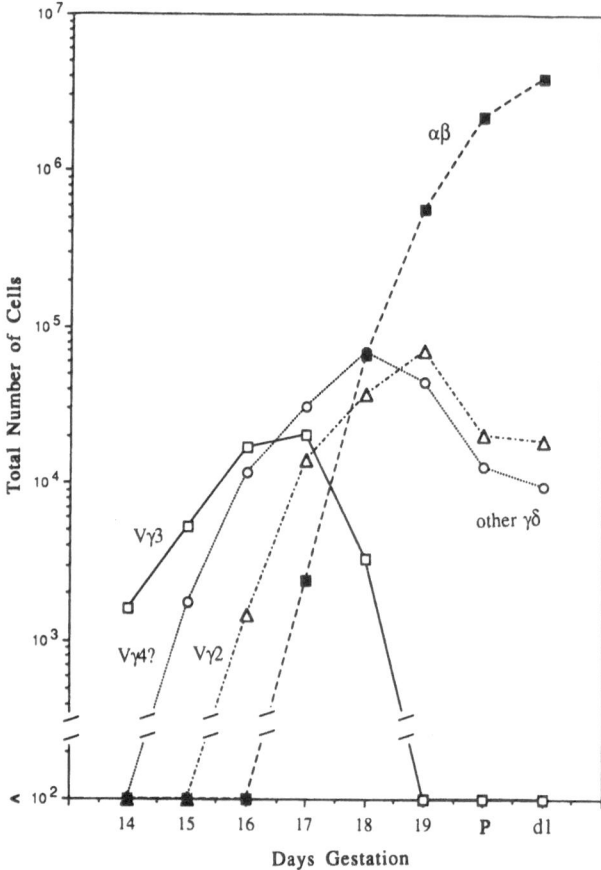

Fig. 1.   Appearance of thymocytes bearing different
TCR during ontogeny.

V GENE USAGE BY γδ T CELLS IN DIFFERENT EPITHELIA

In an analysis of TCR gene usage by a panel of Thy$^-$1$^+$ dendritic epidermal cell (DEC) clones isolated from the skin of adult mice, we noted that each expressed high levels of Vγ3 mRNA.[6] We prepared a monoclonal antibody to one of these, and demonstrated that it detected the Vγ3 gene product.[7] In an analysis of several strains of adult mice, we found that Vγ3$^+$ cells were undetectable in the thymus or spleen, but that essentially all of the Thy-1$^+$ cells in the skin were Vγ3.[7] In an analysis of several strains of adult mice, we found that Vγ3$^+$ cells were undetectable in the thymus or spleen, but that essentially all of the Thy$^-$1$^+$ cells in the skin were Vγ3.[7] There was no evidence of any significant fraction of T cells expressing other Vγ segments or αβ TCR in the skin. Thus, it appears that mouse skin is exclusively the province of cells bearing Vγ3 TCR. As will be discussed in more detail below, the δ chain of the skin-associated T cells is derived from the Vδ1 gene segment.[8,9]

In an examination of intraepithelial lymphocytes (IEL) of the intestine, LeFrancois and his colleagues demonstrated that a large fraction of the T cells expressed γδ TCR.[10] Using oligonucleotide primers specific for different Vγ and Vδ gene segments to amplify cDNA, we found that IEL employ Vγ5 and Vδ4, Vδ5, and Vδ6 in their TCR, a result in agreement with the findings of Bonneville et al.[11]

We have recently isolated and characterized T cells from the epithelial lining of the vagina and uterus. Both CD4$^+$ and CD8$^+$ αβ$^+$ cells and CD4$^-$8$^-$ γδ cells were obtained.

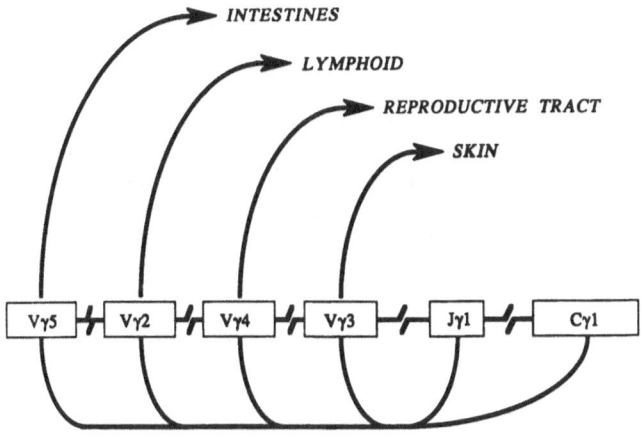

Fig. 2. T cells bearing different Vγ segments seed different epithelial tissues.

PCR and sequence analysis revealed that the γδ cells exclusively employed the Vγ4 and Vδ1 segments in their TCR. These results are consistent with those of Itohara et al.[12]

Together, these findings indicate that there is a strong correlation between Vγ usage and the epithelial tissues with which the γδ T cells home. As shown in Figure 2, these V gene segments are associated with the Jγ1[-]/Cγ1 locus. During ontogeny rearrangement and expression of these V segments appears to begin with the most proximal element 5' to Jγ1, Vγ3, and then proceeds to the more 5' V elements, giving rise successively to cells which may home to the indicated epithelia.

FETAL THYMIC ORIGIN OF EPITHELIAL γδ T CELLS

As will be discussed below, sequence analysis of the rearranged γ and δ genes in the T cells derived from the adult skin and reproductive tract show similarities with rearrangements in the fetal thymus. These features, in addition to the use of fetal specific V segments, include minimal use or absence of non-germline encoded nucleotides, perhaps a result of the low activity of terminal deoxynucleotidyl transferase in the fetal thymus, and the use of a single D element in the δ chains.[3] These observations provide support for the concept that the early fetal waves emigrate to seed the epithelial tissues.

Several lines of direct evidence provide direct confirmation for this concept, at least in the case of the DEC of adult skin. The fact that the skin of athymic nude mice contains normal numbers of Thy-1[+] cells was initially taken as evidence for an extrathymic origin of the cells.[13] However, we[14] and others[15] have shown that the Thy-1[+] cells do not express γδ TCR. Implantation of fetal thymic lobes from normal 14 day embryos or injection of Vγ3[+] cells purified from normal day 16 fetal thymus rapidly give rise to Thy-1[+], Vγ3[+] cells in the adult nude mouse skin,[14] confirming that the fetal thymic cells can seed the skin. In order to determine whether the capacity to generate DEC precursors is limited to the period in fetal development, we treated mice *in utero* by injection of anti-Vγ3 antibody into pregnant females from day 12 to day 19 of gestation, and then analyzed the skin of mice 2, 3, and 4 months after birth. No Vγ3[+] cells were detected in the skin of treated mice, but normal numbers were obtained from the skin of mice treated with control antibody.[2] Taken together, these results indicate that DEC precursors arise in the thymus early in ontogeny, and that the capacity to generate them is lost later in development. This demonstrates that at least the first wave of TCR[+] cells to arise during development, those expressing Vγ3, emigrate to seed an epithelial tissue, the epidermis.

THE ROLE OF THE TCR IN TISSUE SPECIFIC HOMING

The correlation between Vγ usage and tissue association raises the possibility that the γδ TCR is itself responsible for tissue-specific homing. In order to examine this possibility, we constructed mice transgenic for DEC-derived Vγ3 and Vδ1 genes.[16] In these mice all the γδ T cells expressed the Vγ3 transgene as a result of allelic exclusion. These mice contained normal numbers of the gut-associated IEL, but these now expressed Vγ3 rather than Vγ5. Thus it appears that the TCR itself is not involved in tissue localization, but that there may be distinct homing receptors whose expression is coordinately regulated with TCR Vγ gene usage.

TCR REPERTOIRE OF EPITHELIUM-ASSOCIATED γδ CELLS

The restricted V gene usage in the γδ cells of the different epithelia indicated that the TCR repertoire would be restricted to events occurring at the junctions during rearrangement of the gene segments. Largely due to the fact that the δ locus has two D elements, both of which can be read in all three frames and both of which are often used together, junctional diversity can make an extremely large contribution to the γδ TCR repertoire.[17] We therefore examined the junctional sequences of the γ and δ genes in each of the three epithelial T cell populations (unpublished results).[8,9] The results revealed striking features of each of the populations.

In skin derived DEC, the vast majority of in frame rearrangements involving Vγ3 and Jγ1 were identical at the junctions. Similarly, the in frame rearrangements involving Vδ1 almost invariably involved Dδ2 and Jδ2, and contained identical junctions. In both the γ and δ genes, non-germline encoded nucleotides were absent at the junctions. Thus the DEC repertoire is extremely linked, with canonical γ and δ junctional sequences.[8,9] The Vγ3-Jγ1 and Vδ1-Dδ2-Jδ2 junctional sequences are the same that have been reported for the vast majority of these rearrangements in the fetal thymus (unpublished results).[18] This is consistent with our demonstration of the fetal thymic origin of adult DEC.[14] Analysis of reproductive tract γδ T cells revealed a similar lack of variability in both the γ and δ junctions (unpublished observations). All the V4-Jγ1 junctions were identical, and contained no non-germline encoded nucleotides. Similarly, all the in-frame genes used Vδ1-Dδ2-Jδ2, and were identical at the junctions. Strikingly, the sequence of the δ gene used in the reproductive tract cells was identical to that of the DEC cells. The Vγ4-Jγ1 junctions were identical to those described in the fetal thymus,[18] suggesting that the reproductive tract γδ T cells arose from the second wave of fetal thymic emigrants.

The situation is strikingly different in the γδ IEL of the gut epithelia.[9] The Vγ5-Jγ1 junctions exhibit extensive trimming and insertion of non-germline encoded nucleotides. Productive δ rearrangements were largely limited to three V segments, Vδ4, Vδ5, and Vδ6, but exhibited all the features of generating diversity, including the use of both Dδ1 and Dδ2 and extensive trimming and insertion of non-germline encoded nucleotides. Productive δ rearrangements were largely limited to three V segments, Vδ4, Vδ5, and Vδ6, but exhibited all the features of generating diversity, including the use of both Dδ1 and Dδ2 and extensive trimming and insertion of non-germline elements at each junction. Similar results were obtained by Bonneville et al.[11] Thus, although the IEL repertoire is somewhat restricted in V gene usage, compensation by junctional diversity results in a very large potential repertoire.

FUNCTION OF EPITHELIAL γδ T CELLS

The strikingly different repertoires of γδ cells found in the skin or reproductive tract as compared to the gut IEL suggest that the cells have fundamentally different roles in the immune response. Given that the TCR of the skin and reproductive tract γδ cells are essentially monomorphic, it is unlikely that these cells would be effective in conventional immunological surveillance against the myriad of foreign antigens likely to be encountered in these tissues. We have therefore proposed that these cells function in "trauma signal surveillance" for evidence of cellular damage, rather than for the agent inducing that damage.[8,9] Heat shock proteins, which are induced or increased in response to a variety of cellular insults, including viral infection and transformation, would be likely candidates for target antigens.[19] There is evidence that lymphoid $\gamma\delta$ T cells recognize heat shock proteins.[20]

In order to examine the possibility that DEC might recognize stress-induced antigens in the skin, we examined the ability of a cloned DEC line to recognize a keratinocyte cell line and responded by IL-2 secretion and proliferation. The DEC cells do respond, and this response is enhanced by treatment of the keratinocytes with heat or sodium arsenate, both known to increase expression of heat shock proteins. This response was blocked by Fab fragments of anti-Vγ3 or anti-CD3 antibodies, indicating that the TCR was involved in the response. Hybridomas expressing δ TCR other than the DEC Vγ3/Vδ1 did not respond to keratinocytes, nor did DEC respond to treated splenocytes, macrophages, or fibroblasts (unpublished results). Thus it appears that the DEC do have the capacity to specifically recognize trauma-induced antigens in their neighbors in the skin, keratinocytes. We are currently attempting to identify the target antigen. Trauma signal surveillance may represent a primitive form of T cell immunity which may provide a first line of defense in the epi-thelia of young animals.

ROLE OF SELECTION IN DETERMINATION OF EPITHELIAL $\gamma\delta$ TCR REPERTOIRE

The fact that the junctions of the γ and δ genes of T cells associated with the skin and reproductive tract have canonical junctional sequences raises important questions concerning the role of selection in the shaping of the repertoire of these cells. It may be that elements in the thymus provide a positive selection for TCR containing the canonical junctional sequences observed in these cells. Alternatively, it may be that the state of the recombinase apparatus in the early embryonic thymus restricts the generation of functional rearrangements. For example, if TDT, the enzyme thought to be responsible for insertion of non-germline encoded nucleotides, is not expressed at the time the rearrangements are taking place, the degrees of freedom allowed for functional rearrangements would be greatly limited.

In order to examine the role of selection in the generation of $\gamma\delta$ cells, we used agents known to interfere with selection of positive and negative selection of $\alpha\beta$ T cells. It has been demonstrated, for example, that $\alpha\beta$ cells thymocytes can be deleted at the CD4$^+$8$^+$ stage by anti-CD3 antibodies in organ culture,[21] a process which closely mimics negative selection *in vivo*.[22] However, inclusion of anti-CD3 or anti-Vγ3 in fetal thymic organ culture has no effect on the generation of Vγ3$^+$ cells, or even total $\gamma\delta^+$ cells (unpublished observations). Cyclosporin A has been demonstrated to interfere with the production of mature thymocytes *in vivo*, probably as a result of blockage of signals involved in positive selection.[23] We found that addition of cyclosporin A to fetal thymic organ cultures resulted in a virtually complete loss of $\alpha\beta^+$ cells, but that the appearance of Vγ3$^+$ and total $\gamma\delta^+$ cells was unaffected (unpublished observations). These data suggest that at least the early waves of $\gamma\delta$ thymocytes, including the Vγ3$^+$ DEC precursors, do not pass through a developmental stage where they are susceptible to receptor-mediated negative selection, nor do they require cyclosporine a-susceptible positive signals for their generation. Thus, if these $\gamma\delta$ cells do undergo selection in the thymus, the mechanisms must be different from those involved in the positive and negative selection of $\alpha\beta$ T cells.

SUMMARY

The earliest TCR$^+$ cells to appear during fetal development express products of the γ and δ loci, and emerge as successive waves of cells bearing different Vγ gene products. These appear to emigrate and seed different epithelia. The TCR repertoire of the first two of these waves, Vγ3 and Vγ4, respectively, is extremely restricted. Whether the repertoire of these cells is restricted by selective processes or is shaped by developmental restrictions on rearrangements remains to be determined. These cells may function in surveillance for signals of trauma by recognizing self products induced by cellular stress.

ACKNOWLEDGMENTS

This work was supported by grants from the N. I. H. (J. P. A.) and the Lucille P. Markey Charitable Trust (W. L. H.).

REFERENCES

1. D. H. Raulet, The structure, function, and molecular genetics of the δ T cell receptor, *Ann. Rev. Immunol.* 7:175 (1989).
2. R. D. Garman, P. J. Doherty, and D. H. Raulet, Diversity, rearrangement, and expression of murine T cell γ genes, *Cell* 45:733 (1986).
3. Y. H. Chien, M. Iwashima, D. A. Wettstein, K. B. Kaplan, J. F. Elliott, W. Born, and M. M. Davis, T-cell receptor δ gene rearrangements in early thymocytes, *Nature* 330:722 (1987).
4. W. L. Havran and J. P. Allison, Developmentally ordered appearance of thymocytes expressing different T cell antigen receptors, *Nature* 335:443 (1988).
5. S. Itohara, N. Nakanishi, O. Kanagawa, R. Kubo, and S. Tonegawa, Monoclonal antibodies specific to native murine T-cell receptor γδ: Analysis of γδ T cells during thymic ontogeny and in peripheral lymphoid organs, *Proc. Natl. Acad. Sci. USA* 86:5094 (1989).
6. W. A. Kuziel, A. Takashima, M. Bonyhadi, P. R. Bergstresser, J. P. Allison, R. E. Tigelaar, and P. W. Tucker, Regulation of T-cell receptor γ-chain RNA expression in murine Thy-1$^+$ dendritic epidermal cells, *Nature* 328:263 (1987).
7. W. L. Havran, S. C. Grell, G. Duwe, J. Kimura, A. Wilson, A. M. Kruisbeek, R. L. O'Brien, W. Born, R. E. Tigelaar, and J. P. Allison, Limitied diversity of TCR γ chain expression of murine Thy-1$^+$ dendritic epidermal cells revealed by Vγ3-specific monoclonal antibody, *Proc. Natl. Acad. Sci. USA* 86:4185 (1989).
8. D. M. Asarnow, W. A. Kuziel, M. Bonyhadi, R. E. Tigelaar, P. W. Tucker, and J. P. Allison, Limited diversity of γδ antigen receptor genes of Thy-1$^+$ dendritic epidermal cells, *Cell* 55:837 (1988).
9. D. Asarnow, T. Goodman, L. LeFrancois, and J. P. Allison, Distinct antigen receptor repertoires of two classes of murine epithelium-associated T cells, *Nature* 341:60 (1989).
10. T. Goodman and L. LeFrancois, Expression of the gamma/delta T-cell receptor on intestinal CD8$^+$ intraepithelial lymphocytes, *Nature* 333:855 (1988).
11. M. Bonneville, C. A. Janeway, Jr., K. Ito, W. Haser, I. Ishida, N. Nakanishi, and S. Tonegawa, Intestinal intraepithelial lymphocytes are a distinct set of δ T cells, *Nature* 336:479 (1988).
12. S. Itohara, A. G. Farr, J. J. Lafaille, M. Bonneville, Y. Takagaki, W. Haas, and S. Tonegawa, Homing of a δ thymocyte subset with homogenous T-cell receptors to mucosal epithelia, *Nature* 343:754 (1990).
13. P. R. Bergstresser, S. Sullivan, J. W. Streilein, and R. E. Tigelaar, Origin and function of Thy-1$^+$ dendritic epidermal cells in mice, *J. Invest. Derm.* 85:85 (1985).
14. W. L. Havran and J. P. Allison, Origin of Thy-1$^+$ dendritic epidermal cells of adult mice from fetal thymic precursors, *Nature* 344:68 (1990).
15. J. L. Nixon-Fulton, W. A. Kuziel, B. Santerse, P. R. Bergstresser, P. W. Tucker, and R. E. Tigelaar, Thy-1$^+$ epidermal cells in nude mice are distinct from their counterparts in thymus-bearing mice. A study of morphology, function, and T cell receptor expression, *J. Immunol.* 141:1897 (1988).
16. M. Iwashima, A. Green, M. Bonyhadi, M. M. Davis, J. P. Allison, and Y.-H. Chien, Thymic and post-thymic selection of T cells in δ T cell receptor transgenic mice. Submitted.
17. M. M. Davis and P. J. Bjorkman, T cell antigen receptor genes and T cell recognition, *Nature* 334:395 (1988).
18. J. J. Lafaille, A. DeCloux, M. Bonneville, Y. Takagaki, and S. Tonegawa, Junctional sequences of T cell receptor δ genes: Implications for δ T cell lineages and for a novel intermediate of V-(D)-J joining, *Cell* 59:859 (1990).
19. R. A. Young and T. J. Elliott, Stress proteins, infection, and immune surveillance, *Cell* 59:5 (1989).

20. W. Born, M. P. Happ, A. Dallas, C. Reardon, R. Kubo, T. Shinnick, P. Brennan, and R. O'Brien, Recognition of heat shock proteins and gamma/delta cell function, *Immunol. Today* 11:40 (1990).

21. C. A. Smith, G. T. Williams, R. Kingston, E. J. Jenkinson, and J. J. T. Owen, Antibodies to CD3/T-cell receptor complex induce death by apoptosis in immature T cells in thymic cultures, *Nature* 337:181 (1989).

22. B. J. Fowlkes and D. M. Pardoll, Molecular and cellular events of T cell development, *Adv. Immunol* 44:207 (1989).

23. M. K. Jenkins, R. H. Schwartz, and D. M. Pardoll, Effects of cyclosporine A on T cell development and clonal deletion, *Science* 241:1655 (1988).

ACQUISITION OF MATURE FUNCTIONAL RESPONSIVENESS IN T CELLS:

PROGRAMMING FOR FUNCTION VIA SIGNALING

Ellen V. Rothenberg, Dan Chen, Rochelle A. Diamond, Mariam Dohadwala,
Thomas J. Novak, Patricia M. White, and Julia A. Yang-Snyder

Division of Biology
California Institute of Technology
Pasadena, California

SPECIFICATION OF FUNCTIONAL LINEAGES OF T CELLS: THE PROBLEM

Function and Its Conditionality

Over fifteen years ago it was recognized that peripheral T lymphocytes are heterogeneous in their abilities to carry out particular functions. Distinct functional activities are associated with distinct cell-surface phenotypes.[1] Thus, CD4[+] cells are greatly enriched for the ability to provide growth and differentiation factors for other T and B cells, whereas CD8[+] cells are correspondingly enriched for the ability to kill foreign or pathologically altered target cells. At the molecular level, we now understand that each of these functions reflects the transcriptional activation of particular sets of "response" genes, i.e., those encoding lymphokines and/or cytolytic molecules, when triggered by recognition of a foreign antigen. Thus, different T-cell subsets are defined by the fact that they respond to antigen by induction of different sets of genes. As all of these subsets are derived from common precursors, developing T cells must not only mature but also diverge in their properties in a regulated way. In this paper, we will consider how different programs of transcriptional inducibility become allocated to different sets of cells.

T-cell functional lineage commitment may provide an interesting model for understanding other kinds of committed progenitor states. At least in the helper lineages, genes encoding the potent lymphokines IL-2 and IL-4 are generally silent even in cells whose primary function is to express them. These genes are activated abruptly and reproducibly, however, in response to stimulation with the cell's target antigen. While expression is transient on any occasion, the fact that clonal helper T-cell lines can make essentially identical responses to experimental stimulation over a period of months or years justifies the notion that such cells are committed. The question is how cells can maintain the preferential inducibility of certain loci over many cell generations, often in the absence of any detectable transcription from those genes, without permitting induction of any of the genes that respond to the same stimuli in other T-cell types.

The induction-dependence of the T-cell function has an important consequence, however, which colors the issue of commitment. The ability of a T cell to make a particular response can only be assessed in the context of a particular experimental stimulus, not in an absolute way. A cell may therefore fail to act "as a helper" or "as a killer" for a variety of reasons: the response gene may be inacccessible in chromatin (e.g., heavily methylated), or the cell may lack crucial transcription factors, or the necessary transcription factors may fail to be activated because an inadequate stimulus was used. Under other conditions of stimulation, the cell may respond quite well, raising the question of which condition is physiologically

Mechanisms of Lymphocyte Activation and Immune Regulation III
Edited by S. Gupta *et al.*, Plenum Press, New York, 1991

71

relevant. Since commitment is defined not only by the ability to express one kind of response, but equally by the *inability* to make another kind of response, the signaling require ments for induction of different response genes are a major component of the mechanism underlying the committed state. Thus, if distinct response genes require distinct pathways to be activated, then some crucial differences between T-cell subsets may reside not in the configuration of those response genes in chromatin, but rather in the coupling of common membrane receptors to distinct mediator cascades.

Positive Selection and Functional Maturation

In the thymus, the ultimate stage of phenotypic maturation is the conversion from a $CD4^+CD8^+TcR^{low}$ phenotype to a $CD4^+$ or $CD8^+$ ("single-positive") TcR^{high} phenotype. Cells that have undergone this conversion account for the great majority of effector function in the thymus that is inducible "normally" by TcR ligands. Early studies[2,3] indicated that such thymocytes as a whole display the same association of function with phenotype as peripheral T cells, with the IL-2 producers found in the $CD4^+$ population and the killer-cell precursors (CTL-p) in the $CD8^+$ population. This picture suggests that acquisition of certain kinds of functional competence, exclusion of other kinds of functional competence, and maturation of CD4/CD8 phenotypes all occur at roughly the same time in the thymus.

On closer examination, however, this coordination seems difficult to explain. Recent work has shown that a cell's ultimate phenotypic class ($CD4^+$ or $CD8^+$) is determined intrathymically by the major histocompatibility complex (MHC) restriction specificity of its TcR, and mediated by TcR-MHC contacts at a relatively late stage of differentiation.[4-9] This process rescues a cell from death and is termed positive selection. The TcR interaction which dictates loss of CD4 or CD8 from the $CD4^+CD8^+$ precursor does not appear to induce expression of IL-2, IL-4, or any known killer-specific genes. The properties demonstrably affected by positive selection are not coordinately regulated with response genes. The CD4 or CD8 genes themselves are constitutively expressed at fixed levels in cells that have under-gone positive selection and all their effector and memory cell progeny, in sharp contrast to the inducible and largely transient expression of the response genes. Furthermore, as dis-cussed below and elsewhere, there is significant evidence showing that high-level inducibil-ity of various response genes need not await positive selection. When TcR-independent stimuli are used, IL-2 and killer function can be induced efficiently in TcR^- immature thymocytes.[10-16] Thus, acquisition of these elements of competence can be uncoupled from positive selection. In addition, increasing evidence suggests that the exclusion of inappropriate functions is far from complete at the single-positive thymocyte stage, or even afterwards in fresh peripheral T cells. In thymocytes fractionated without prior glucocorticoid lysis, evidence for IL-2 production from $CD8^+$ cells was noted even in response to TcR ligands,[17] and numerous reports have confirmed the ability of fresh $CD8^+$ peripheral cells to express IL-2.[18-21] This is inconsistent with a tight mechanistic coupling between programming to become a $CD4^+$ or $CD8^+$ cell and programming for helper vs. killer function. The elegant demonstrations of positive selection thus leave unresolved how thymocytes acquire particular aspects of responsiveness and how they block acquisition of others.

Models for Lineage Specification

The result of the divergent programming for function of developing T cells is the emergence of at least three distinct, terminal cell types: the type 1 helper, which makes IL-2 and interferon-γ (IFN-γ); the type 2 helper, which makes IL-4 and IL-5; and the cytolytic T cells, which expresses granzymes A through H, IFN-γ, and perforin. Prior to stimulation, the IL-2 producer, IL-4 producer, and CTL-p do not express these genes, but can do so with predictable kinetics following encounter with their respective target antigens. To understand how these three end states arise we must formulate an explanation for how they differ at the molecular level. Due to the induction-dependence of function and the long proliferative lifespan of T cells, however, several models are possible. We will here discuss three extreme models which stress the centrality of different mechanisms that distinguish the responses of these three terminal cell types.

Figure 1 shows the first model, perhaps the most straightforward and appealing. In this model, the three functional lineages are distinguished only in that different sets of genes are susceptible to activation in these cell types. Activation, in this model, is the *same*

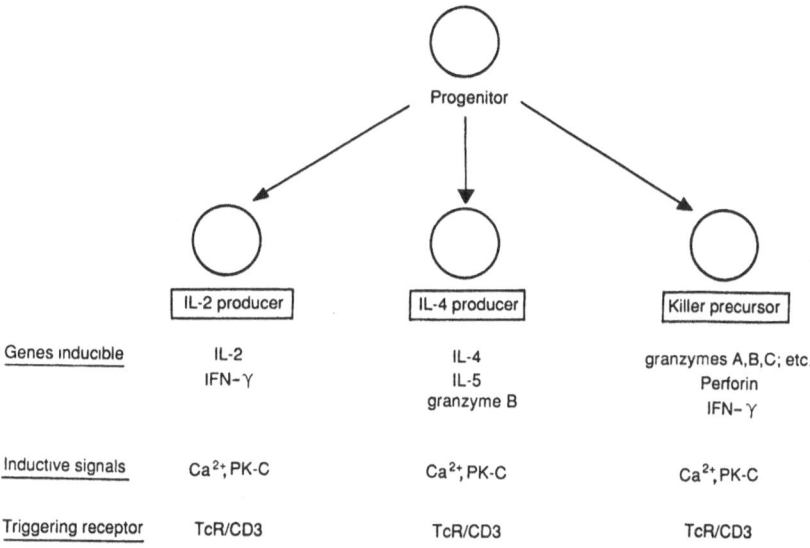

Figure 1. Model 1 for T-cell commitment to different effector lineages. Differential inducibility of response genes (see text).

for all the cells, namely a TcR-generated signal probably operating through the elevation of intracellular Ca^{2+} and the mobilization of protein kinase C. The different subtypes are all terminally differentiated in their various lineages and thus developmentally equivalent. This model implies that the responses of different cell types are inflexibly distinct, and that the essential result of their programming for function is the differential accessibility of their characteristic response genes to a common inductive stimulus. The questions this model raises are, e.g., how the IL-4 gene is kept noninducible in IL-2 producing cells, and vice versa. It implies either that such genes are kept in locked configurations in chromatin, or that they depend on the synthesis of unique transcription factors which themselves are permanently blocked from expression in the inappropriate cell types.

Figure 2 shows the second extreme model which could explain differential function of T-cell subsets. In this model, all response genes in principle are inducible in all T cells, but each requires somewhat different inductive signals to be activated. Such signals would include kinase cascades, cyclic nucleotides, and ions. The functional subsets then would differ in their abilities to generate the appropriate inductive signals for different response genes. If all response genes need not be activated by a common signaling pathway, there are complex possibilities for regulating cell function. Costimulating combinations of factors might be required for some responses. Also, negative regulatory pathways may suppress some responses but not others, as will be discussed further. Note that this model offers a way for hypothetically distinct signals from CD4 and CD8 themselves to participate in directing the response. Furthermore, although the cells are presumed to be developmentally equivalent, their capacity to generate particular signals could vary according to other factors such as cell cycle state. This model is inspired by reports in the literature of possibly complex triggering requirements, such as an accessory cell factor apparently needed for IL-2 expression.[22]

Another relevant case may be the induction of killer function, which lags behind lymphokine production by 2-3 days in murine cells and appears to be highly dependent on particular lymphokines.[23-26] Conceivably the proximate inducing signal for the induction of perforin and certain granzymes is the lymphokine/lymphokine receptor interaction, not antigen recognition *per se*. The observation that antigen *initiates* the conversion of CTL-p to CTL may thus reflect a prerequisite for induction of the relevant lymphokine receptor, if it is not expressed on the resting cell. Alternatively, lymphokine signals may act combinatorially with a delayed effect of TcR stimulation to induce these genes.

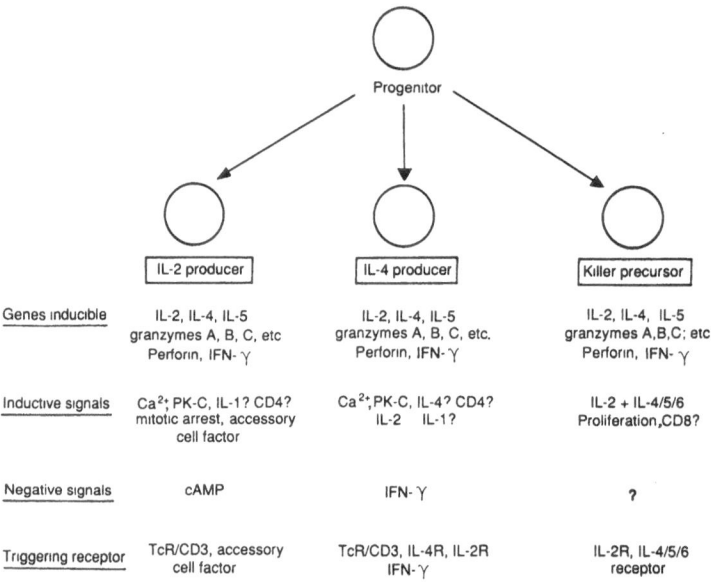

Figure 2. Model 2 for T-cell commitment to different effector lineages. Cells distinguished by signaling cascades (see text).

Overall, an important difference between this model and the first is its prediction that even a committed cell will not respond to all modes of stimulation in the same way. In the extreme, this model predicts that certain conditions of stimulation might be able to blur distinctions between different subsets altogether.

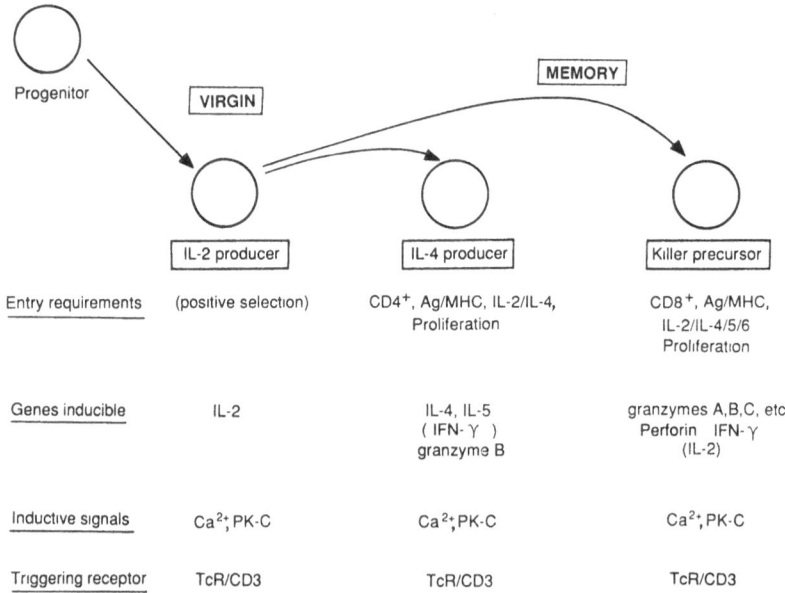

Figure 3. Model 3 for T-cell commitment to different effector lineages. IL-2 producers differentiating irreversibly into CTL or IL-4 producers. This version is an adaptation of Model 1; a similar adaptation of Model 2 is also possible (see text).

The third extreme model, shown in Figure 3, introduces the possibility that different functional subtypes are not in fact developmentally equivalent. This is based on an accumulating body of evidence that IL-2 responses are stronger in virgin cells and weaker in memory cells of both CD4[+] helper and CD8[+] killer types.[27-30] In this view, the classical Type 1 helper cell is not a separate lineage, but an artifact of *in vitro* selection; the unquestionably vital function of IL-2 production is performed by *in vivo* by a developmentally intermediate cell type. Overall, a critical consideration for predicting cell function thus becomes whether or not a T cell has yet fulfilled the requirements for its irreversible "promotion" into one of the two memory cell states.

For simplicity, we have presented this third model as a modification of Model 1, in which memory-type differentiation alters the battery of genes to which a simple activating stimulus has access. It is important to note, however, that this model can also be merged seamlessly with model 2, as further differentiation may be considered as a special case of pre-exposure to a subset of the required co-stimuli. In the following, we will provide evidence consistent with such a fusion.

Model 1 provides a formal explanation of T-cell subsets as committed states. The other two models, however, are based on the proposition that cells can carry out certain differentiated functions without losing the potential to carry out others. In Model 2, there is always the possibility that some signal, encountered someday, might elicit an unexpected response, e.g., the synthesis of perforin and the whole cytolytic apparatus in a cell that more commonly expressed IL-4. In model 3, the essential IL-2-producer subset is destined to be converted into cells of wholly different response capability. These scenarios are not commitment to a functional subset in any familiar sense.

Another developmental concept may thus be of value here, namely that of *specification*.[31] This refers to the condition of a cell which will proceed along a fixed line of development if unperturbed, but which has not yet lost the capacity to change course if exposed to other stimuli. Specification of a lineage commonly precedes commitment proper. The important difference between commitment and specification, in embryological terms, sheds an oblique light on the relationship between T-cell lines and normal T cells *in vivo*. .While T-cell lines in common use have been selected for phenotype stability, many of their normal counterparts may be far more plastic, at least initially. We will now consider evidence for the levels at which subset-specific T-cell functions may be specified.

EVIDENCE FOR DEVELOPMENTAL REGULATION OF SIGNALING

In the past several years, we have explored developmental changes in the capacity of different T-cell populations to express IL-2 on induction, as an accessible model for acquisition of competence in general. In these studies, we took advantage of the findings of Weiss and Stobo[32] and Truneh et al.[33] that the critical TcR-generated signals for IL-2 induction were simply $[Ca^{2+}]_i$ elevation and activation of protein kinase C. This allowed us, in principle, to use Ca^{2+} ionophores such as A23187 and phorbol esters such as 12-0-tetradecanoyl phorbol 13-acetate (TPA) to induce IL-2 in developmentally immature cells irrespective of the presence or absence of a complete TcR complex. Since TcR expression and specific interactions are apparently required for the emergence of mature CD4[+] or CD8[+] phenotypes, we critically examined whether TcR expression was also required for the acquisition of competence to make this CD4-associated response. Our findings have been reported recently.[15,16] They demonstrate that maturation-associated differences in IL-2 inducibility clearly exist, but that an important component of the difference may simply be a developmental shift in the signaling requirement for IL-2 induction.

Briefly, immature thymocytes at a stage lacking TcR expression (CD4[-]CD8[-]CD5[low]TcR/CD3[-]) made little or no IL-2 RNA following stimulation with A23187 and TPA. What IL-2 or IL-2 mRNA could be detected under these conditions was due to rare individual contaminants, rather than to low-level expression throughout the population, as established by *in situ* hybridization. The cells capable of making IL-2 even with this TcR-bypassing stimulus appeared to be drawn exclusively from TcR[+] populations, namely the CD4[+] and CD8[+] "single positives" and the TcR[+] subset of CD4[-]CD8[-] cells. However, the immature cells were prolific IL-2 producers if the stimulus was changed slightly. At least 10% of the TcR⁻ population rapidly accumulated high levels of IL-2 mRNA when stimulated with A23187 and TPA in the presence of IL-1α. Thus, their inability to make IL-2 in response to TcR-type signaling mediators did not reflect intrinsic noninducibility of the IL-2 gene in these cells, but rather a need for an additional signal.[15]

The synergy between IL-1 and the A23187/TPA stimuli itself appeared to be developmentally regulated. Not only was IL-1 unnecessary to stimulate detectable IL-2 production from TcR⁺ subsets of thymocytes and peripheral T cells, but it was even ineffective at providing any enhancement over the levels stimulated by A23187 and TPA. This was ascertained both by RNase protection assays and by titration of secreted IL-2 activity. Even when peripheral T cells were submaximally induced by anti-CD3 antibody and TPA, IL-1 had no adjuvant effect.[16] This is in accord with previous conclusions that IL-1 can be fully replaced by TPA in the stimulation of peripheral T cells. The pathway by which IL-1 complements A23187/TPA in immature cells is therefore likely either to be constitutively activated or to be superseded with further development.

To examine the mechanisms responsible for these effects, we have undertaken the analysis of the molecular targets of IL-1 costimulation. As a point of departure, we have taken advantage of a tumor-cell system in which IL-1 shows synergy with A23187 and TPA. The EL4.E1 thymoma line is not the equivalent of any normal cell type, as it can make IL-2 when stimulated with TPA alone.[34] However, addition of A23187 increases this level of expression, and the further addiition of IL-1, which is ineffective alone, causes an addi tional early increase (at 5 hr) in IL-2 RNA expression in these cells. The distinctive kinetics of the IL-1 effect here recall the kinetics of IL-1-dependent IL-2 production in immature thymocytes. Thus, while far from proven, these observations raise the possibility that IL-1 exerts its enhancing effect on EL4.E1 cells through a similar pathway to that of its obligatory costimulating effect on immature thymocytes.

The IL-2 promoter contains at least four inducible elements: two binding sites for AP-1-like factors, one for an NF-κB-like factor, and an NFAT-1 binding site.[35-38] Using oligo nucleotides derived from these sequences in the murine IL-2 gene, we have assayed nuclear extracts of variously stimulated EL4.E1 cells for the corresponding factors. The results, which will be presented elsewhere,[39] are summarized here in Table 1.

While the EL4.E1 cells constitutively express NFAT-1 activity and activity specific for binding the upstream AP-1 site, expression of the NF-κB-like factor and of factors binding the downstream AP-1 site are highly induction-dependent. IL-1 alone induces neither, but enhances the ability of A23187 and TPA to induce binding to the downstream AP-1 site. Furthermore, in combination with A23187 and TPA it also significantly increases the NF-κB-like binding activity over that observed with A23187/TPA alone. By contrast, the activity of NFAT-1 appears unchanged whether the cells are stimulated with any one, or all, of these compounds. These results cannot identify the cause of the IL-1 *requirement* in immature cells, since EL4.E1 cells do not share this requirement; but they nevertheless suggest that cells which are sensitive to IL-1 costimulation may react by coupling the IL-1 receptor to a pathway leading to higher expression both of NF-κB and of the AP-1 variant that specifically binds the downstream site. The mediators in this pathway are as yet unknown; we argue elsewhere that they are unlikely to be the same as those mobilized by increased levels of cAMP.[40] It will be interesting to determine whether the immature cell response may reflect the hyperinduction of these two factors in the *absence*, say, of NFAT-1.

To summarize, while the molecular mechanisms are still under study, it seems clear that cells of different developmental stages generate different signals in response to exogenous stimuli, so that they differ in the stimuli needed for transcriptional induction of the IL-2 gene. Perceived competence to exercise this effector function is therefore a function of the stimulus used to elicit it. During the initial appearance of IL-2 producer effector function, the phosphoinositide pathway mimicking agents A23187 and TPA can be insufficient to induce IL-2 expression, even in cells intrinsically competent to make this response. This is a result more in accord with Model 2 than with Model 1.

We have previously reported an example of the converse, where A23187 and TPA may

Table 1. Modulation of IL-2 Expression by Differential Induction of DNA-Binding Proteins.

Stimulation Conditions	DNA Binding Activities in EL4.E1 Cells[a]			
	NF-AT[b]	"NF-κB"	"AP-1$_D$"[d]	"AP-1$_P$"[e]
None	±/+	-	+	-
TPA	+	+/++	+	+
IL-1	+	-	+	-
TPA/A23187	+	++	+	++
TPA/A23187/IL-1	+	+++	+	+++
TPA/A23187/Forskolin	+	+/++	+	+++

[a]Nuclear extracts were made by the methods of Dignam et al.[50] or Crabtree[51] (personal communication, G. Crabtree), at 3-4 hr of stimulation under the indicated conditions. Results in the table are those obtained with both types of extracts. For binding assays, 2.5 μg of nuclear protein were incubated with ^{35}S-labeled or ^{32}P-labeled double-stranded oligonucleotides (0.1 ng-0.2 ng per reaction in different experiments) in the presence of 0.25 μg of poly(DI·dC) (for "NF-κB", AP-1$_D$", and "AP-1$_P$") or 2.5 μg of poly(dA·dT) (NF-AT binding). Complexes were separated from free oligonucleotides on 8% acrylamide nondenaturing gels. The numbers of +'s gives a semiquantitative measure of band intensity, in which differences between levels are at least twofold. In cases where different results are obtained in different experiments, both values are given, separated by a slash.
[b]The NF-AT oligonucleotide spanned residues from -260 to -289 in the 5' flanking region of the murine IL-2 gene.[48] The constitutive expression of a factor binding this sequence was verified in three sets of extracts from our line of EL4.E1 cells. A factor with similar mobility and indistinguishable binding specificity was found to be highly inducible, however, in Jurkat cells.[39]
[c]The "NF-κ-B" oligonucleotide spans from -192 to -211. This oligonucleotide can be associated with at least two discrete complexes. The table shows only data for the slower-migrating complex, which is more specific by the criterion of homologous fragment competition (D. Chen and E. V. Rothenberg, unpublished). This may not be identical with the NF-κB complexes described for other genes.
[d]"AP-1$_D$", so called because it includes the site TTCAGTCAG, extends from -178 to -195 and overlaps the 3' end of the "NF-κB" oligonucleotide. It is bound by a complex of distinct mobility and binding specificity from that which binds the "AP-1$_P$" site, although both can be competed by a canonical SV40-derived AP-1 binding site.
[e]"AP-1$_P$", which includes the sequence AGAGTCA, extends from -143 to -161.

be excessive as stimuli for IL-2 production.[41] This is in the case of splenic CD4$^+$ and CD8$^+$ T cells, which exhibit marked differences in IL-2 mRNA expression and protein secretion when stimulated via the TcR. When the same cells are activated with A23187 and TPA, both express IL-2 at high and virtually indistinguishable levels. The cells thus differ in their coupling of TcR signaling to IL-2 induction, more than in their intrinsic IL-2 inducibility *per se*. The inducibility of the IL-2 gene in splenic CD8$^+$ cells itself violates a key prediction of Model 1, namely that the characteristic response genes of one subset are not inducible in other subsets. More generally, the sensitivity of cellular responses to different modes of stimulation are more consistent with Model 2 than with Model 1.

ORIGIN OF IL-4 PRODUCERS AND CTL: NONCOORDINATE SHIFTS IN GENE INDUCIBILITY

The ability to express IL-2 is readily demonstrable in fresh cell populations as described above, but the abilities to kill and to secrete IL-4 are not. Each of the latter two types of function is dramatically amplified by antigen priming and incubation with lymphokines for several days, whether *in vivo* or *in vitro*, as discussed above. This experience provides the basis for Model 3, in which the T_H2 and CTL effector types are proposed to be derived from a precursor which is neither, and instead can express IL-2. *A priori*, the precursors of T_H2 and CTL might be functionally silent. Many investigators have provided evidence, however, for dual-function producers of IL-2 and IL-4, or dual-function clones that secrete their own IL-2 and kill.[27,28,42,43] The common observation that such cells lose the ability to make IL-2 on further passage supports the interpretation that these are transitional forms. Examination of CTL-specific gene expression provides further evidence for the detailed mechanisms proposed to operate in Model 3.

The genes used by cytolytic T cells for their functions fall into two functional groups. One is a set of moderately toxic lymphokines, largely shared with T_H1-type cell lines, and apparently regulated similarly in both the "helpers" and the "killers". The second group encodes the constituents of the cytolytic granules that are uniquely characteristic of mature CTL's and their TcR⁻ functional counterparts, the natural killer cells. These genes are not detectably expressed in fresh splenic T cells, but show constitutive activity in established

Table 2. Noncoordinate Regulation of CTL-Specific Genes

Cell Type	Stimulation	mRNA expressed for[a]				
		IL-2	IL-4	granzyme B	granzyme C	perforin
CTLL-2	none	-	-	+++	++	++
	A23187/TPA	-	-	+++	++	++
MTL2.8.2	none	-	-	+++	++	++
	A23187/TPA	-	-	+++	++	++
EL4.E1	none	-	-	-	-	-
	A23187/TPA	+++	++	-	-	-
A.E7	none	-[b]	-[b]	++	(ND)	+
	A23187/TPA	++[b]	-[b]	+++	(ND)	++
D10.G4.1	none	-	-	++	-	-
	A23187/TPA	-	+++	++	-	-
CDC-25	none	-	-	+	-	-
	A23187/TPA	-	+++	++	-	-
HT-2	none	-	-	++	(ND)	+
	A23187/TPA	-	+	++	(ND)	+
Normal splenocytes	none	-	-	-	-	-
	A23187/TPA (24 hr)	++	±	+	-	+
CD4⁻CD8⁻ thymocytes	none (19 hr)	-	-	-	-	(ND)
	A23187/TPA	(+)[c]	±	+	-	(ND)
	A23187/TPA/ IL-1 (19 hr)	(++)[c]	±	++	-	(ND)

[a]Cytoplasmic RNA analyzed by Northern blot (5 μg/lane), using actin mRNA level for normalization. -, undetectable, ± to +++, semiquantitative estimate of level of expression of a given RNA species relative to the range observed in other cell types.
[b]See reference 22.
[c]At 5 hr, these cells express IL-2 well, but at 19 hr, IL-2 levels have dropped.[15]

CTL lines.[24,44-47] Among this class are the genes encoding granzymes B and C, and perforin. None of these genes are expressed detectably in EL4.E1 cells. We have therefore considered whether these "killer-specific" genes constitute a coordinately regulated gene battery and what conditions suffice for their induction in fresh T cells.

The results of these studies are summarized in Table 2. Two general points emerge. First, it is clear that neither granzyme B nor perforin is completely CTL-specific. Both mRNAs are readily detectable in at least some helper T-cell lines including $T_H 1$ and $T_H 2$. Actual killing activity has not been reported for these lines, so that their expression of granzyme and perforin genes is unlikely to reflect commitment to a specialized dual helper-killer lineage. Instead, it appears to indicate that the gene expression programs of helper cells can routinely overlap with those of killers.

Second, the CTL-specific genes are not coordinately regulated. All the helper lines examined expressed granzyme B mRNA (A.E7, D10.G4.1, CDC-25, and HT-2), but only two expressed mRNA for perforin (A.E7 and HT-2). Furthermore, whereas activated splenocytes expressed both granzyme B and perforin, their level of granzyme C mRNA was below the threshold of detection, at least an order of magnitude lower relative to actin than in established killer lines or in an IL-2 dependent large granular lymphocyte line (T. J. N. and M. D., unpublished results). This is consistent with other reports, in which granzyme C only becomes expressed at high levels in murine splenocytes in response to lymphokines, after several days of CTL-p stimulation.[26]

These data thus segregate CTL-specific gene expression into components which are either inducible by different signals or inducible in different developmental states. The inducibility of perforin and granzyme B in committed helper cell lines is inconsistent with a simple form of Model 1, as is the apparent inaccessibility to induction of granzyme C in most fresh splenic $CD8^+$ cells. On the other hand, both Model 2 and Model 3 may explain these results. In particular, the irreversible differentiation component of Model 3 may be an important feature of CTL induction. Since all the CTL genes, once activated, are expressed constitutively, even the onset of granzyme B RNA expression defines a new developmental state. Yet the results with the A.E7 and HT-2 helper lines show that induction of granzyme B and perforin mRNAs is probably not sufficient to confer cytolytic activity on a cell. Additional events, including the upregulation of granzyme C, appear to be necessary to complete the CTL differentiation program over the time course of several days.[23-26] We have been unsuccessful thus far in accelerating the induction of granzyme C by any combination of IL-2 and IL-4 with TcR ligands (M. D. and E. V. R., unpublished data). Thus, even if the inductive signal for the late events is a lymphokine rather than a TcR ligand, it seems unlikely that the $CD8^+$ cell is competent to respond to it until initially primed to a novel state. Thus, the inducible IL-2 producer and the penultimate pre-CTL are not developmentally equivalent.

If cells initially expressing IL-2 give rise to cells with other kinds of functional competence, how do they lose inducibility of IL-2? This question is posed most acutely with respect to cells which become IL-4 producers, because perhaps unlike key CTL genes, but like IL-2, the IL-4 gene can be induced directly by signaling pathways coupled to the TcR. In committed IL-4 producers, signals would appear to be generated which in principle could induce IL-2 as well as IL-4 expression, if the IL-2 locus were accessible in chromatin.

Transient transfection studies using the IL-2 promoter to drive a reporter gene in fact suggest that the signaling environment within an IL-4 producer need not be permissive for IL-2 inducibility.[48] A series of IL-2 promoter-chloramphenicol acetyl transferase (CAT) constructs which were readily inducible in EL4.E1 and Jurkat cells could not be induced by A23187 and TPA in the $T_H 2$ lines D10.G4.1 or HT-2, nor in the IL-4 inducible pre-mast cell line 32Dcl5. This result specifically establishes a block to IL-2 expression in the signaling environment, because the exogenous IL-2 regulatory DNA is presumably neither methylated nor sequestered in chromatin, and yet fails to respond to inductive treatments.

An example of a mechanism which could selectively restrict inducibility of IL-2 is provided by the effects of combining A23187 and TPA stimuli with cAMP increases.[49] Elevation of intracellular cAMP levels by any of a variety of routes sharply inhibits the ability of A23187 and TPA to induce IL-2 expression. Thus, the cAMP pathway can abort IL-2 induction even at a level distal to the generation of Ca^{2+} and protein kinase C signals. By contrast, cAMP agonists have little or no effect on the induction of IL-4 under similar conditions. The difference is not a function of sensitivity or resistance to cAMP at the cellular level, for at least one case exists, that of EL4.E1 cells, where the same cells can express both lymphokines. In these cells, elevation of cAMP blocks induction of IL-2 at the same time that IL-4 induction proceeds normally. This case suggests the plausibility of using amplified cAMP generation as a developmental switch in T_H2 cell differentiation. While this would not account for the acquisition of IL-4 inducibility by IL-2 producers, any subsequent trends toward increasingly efficient generation of cAMP itself, of A-kinase activity or of A-kinase substrates, could indeed play a role in silencing the earlier effector function.

SUMMARY

The results discussed here provide strong evidence that different T-cell effector gene programs are activated by different signals, and that in several cases their responses to the same exogenous stimuli shift during the development and antigen responses of the cells. T-cell responses are thus conditional and plastic at the individual cell level. In the formalism of the introductory section, the results support elements of Models 2 and 3, and suggest a fusion between them as differentiation is explained in terms of alteration in the relative strengths of different intracellular signaling pathways.

Returning to an initial question, how are different functional capabilities assigned nonrandomly to cells with different antigen recognition specificities? This question has not been answered, but it can be reformulated. If all virgin T cells can transiently make IL-2, then we must ask what features of cell biology explain the preferential preservation of IL-2 inducibility in $CD4^+$ cells as opposed to $CD8^+$ cells. If the capacity to induce IL-4 expression is not acquired in the thymus, then we may ask whether the initial opening of this locus depends on a CD4-transmitted signal. Similarly, the CD8 molecule itself might participate in inducing the initial differentiation events that render CTL-p inducible for granzyme C and perforin. This would be in accord with a large literature showing that CD8 engagement is much more important in the initial induction of CTL activity than in the exercise of function by pre-primed CTL effectors. The subtext of each of these "questions", however, is that intrathymic events may not directly affect the genes used by terminal effectors for function at all. They may instead bias a cell's complement of triggering receptors, thus rendering it differentially sensitive to particular signals generated during antigen reception.

This view is extreme, and will probably turn out to be an overstatement. But it does inspire a unique set of investigations into the basis of T-cell function. It lends urgency to the question of whether $CD4^+$ and $CD8^+$ cells differ in their G proteins, kinases, or inducible proto-oncogenes. If they do, we can then ask whether such differences themselves arise in the periphery, or whether they can be traced back to thymocytes fresh from positive selection—or before.

REFERENCES

1. J. Sprent and S. R. Webb, Function and specificity of T-cell subsets in the mouse, *Adv. Immunol.* 41:39 (1987).
2. R. Ceredig, A. L. Glasebrook, and H. R. MacDonald, Phenotypic and functional properties of murine thymocytes. I. Precursors of CTLs and interleukin-2 producing cells are all contained within a subpopulation of "mature" thymocytes as analyzed by monoclonal antibodies and flow microfluorometry, *J. Exp. Med.* 155:358 (1982).
3. R. Ceredig, D. P. Dialynas, F. W. Fitch, and H. R. MacDonald, Precursors of T-cell growth factor producing cells in the thymus: Ontogeny, frequency, and quantitative recovery in a subpopulation of phenotypically mature thymocytes defined by monoclonal antibody GK-1.5, *J. Exp. Med.* 158:1654 (1983).
4. H. S. Teh, P. Kisielow, B. Scott, H. Kishi, Y. Uematsu, H. Bluthmann, and H. von Boehmer, Thymic major histocompatibility complex antigens and the $\alpha\beta$ T-cell receptor determine the CD4/CD8 phenotype of T cells, *Nature* 335:229 (1988).

5. W. C. Sha, C. A. Nelson, R. D. Newberry, D. M. Kranz, J. H. Russell, and D. Y. Loh, Positive and negative selection of an antigen receptor on T cells in transgenic mice, *Nature* 336:73 (1988).

6. B. Scott, H. Blüthmann, H. S. Teh, and H. von Boehmer, The generation of mature T cells requires interaction of the $\alpha\beta$ T-cell receptor with major histocompatibility antigens, *Nature* 338:591 (1989).

7. L. J. Berg, A. M. Pullen, B. Fazekas de St. Groth, D. Mathis, C. Benoist, and M. M. Davis, Antigen/MHC-specific T cells are preferentially exported from the thymus in the presence of their MHC ligand, *Cell* 58:1035 (1989).]

8. H. von Boehmer, H. S. Teh, and P. Kisielow, The thymus selects the useful, neglects the useless and destroys the harmful, *Immunol. Today* 10:57 (1989).

9. E. V. Rothenberg, Death and transfiguration of cortical thymocytes: A reconsideration, *Immunol. Today* 11:116 (1990).

10. J. P. Lugo, S. N. Krishnan, R. D. Sailor, and E. V. Rothenberg, Early precursor thymocytes can produce interleukin 2 upon stimulation with calcium ionophore and phorbol ester, *Proc. Natl. Acad. Sci. USA* 83:1862 (1986).

11. R. Palacios and H. von Boehmer, Requirements for growth of immature thymocytes from fetal and adult mice *in vitro*, *Eur. J. Immunol.* 16:12 (1986).

12. R. C. Howe, J. W. Lowenthal, and H. R. MacDonald, Role of interleukin 1 in early T-cell development: Lyt-2⁻L3T4⁻ thymocytes bind and respond *in vitro* to recombinant IL-1, *J. Immunol.* 137:3195 (1986).

13. K. L. McGuire and E. V. Rothenberg, Inducibility of interleukin-2 RNA expression in individual mature and immature T lymphocytes, *EMBO J.* 7:939 (1987).

14. R. C. Howe and H. R. MacDonald, Heterogeneity of immature (Lyt-2⁻/L3T4⁻) thymocytes. Identification of four major phenotypically distinct subsets differing in cell cycle status and *in vitro* activation requirements, *J. Immunol.* 140:1047 (1988).

15. E. V. Rothenberg, R. A. Diamond, K. A. Pepper, and J. A. Yang, IL-2 gene inducibility in T cells before T-cell receptor expression. Changes in signaling pathways and gene expression requirements during intrathymic maturation. *J. Immunol.* 144:1614 (1990).

16. E. V. Rothenberg, R. A. Diamond, T. J. Novak, K. A. Pepper, and J. A. Yang, Mechanisms of effector lineage commitment in T-lymphocyte development, *Devel. Biol.*, *UCLA Symp. in Molec. & Cell Biol.* 125:225 (1990).

17. B. Caplan and E. V. Rothenberg, High-level secretion of IL-2 by a subset of proliferating thymic lymphoblasts, *J. Immunol.* 133:1983 (1984).

18. K. Heeg, C. Steeg, J. Schmitt, and H. Wagner, Frequency analysis of class I MHC-reactive Lyt2⁺ and class II MHC-reactive L3T4⁺ IL2-secreting lymphocytes, *J. Immunol.* 138:4121 (1987).

19. D. E. Kehn, L. B. Lachmann, and P. D. Greenberg, Lyt-2⁺ cells. Requirements for concanavalin A-induced proliferation and interleukin 2 production, *J. Immunol.* 139:2880 (1987).

20. T. Mizuochi, D. J. McLean, and A. Singer, IL-2 as a co-factor for lymphokine-secreting CD8⁺ murine T cells, *J. Immunol.* 141:1571 (1988).

21. Y. Samstag, F. Emmrich, and T. Staehelin, Activation of human T lymphocytes: Differential effects of CD3- and CD8-mediated signals, *Proc. Natl. Acad. Sci. USA* 85:9689 (1988).

22. D. L. Mueller, M. K. Jenkins, and R. H. Schwartz, Clonal expansion versus functional clonal activation: A costimulatory signaling pathway determines the outcome of T-cell antigen receptor occupancy, *Ann. Rev. Immunol.* 7:445 (1989).

23. H. Wagner, K. Heeg, and C. Hardt, Multiple signals required in cytolytic T-cell responses, *Prog. Immunol.* VI:386 (1986).

24. J. A. Garcia-Sanz, G. Plaetinck, F. Velotti, D. Masson, J. Tschopp, and H. R. MacDonald, Perforin is present only in normal activated Lyt2⁺ T lymphocytes and not in L3T4⁺ cells, but the serine protease granzyme A is made by both subsets, *EMBO J.* 6:933 (1987).

25. J. D. Pfeifer, D. T. McKenzie, S. L. Swain, and R. W. Dutton, B-cell stimulatory factor 1 (interleukin 4) is sufficient for the proliferation and differentiation of lectin-stimulated cytolytic T-lymphocyte precursors, *J. Exp. Med.* 166:1464 (1987).

26. J. W. L. Hooton, C. L. Miller, C. D. Helgason, R. C. Bleackley, S. Gillis, and

V. Paetkau, Development of precytotoxic T cells in cyclosporine-suppressed mixed lymphocyte reactions, *J. Immunol.* 144:816 (1990).

27. H. von Boehmer, P. Kisielow, W. Leiserson, and W. Haas, Lyt-2⁻ T-cell independent functions of Lyt-2⁺ cells stimulated with antigen or ConA, *J. Immunol.* 133:59 (1984).

28. G. D. Powers, A. K. Abbas, and R. A. Miller, Frequencies of IL-2 and IL-4-secreting T cells in naive and antigen-stimulated lymphocyte populations, *J. Immunol.* 140:3352 (1988).

29. K. Hayakawa and R. R. Hardy, Murine CD4⁺ T-cell subsets defined, *J. Exp. Med.* 168:1825 (1988).

30. K. Hayakawa and R. R. Hardy, Phenotypic and functional alteration of CD4⁺ T cells after antigen stimulation. Resolution of two populations of memory T cells that both secrete interleukin 4, *J. Exp. Med.* 169:2245 (1989).

31. E. H. Davidson, How embryos work: A comparative view of diverse modes of cell fate specification, *Development* 108:365 (1990).

32. A. Weiss and J. D. Stobo, Requirement for the coexpression of T3 and the T-cell antigen receptor on a malignant human T-cell line, *J. Exp. Med.* 160:1284 (1984).

33. A. Truneh, F. Albert, P. Golstein and A. M. Schmitt-Verhulst, Early steps of lymphocyte activation bypassed by synergy between calcium ionophores and phorbol ester, *Nature* 313:318 (1985).

34. J. J. Farrar, J. Fuller-Farrar, P. L. Simon, M. L. Hilfiker, B. M. Stadler, and W. L. Farrar, Thymoma production of T-cell growth factor (interleukin 2), *J. Immunol.* 125:2555 (1980).

35. G. R. Crabtree, Contingent genetic regulatory events in T-lymphocyte activation, *Science* 243:355 (1989).

36. B. Hoyos, D. W. Ballard, E. Böhnlein, M. Siekevitz, and W. C. Greene, Kappa B-specific DNA binding proteins: Role in the regulation of human interleukin-2 gene expression, *Science* 244:457 (1989).

37. K. Muegge, T. M. Williams, J. Kant, M. Karin, R. Chiu, A. Schmidt, U. Siebenlist, H. A. Young, and S. K. Durum, Interleukin-1 costimulatory activity on the interleukin-2 promoter via AP-1, *Science* 246:249 (1989).

38. E. Serfling, R. Barthelmas, I. Pfeuffer, B. Schenk, S. Zarius, R. Swoboda, F. Mercurio, and M. Karin, Ubiquitous and lymphocyte-specific factors are involved in the induction of the mouse interleukin-2 gene in T lymphocytes, *EMBO J.* 8:465 (1989).

39. T. J. Novak, D. Chen, and E. V. Rothenberg, Interleukin 1 synergy with phosphoinositide pathway agonists for induction of interleukin 2 gene expression: molecular basis of costimulation. (1990, submitted for publication).

40. F. Shirakawa, M. Chedid, J. Suttles, B. A. Pollok, and S. B. Mizel, Interleukin-1 and cyclic AMP induce κ immunoglobulin light-chain expression via activation of an NF-κB-like DNA-binding protein, *Mol. Cell. Biol.* 9:959 (1989).

41. K. L. McGuire, J. A. Yang, and E. V. Rothenberg, Influence of activating stimulus on functional phenotype: Interleukin-2 mRNA accumulation differentially induced by ionophore and receptor ligands in subsets of murine T cells, *Proc. Natl. Acad. Sci. USA* 85:6503 (1988).

42. X. Paliard, R. de Waal Malefijt, H. Yssel, D. Blanchard, L. Chretien, J. Abrams, and J. de Vries, Simultaneous production of IL-2, IL-4, and IFN-γ by activated human CD4⁺ and CD8⁺ T-cell clones, *J. Immunol.* 141:849 (1988).

43. G. S. Firestein, W. D. Roeder, J. A. Laxer, K. S. Townsend, C. T. Weaver, J. T. Hom, J. Linton, B. E. Torbett, and A. L. Glasebrook, A new murine T cell subset with an unrestricted cytokine profile, *J. Immunol.* 143:578 (1989).

44. H. K. Gershenfeld and I. L. Weissman, Cloning of a cDNA for a T cell-specific serine protease from a cytotoxic T lymphocyte, *Science* 232:854 (1986).

45. C. L. Lobe, B. B. Finlay, W. Paranchych, V. H. Paetkau, and R. C. Bleackley, Novel serine proteases encoded by two cytotoxic T lymphocyte-specific genes, *Science* 232:858 (1986).

46. J.-F. Brunet, M. Dosseto, F. Denizot, M.-G. Mattei, W. R. Clark, T. M. Maqqi, and P. Golstein, The inducible cytotoxic T-lymphocyte-associated gene transcript CTLA-1 sequence and gene localization to mouse chromosome 14, *Nature* 322:268 (1986).

47. D. Masson and J. Tschopp, A family of serine esterases in lytic granules of cyto-lytic T lymphocytes, *Cell* 49:679 (1987).

48. T. J. Novak, P. M. White, and E. V. Rothenberg, Regulatory anatomy of the murine interleukin-2 gene, *Nucl. Acids Res.* 18 (1990, in press).

49. T. J. Novak and E. V. Rothenberg, cAMP inhibits induction of interleukin 2 but not of interleukin 4 in T cells (1990, submitted for publication).

50. J. D. Dignam, R. M. Lebowitz, and R. G. Roeder, Accurate transcription initiation by RNA polymerase II in a soluble extract from isolated mammalian nuclei, *Nucl. Acids Res.* 11:1475 (1983).

51. J.-P. Shaw, P.-J. Utz, D. B. Durand, J. J. Toole, E. A. Emmel, and G. R. Crabtree, Identification of a putative regulator of early T-cell activation genes, *Science* 241:202 (1988).

MOLECULAR ANALYSIS OF THE INTERACTION OF p56^{lck} WITH THE CD4 AND CD8 ANTIGENS

Christopher E. Rudd,[1,3] Elizabeth K. Barber,[1] Kristine E. Burgess,[1], Julie Y. Hahn,[1] Andreani D. Odysseos,[1,3] Man Sun Sy,[2,3] and Stuart F. Schlossman[1,3]

[1]Division of Tumor Immunology
 Dana-Farber Cancer Institute
[2]Massachusetts General Hospital
[3]Harvard Medical School
 Boston, Massachusetts

ABSTRACT

The CD4 and CD8 antigens on the surface of T cells appear to bind to major histocompatibility complex (MHC) class II and I antigens, respectively. These antigens also synergize with the Ti(TcR)/CD3 complex in the potentiation of T-cell proliferation. Our earlier work demonstrated that the CD4 and CD8 receptors are coupled to a protein-tyrosine kinase termed p56^{lck} from normal and transformed T lymphocytes. The p56^{lck} protein is a member of the src family and its homology with receptor-kinases such as the epidermal growth factor receptor (EGFR) make it an important candidate in signal transduction. In this paper, we show in transfectants that p56^{lck} interacts with the cytoplasmic tail of the CD4 antigen. Murine p56^{lck} can interact across species with the human CD4 receptor. Furthermore, peptide competition studies showed that a specific sequence within the cytoplasmic tail of CD4 interacts with the kinase. Cysteine residues also appear to play key roles in this interaction. Lastly, we show biochemically that the CD4:p56^{lck} complex can physically associate with the ϵ chain of the CD3 complex on HPB-ALL transformed T cells. This interaction may provide a bridge by which events related to ligand binding to Ti(TcR)/CD3 may trigger T cells via the CD4/CD8:p56^{lck} complex.

INTRODUCTION

The CD4 and CD8 antigens are surface glycoproteins expressed on reciprocal subsets of T cells which recognize non-polymorphic regions of the major histocompatibility complex class I and class II antigens, respectively.[1-3] The CD4 antigen also acts as a receptor for the human immunodeficiency virus-1 (HIV-1).[4,5] Both antigens synergize with the T cell receptor complex in the generation of intracellular signals linked to T-cell growth.[6,7] A major issue has been to understand the molecular mechanism by which these antigens regulate T cell function. We recently showed that the CD4 and CD8 antigens are associated with a protein-tyrosine kinase termed p56^{lck} in resting and transformed T lymphocytes.[8-10] The interaction has been verified in mice[11] and humans[12-14] and has provided an important clue as to the molecular basis of CD4 and CD8 function in T cells. The purified CD4:p56^{lck} complex is catalytically active as observed by its ability to autophosphorylate and to phosphorylate members of the CD3 complex.[9] The homology of p56^{lck} with other members of the src family and receptor-kinases such as the epidermal growth factor receptor (EGF) make this an important candidate in signal transduction.[15-18]

Mechanisms of Lymphocyte Activation and Immune Regulation III
Edited by S. Gupta *et al.*, Plenum Press, New York, 1991

The CD4/CD8 associated protein-tyrosine kinase p56[lck] is a member of the src family of protein-tyrosine kinase genes which includes the c-src, c-yes, c-fgr, fyn, hck, tkl and lyn proto-oncogenes.[15-18] Each of these proto-oncogene products is associated with the inner face of the plasma membrane by means of a myristic group. Altered forms of certain kinases (v-src, v-yes, v-fgr) found in various retroviruses such as Rous sarcoma virus have an ability to transform mammalian cells.[15] In the case of p56[lck], the overexpression and enhanced activity of the kinase appears causally linked to the development of the murine thymic lymphoma, LSTRA.[16-17] However, its role in the regulation of normal T-cell growth and proliferation is unclear. T cell stimulation is accompanied by the phosphorylation of the kinase by protein kinase C[19] and by a dramatic decrease in the abundance of lck mRNA transcripts in response to mitogen and phorbol ester.[20] Crosslinking of CD4 is accompanied by a transient increase in p56[lck] activity; however, these events do not lead to the proliferation of T cells.[21,22] In contrast, engagement of the Ti(TcR)/CD3 complex does not appear to induce a detectable increase in p56[lck] activity as assessed in blotting assays using anti-phosphotyrosine antibodies[23] (unpublished data). The target substrates of the CD4/CD8:p56[lck] complex and the question of whether tyrosine phosphorylation is an obligatory step in a cascade of events linked to proliferation is presently the subject of keen interest.

In addition to a potential involvement in T-cell activation, the CD4/CD8:p56[lck] complex serves as a model by which other members of the src family may be found to interact with growth receptors. Emerging data have begun to identify regions on CD4/CD8 and p56[lck] proteins that are involved in the interaction. A comparison of cytoplasmic sequences has revealed only limited homology between CD4 and CD8. The most homologous region is encoded by the CD4 sequence KKTCQCPHRFQKT and the CD8 sequence RRVCKCPRPVV-KS.[10] From this, a possible binding sequence reads K/RK/RXCXCPXXXXKT/S with the K to R and T to S representing conservative substitutions. By contrast, the CD8β subunit does not appear to have any sequence homology to CD4.[24,25]

In the present paper, we extend our studies by showing that p56[lck] interacts with the cytoplasmic tail of the CD4 antigen. Murine p56[lck] can interact across species with the human CD4 receptor. Peptide competition studies further showed that a specific sequence within the cytoplasmic tail of CD4 interacts with kinase. Cysteine residues appear to play key roles in this interaction. Lastly, we show that the CD4:p56[lck] complex can physically associate with the ϵ chain of the CD3 complex on HPB-ALL transformed T cells. This interaction may provide a bridge by which ligand binding to Ti(TcR)/CD3 may generate intracellular signals via the CD4/CD8:p56[lck] complex.

MATERIALS AND METHODS

Monoclonal Antibodies, Antisera and Cells

The production and characterization of the monoclonal antibodies to the CD4 antigen (19Thy5D7;IgG2) and the CD8 antigen (21Thy2D3;IgG1) have been described elsewhere.[1] The anti-PTK antisera was generated against a synthetic peptide of the sequence CKERPEDR-PTFDYLRSVLEDFFT ATEGQYQPQP, as described.[18] The anti-CD3 antibodies sp34 (a kind gift from Dr. Cox Terhorst) and T1A-2 have been reported elsewhere.[26,27] The transformed human and mouse T cell lines REX, MOLT 4, HPB-ALL and 5D5.63, 63.CD4.33, Δ19 (a gift from Dr. B. Sleckman) were cultured in RPMI 1640 with 10% (v/v) fetal calf serum and 1% (w/v) penicillin streptomycin at 37°C in an atmosphere containing 5% CO_2.

Immunoprecipitation and Kinase Analysis

Cells at 50×10^6/ml were solubilized in 1% (v/v) Nonidet P-40 lysis buffer in 20mM Tris-HCl buffer, pH 8.0 containing 150 mM NaCl and 1 mM phenylmethyl sulphonyl fluoride (PMSF) for 30 min at 4°C, as previously described.[8,9] The immunoprecipitates were washed three times with Nonidet P-40 lysis buffer, prior to incubation with 30 μl of 25mM Hepes containing 0.1% (v/v) NP-40, 10mM $MnCl_2$ and 1-10μCi of [γ-^{32}P] ATP (ICN Chemicals Ltd.). Occasionally, cells were pre-incubated with anti-CD4 antibody (10μl of ascites fluid/50 x 10⁶ cells/10ml of RPMI 1640 media) at 4° C for 1 hour prior to washing and solubilization in lysis buffer. After an incubation of 15 minutes to 30 minutes at 25°C,

the reaction mixture was subjected to SDS-PAGE electrophoresis and autoradiography.[28]
Two dimensional gel electrophoresis (NEPHGE/SDS) was conducted as previously described.[8]

Peptide Competition Assays

Immunoprecipitates formed by anti-CD4 antibodies were labeled in the phosphotransferase assay with $[\gamma\text{-}^{32}P]$-ATP as described.[8,9] The reaction mixture (termed supernatant) was then removed from the immune complexes on beads. The pellet was then exposed to a solution of 10mM Hepes, pH 7.4 to 10-20 percent dimethylformamide (DMF) supplemented with various concentrations of peptide. The peptides corresponded to the various sequences: CD4: KKTCQCPHRFQKT; CD2: TLTCEVMNGT and LCK: CKERPEDRPTFDYLRSVLEDFFTA-TEGQYQPQP. After a 10 minute incubation at room temperature, the supernatant was separated from the pellet. The pellet and the supernatant were then supplemented with 3X SDS sample buffer, boiled and subjected to SDS-PAGE electrophoresis.[28]

RESULTS

CD4 and CD8 are Associated with the Protein-Tyrosine Kinase p56^lck

Figure 1A illustrates the finding that the CD4 and CD8 antigens from T cells are physically coupled to the protein-tyrosine kinase p56^lck at 55-60Kd (lanes 2, 3, 6). The 55-60Kd protein was found to co-migrate with p56^lck as directly precipitated by an antisera to the kinase (lanes 4, 8) and was found to undergo autophosphorylation at tyrosine residues

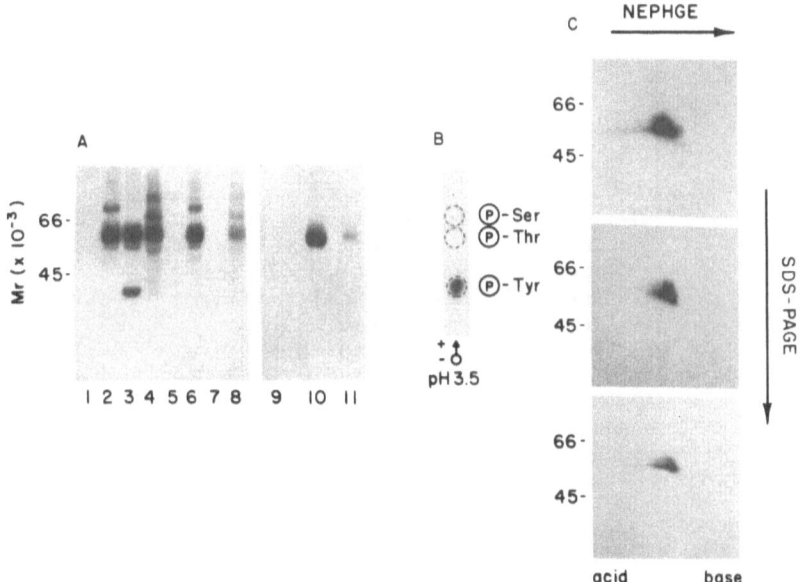

Figure 1. p56^lck is associated with the CD4 and CD8 antigens. (A) Phosphotransferase activity of immunoprecipitates derived from REX (lanes 1-4) and MOLT 4 cells (lanes 5-8). (1,5) rabbit anti-mouse antibody (2,6) anti-CD4 antibody; (3,7) anti-CD8 antibody; (4,8) anti-p56^lck antiserum; (9) anti-CD8 reprecipitated with the W6/32 antibody; (10) anti-CD8 reprecipitated with the anti-p56^lck serum; (11) anti-CD8 reprecipitated with an anti-phosphotyrosine antiserum. (B) Phosphoamino acid analysis of the 55-60KDa band. (C) Two dimensional NEPHGE/SDS-PAGE of p56^lck labeled in the phosphotransferase assay. Upper panel: anti-p56^lck antiserum; middle panel: anti-CD8 antibody; lower panel: anti-CD4 antibody.

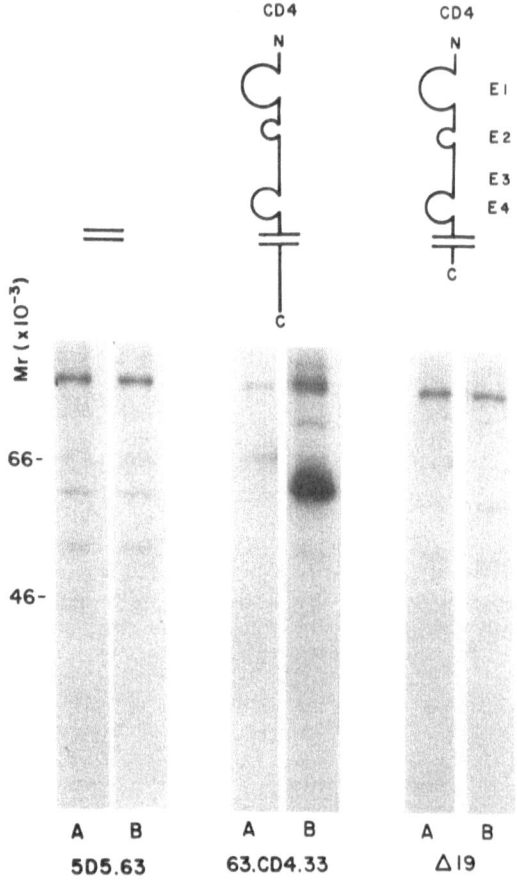

Figure 2. The C-terminal 31 amino acids of human
CD4 are required for the binding of p56^{lck}.
The parental T cell hybridoma 5D5.63 (lpr/lpr)
was transfected with a full-length CD4 and
cytoplasmically-deleted CD4 and assayed for
the presence of associated p56^{lck} activity.
Left panel: parental cell 5D5.63; middle panel:
transfectant expressing full-length CD4,
63.CD4.33; right panel: transfectant expressing
cytoplasmically-truncated CD4, Δ19. A, rabbit
anti-mouse control; B, anti-CD4 immunoprecipitate.

immunoprecipitates, followed by re-precipitation with anti-p56^{lck} antisera identified the kinase
as p56^{lck} (Figure 1A, lane 10). Expression of CD4 did not prevent p56^{lck} from associating
with CD8 and vice versa (Figure 1A, lane 3, 4).
Further, identification of the complex on the Ti(TcR)/CD3 negative T-cell line MOLT-4
showed that the expression of the antigen-receptor is not required for the association between
CD4 and p56^{lck} to occur (Figure 1A, lane 6). Lastly, two-dimensional NEPHGE/SDS-PAGE
further showed that the kinase associated with the CD8 antigen had the same Mr and charge
(approximate pI of 4.8 to 5.8) as that associated with CD4 antigen (Figure 1C, middle panel
and lower panel, respectively), and with that recognized directly by the anti-p56^{lck} antiserum (Figure 1C, upper panel). Taken together, these data revealed that both the CD4 and
CD8 antigens are associated with a catalytically active form of p56^{lck} that is capable of
undergoing autophosphorylation at a tyrosine residue(s).

In order for the interaction to be of functional significance, $p56^{lck}$ must interact with the cytoplasmic tail of the CD4 antigen. To test this directly, we assessed whether $p56^{lck}$ could associate with human CD4 transfected into murine T cells using cDNA encoding for full-length CD4 or truncated CD4 lacking the C-terminal 31 amino acids of the cytoplasmic tail. Figure 2 shows the results of an assay in which an anti-CD4 antibody (19Thy5D7) was used to precipitate CD4 from the mouse parental cell line 5D5.63 (left panel), parental cells expressing the full-length human CD4 antigen (63.CD4.33) (middle panel) and parental cell expressing the CD4 with the truncated cytoplasmic tail (Δ19) (right panel). Immune complexes were assessed for their ability to undergo autophosphorylation in the presence of $[\gamma-^{32}P]$-ATP. Figure 2 shows that anti-CD4 was unable to precipitate $p56^{lck}$ from the parental cell line lacking CD4 (5D5.63: left panel, lane B), but did precipitate $p56^{lck}$ from the hybridoma 63.CD4.33 that expressed the full-length CD4 receptor (middle panel, lane B). These data indicated that the transfected human CD4 antigen can associate with the endogenous murine $p56^{lck}$ from the T-cell hybridoma. By contrast, anti-CD4 failed to precipitate $p56^{lck}$ from the T-cell hybridoma expressing the truncated CD4 antigen (right panel, lane B). These data indicate that $p56^{lck}$ binds a region within the cytoplasmic tail of the CD4 antigen.

Peptide Competition Analysis

A comparison of cytoplasmic sequences of CD4 and CD8 has revealed a common sequence corresponding to the sequence KKTCQCPXXXXKT.[10] In order to test whether this sequence may be involved in binding to the $p56^{lck}$ kinase, various peptides were synthesized and used in a peptide competition assay. $p56^{lck}$ was found to be stably associated with the CD4 antigen following the *in vitro* kinase assay. As demonstrated in Figure 3, greater than 95 percent of the labeled kinase is found associated with the beads after labeling. This finding allowed a determination of whether the addition of various peptides could dissociate $p56^{lck}$ from the CD4 antigen. Peptides corresponding to an extracellular region of the CD2 antigen and the C-terminus of $p56^{lck}$ consistently failed to dissociate the kinase as measured by the appearance of the $p56^{lck}$ in the supernatant (Figure 3, CD2 peptide and lck peptide). However, a peptide corresponding to the cytoplasmic region of CD4 was able to dissociate $p56^{lck}$, resulting in its appearance in the supernatant (Figure 3, CD4 peptide). Approximately

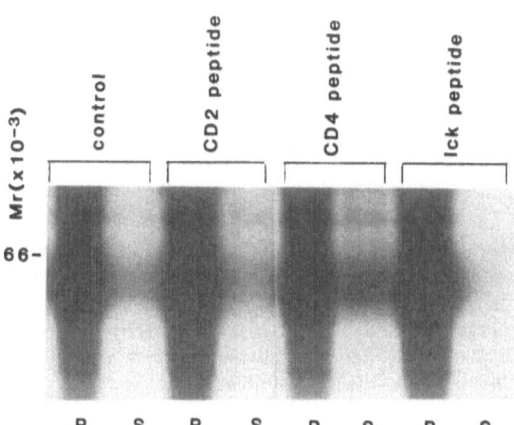

Figure 3. $p56^{lck}$ binds to a specific region within the CD4 cytoplasmic tail. Anti-CD4 immunoprecipitates were labeled in a phosphotransferase assay with $[\gamma-^{32}P]$ATP, incubated with peptides for 10 min at RT prior to separation of the supernatant from the pellet and analysis by SDS-PAGE. Each of the lanes is labeled with respect to peptide. P = pellet; S = supernatant.

Figure 4. Effect of alkylating agents on the detection of CD4: associated p56[lck] activity. Alkylating agents were added to detergent lysates (D-H) or incubated with intact cells for 10 min at 4°C (B, C) prior to immunoprecipitation analysis with anti-CD4 (lanes A-E) or anti-p56[lck] antibodies (lanes F-H). Lane A, anti-CD4 control; lane B, anti-CD4 from cells pretreated with iodoacetic acid; lane C, anti-CD4 from cells pretreated with iodoacetamide; lane D, anti-CD4 from cell lysates containing 10 mM iodoacetic acid; lane E, anti-CD4 from cell lysates containing 10 mM iodoacetamide; lane F, anti-p56[lck] control; lane G, anti-p56[lck] from cell lysates containing 10 mM iodoacetic acid; lane H, anti-p56[lck] from cell lysates containing 10 mM iodoacetamide.

15 to 25 percent of the kinase was dissociated by this peptide. In addition, relatively high concentrations of peptide were required (100 to 1000 μg/ml) for the dissociation to occur. However, the same concentration of the CD2 and lck peptide mixture had no observable effect. None of these peptides were found to effect the binding of antibody to CD4 itself (data not shown). Thus, these data show that p56[lck] binding to the cytoplasmic tail of CD4 is non-covalent and involves the CD4 region encoded by the sequence KKTCQCPHRFQKT.

The CD4:p56[lck] Interaction Involves Cysteine Residues

The putative binding region on the CD4 antigen possesses two cysteine residues of possible importance in the interaction with p56[lck]. In order to test the relevance or this amino acid, cell lysates were treated with the alkylating agents iodoacetic acid of iodacetamide and assessed for CD4 associated p56[lck]. Figure 4 shows that the presence of these agents resulted in a complete loss of detectable p56[lck] activity in anti-CD4 immunoprecipitates (lanes D, E). The same reagents had a negligible to mild effect on kinase activity detected by the anti-p56[lck] antibody (lanes F-H). These data demonstrate that cysteine residues are important in the interaction between the kinase and CD4.

To verify that the cysteines of interest were located in the cytoplasm, intact cells were treated briefly with iodoacetic acid (lane B) or iodoacetamide (lane C) followed by washing and an assessment of the presence of CD4 precipitable p56[lck]. Iodoacetic acid is a charged molecule and relatively impermeable to intact cells at 4°C. In contrast, iodoacetamide is uncharged and permeable to cells. Figure 4 shows that the pretreatment of cells with iodo-acetamide completely prevented the detection of CD4:p56[lck] (lane C), while iodoacetic acid had little if any effect (lane B). These data are consistent with the notion that the cytoplasmic cysteine residues within the CD4 antigen and/or the p56[lck] molecule are crucial to the formation of an active complex.

Phorbol esters are well-established to induce the modulation of CD4, but not CD8 from the surface of T cells.[29] It was therefore of interest to determine whether protein kinase C could also phosphorylate the p56^{lck} associated with CD4 and CD8, and whether this event would alter the association and/or the level of phosphotransferase activity in the complex. As seen in Figure 5 (top left panel), PMA did not change the level of CD8 expression on the surface of the cells. However, PMA did cause a shift in the intensity from the 56Kd to 60Kd bands, similar to that observed for the anti-p56^{lck} pattern (lanes A vs B). The overall intensity of p56^{lck} labeling associated with CD8 was comparable between control and treated cells indicating that PKC-induced phosphorylation did not induce any detectable change in p56^{lck} activity. Similar effects were observed on p56^{lck} precipitated directly with anti-p56^{lck} antisera (lane C, D).

In contrast to CD8, treatment with PMA caused the complete loss of CD4 expression from the surface of REX cells (Figure 5, top right panel). Interestingly, the loss of the surface form of CD4 was accompanied by the loss of p56^{lck} associated activity (lane E vs F). These data demonstrate that the modulation of CD4 from the surface of the T cell results in the dissociation of p56^{lck} from the receptor. Similar findings have been reported by others.[12]

The CD4:p56^{lck} Complex is Physically Linked to the TcR/CD3 Complex

CD4 and CD8 are known to synergize with the CD3/Ti complex in both the generation of signals leading to T-cell activation.[6,7] Although CD4 and Ti/CD3 can be induced to co-modulate,[30,31] it has proven difficult to demonstrate a direct physical interaction between these molecules. Given the sensitivity of CD3 subunits to *in vitro* labeling by the CD4:p56^{lck} complex,[9] we attempted to assess whether any of the CD3 subunits could be identified in direct association with CD4:p56^{lck}. Figure 6A shows that anti-CD4 immuno-

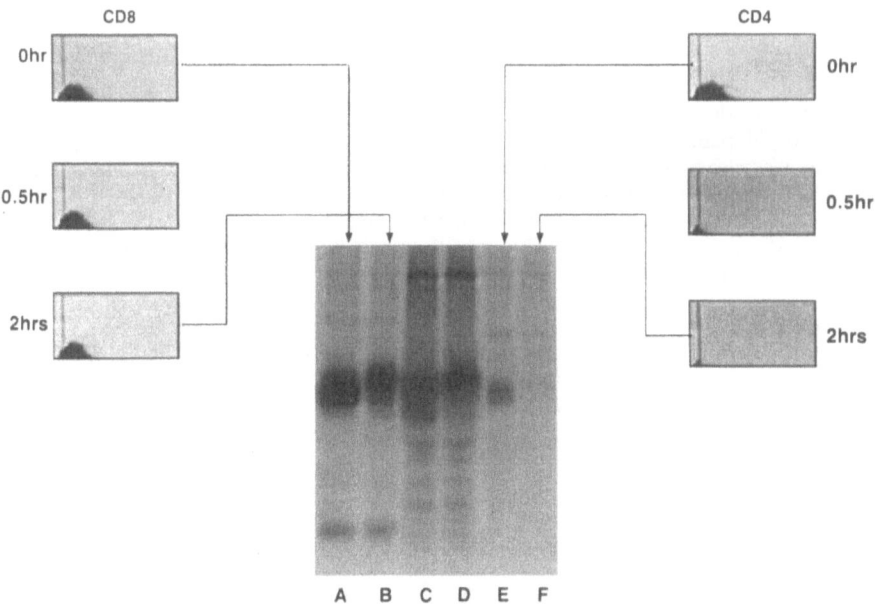

Figure 5. Co-modulation of CD4 and p56^{lck} and the effect of PMA on the CD8:p56^{lck} linkage. Rex cells were treated for various times with 20 ng/ml of phorbol myristic acid (PMA) and assessed for p56^{lck} activity. Lane A, anti-CD8 precipitate at 0 time; lane B, anti-CD8 precipitate at 2 hr of PMA treatment; lane C, anti-p56^{lck} at 0 time; lane D, anti-p56^{lck} at 2 hr of PMA treatment; lane E, anti-CD4 at 0 time; lane F, anti-CD4 at 2 hr of PMA treatment.

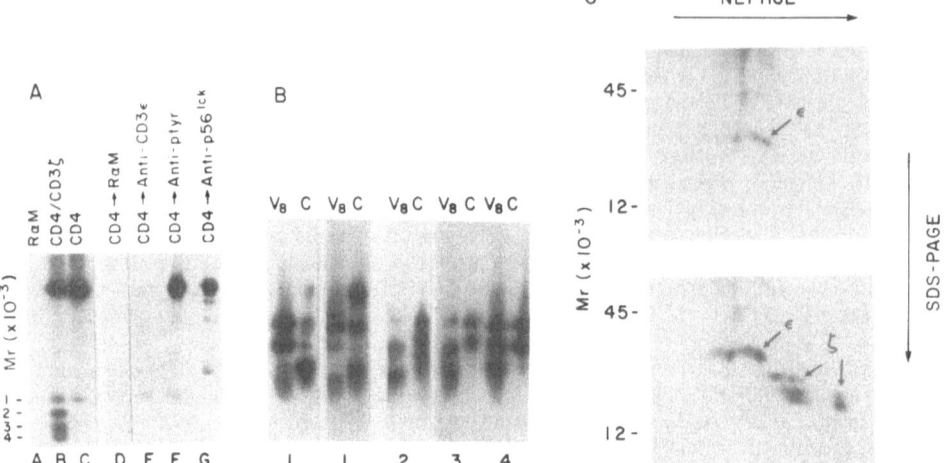

Figure 6. Identification of a physical association between the CD4:p56^{lck} and Ti/CD3 complexes on HPB-ALL cells. (A) HPB-ALL cells were solubilized in NP-40 lysis buffer and immunoprecipitated with either rabbit anti-mouse (lane A), a combination of anti-CD4 and anti-zeta (TIA-2) antibodies (lane B) or an anti-CD4 antibody (lane C). The anti-CD4 immunoprecipitate was then subjected to labeling in the kinase assay and to re-precipitation using rabbit anti-mouse (lane D), anti-CD3ε antibody (sp34) (lane E), anti-phosphotyrosine antibody (lane F) or anti-p56^{lck} antisera (lane G). (B) Peptide map analysis of the individual polypeptides. Individual bands (labeled 1-4 from Figure 5A, lanes B and C) were eluted and subjected to peptide mapping with either V8 protease or chymotrypsin. Panel 1 (left): band 1 from Figure 5A, lane B; panel 1 (right): band 1 from Figure 5A, lane C; panels 2-4 were derived from bands 2-4 from Figure 5A, lane B. (C) Two-dimensional gel electrophoresis of immunoprecipitates. Upper panel: anti-CD4 precipitation. Lower panel: an anti-CD3 immunoprecipitation.

precipitates from the T-cell line HPB-ALL contained a polypeptide at 20Kd in addition to the 55-60Kd polypeptide corresponding to p56^{lck}(lane C, band 1). The position of the 20Kd polypeptide was clearly distinct from a series of three labeled ζ chains found in precipitates formed by a combination of anti-CD4 and anti-ζ chain antibodies (lane B, band 1 vs bands 2-4). Peptide map analysis using both V8 protease and chymotrypsin confirmed the similarity of the individual ζ chains (Figure 6B, panels 2, 3, 4). Additionally, the peptide maps showed that the ζ chain patterns were distinct from the CD4:p56^{lck} co-precipitated 20Kd band (Figure 6B, lane 1) Repvecipitation analysis showed that the 20Kd band could be specifically reprecipitated by an antibody to the C3ε chain (Figure 6A, lane E). An anti-phosphotyrosine antibody re-precipitated both the lck and CD3ε band (Figure 6A, lane G). Two dimensional NEPHGE gels showed that the CD4 associated 20Kd protein (Figure 6C, upper panel) co-migrated in Mr and pI with the CD3ε chain precipitated by anti-CD3 antibodies (Figure 6C, bottom panel). These observations demonstrate that the CD3ε chain can physically associate with the CD4:p56^{lck} complex.

DISCUSSION

The CD4 and CD8 antigens can play crucial roles in the stimulation and effector functions of T lymphocytes. The interaction of the CD4 and CD8 antigens with the protein-tyrosine kinase p56^{lck} provides a compelling model by which these antigens may generate intracellular signals leading to proliferation.[10] In this study, we show by use of transfection analysis that the CD4 antigen requires the terminal 31 amino acids of the 38 amino acid cytoplasmic tail in order to interact with the kinase (Figure 2). Given that both CD4 and CD8 can associate with p56^{lck}, we proposed that a region within the sequence K/RK/RXCXCPXXXXKT/S could serve as the binding region for p56^{lck}.[10] Peptide competition analysis revealed that a peptide corresponding to this region on CD4 can dissociate small amounts of the kinase from the CD4 antigen (Figure 3). Other unrelated peptides failed to dissociate the kinase. Therefore, at least part of the cytoplasmic sequence that is conserved between CD4 and CD8 appears to mediate the binding to the kinase. Cysteine residues also appear to play an important role as shown by the inhibitory effects of alkylating agents on the detection of the CD4 associated kinase (Figure 4). The interaction is non-covalent, not involving di-sulphide bonding between the proteins.[9] The apparent inability of the peptide corresponding to the C-terminus of lck to dissociate the complex argues against this region being involved in the interaction. This is consistent with the recently reported role of the N-terminal region of p56[lck] in binding to CD4.[32]

The identification of the regions of CD4 and p56[lck] which involved in binding to each other has allowed for the development of a model as illustrated in Figure 7. In this model, the CD4 and CD8 receptors make up the ligand binding component of the complex, while the p56[lck] kinase functions in the generation of intracellular signals involved in the regulation of T-cell growth. The kinase itself is attached to the inner face of the membrane by a myris-

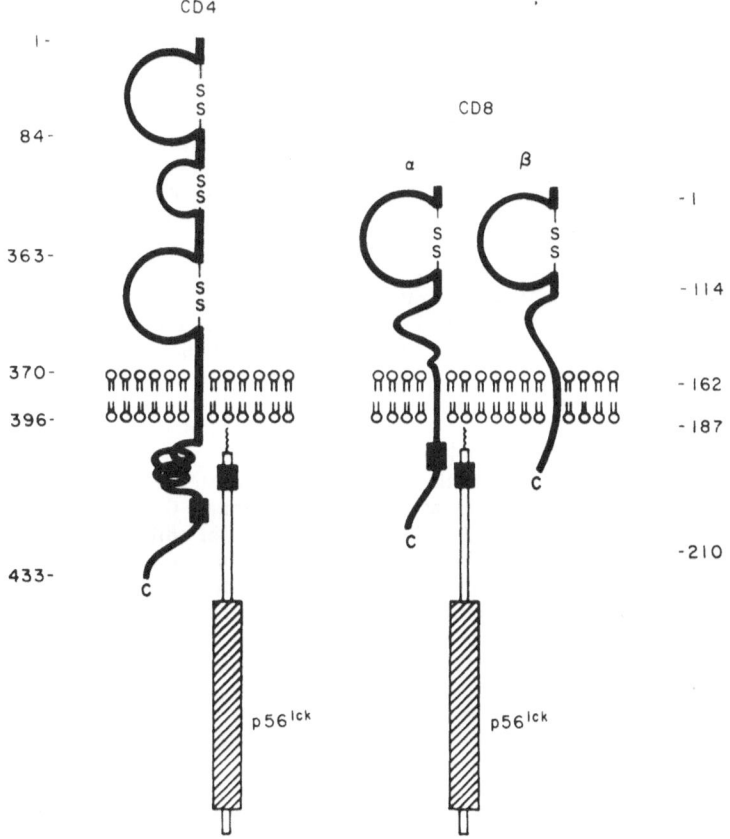

Figure 7. Model of the interaction of p56[lck] with CD4 and CD8 antigens. The CD4 and CD8α chains are physically associated with the protein-tyrosine kinase p56[lck] via a specifc region within their cytoplasmic tails. The region within the cytoplasm is encoded by the sequence KKTCQCPHRFQKT of CD4 and RRVCKCPRPVVKS of CD8. The sequence (black box) is located midway in the cytoplasmic tail of CD4 and proximal to the plasma membrane in the case of CD8α. The second chain CD8β does not possess a binding region for the kinase. The binding region (black box) on p56[lck] is also shown at the N-terminus of the kinase.

tic group, an appropriate location for an interaction with surface receptors. Both CD4 and the α chain of CD8 carry a p56[lck] binding region; however, they are located in different regions of the cytoplasmic tail. The region is located midway in the tail of CD4 (residues 419-431), while in the case of CD8, it is located proximal to the plasma membrane (residues 190-203). In contrast, the CD8β chain lacks the sequence defining the binding region. With respect to p56[lck], Shaw and co-workers (1990) have recently shown that the N-terminal region consisting of residues 10 to 30 is sufficient to allow binding to CD4 and CD8.[32] Within this region, cysteine residues appear to play a crucial role (Turner and Littman, personal communication). It is noteworthy that the binding site is situated in the region of the kinase that varies between members of the src family. This is consistent with the notion that the N-terminus of the members of the src family may specify an interaction with other mammalian receptors.

The exact mechanism by which p56[lck] activity is regulated and its role in activation process is poorly understood. The binding of CD4 antibodies to the CD4 antigen can cause a transient increase in p56[lck] catalytic activity.[21,22,33] However, this effect is not epitope specific since stimulation has been observed using antibodies to various sites on the CD4 antigen.[22] Furthermore, the relationship of this enhanced kinase activity to events involved in the T-cell activation is unclear. Unlike ligand binding to the EGF and PDGF receptors, the addition of antibodies to CD4 on intact T cells inhibits rather than stimulates cell growth.[34] Futhermore, stimulation of thymocytes and peripheral blood lymphocytes through the CD3 pathway appears to stimulate little if any p56[lck] activity.[33] It is therefore of particular interest that we have been able to identify a physical link between the CD3ε chain and the CD4:p56[lck] complex in HPB-ALL cells (Figure 6). The HPB-ALL cells are the only cells examined to date that demonstrate this interaction. This is likely to be related to the exceptionally high levels of CD3 expression on their cell surface. However, the fact that instances of this interaction can be demonstrated is consistent with the notion that these cell surface molecules can physically interact during events such as antigen presentation. Previous data have shown that crosslinking of the CD4/L3T4 antigen with the Ti/CD3 complex potentiates activation of T cells.[6,7] The localization of the CD4/CD8:p56[lck] complex relative to the Ti(TcR)/CD3 complex may therefore be a key factor in understanding the role of p56[lck] in the activation process of the T cell.

Another clue as to the involvement of p56[lck] in the regulation of T cell growth has come from modulation experiments induced by phorbol ester. T-cell activation is accompanied by the initiation of phosphatidylinositol hydrolysis and the activation of protein kinase C.[35] It was thus of interest to analyze the effect of phorbol esters on p56[lck] since they mimic an aspect of T-cell activation by activating protein kinase C. Interestingly, p56[lck] associated with CD8 is phosphorylated by PKC as shown by the shift in bands from 55Kd to 62 Kd (Figure 5). Despite this, no effect was observed on the autophosphorylative potential of the kinase. In contrast to CD8, CD4 is modulated from the cell surface by the phorbol ester (Figure 5). In addition, this modulation is accompanied by the loss of CD4 associated p56[lck] activity. Similar studies have been reported by others.[12] The dissociation of the kinase from the receptor may be integral to altering the location and regulatory function of the kinase within the T cell.

REFERENCES

1. E. L. Reinherz, S. C. Meuer, and S. F. Schlossman, The human T cell receptor: analysis with cytotoxic T cell clones, *Immunol. Rev.*, 74:83 (1983).
2. C. Doyle, and J. L. Strominger, Interaction between CD4 and class II MHC molecules mediates cell adhesion, *Nature*, 330:256 (1987).
3. D. R. Littman, The structure of the CD4 and CD8 genes, *Ann. Rev. Immunol.*, 5:561 (1987).
4. A. Dalgleish, P. C. L. Beverly, P. R. Clapham, D. H. Crawford, M. F. Greaves, and R. A. Weiss, The CD4 (T4) antigen is a essential component of the receptor for the AIDS retrovirus, *Nature (London)* 312:763 (1984).
5. D. Klatzman, E. Champayne, S. Chamaret, J. Gruest, D. Guefard, T. Hercend, J. C. Gluckman, and L. Montagnier, T-lymphocyte T4 molecule behaves as the receptor for human retrovirus LAV, *Nature (London)*, 312:767 (1984).

6. K. Eichmann, J. I. Johnson, I. Falk, and F. Emmrich, Effective activation of resting mouse T lymphocytes by cross-linking submitogenic concentrations of the T cell antigen receptor with either Lyt-2 or L3T4, *Eur. J. Immunol.*, 17:643 (1987).

7. P. Anderson, M. L. Blue, C. Morimoto, and S. F. Scholssman, Crosslinking of T3 (CD3) with T4 (CD4) enhances the proliferation of resting T lymphocytes, *J. Immunol.*, 139:678 (1987).

8. C. E. Rudd, J. M Trevillyan, L. L. Wong, J. D. Dasgupta, and S. F. Schlossman, The CD4 receptor is complexed to a T-cell specific tyrosine kinase (pp.58) in detergent lysates from human T lymphocytes, *Proc. Natl. Acad. Sci. USA*, 85:5190 (1988).

9. E. K. Barber, J. D. Dasgupta, S. F. Schlossman, J. M. Trevillyan and C. E. Rudd, The CD4 and CD8 antigens are coupled to a protein-tyrosine kinase (p56^{lck}) that phosphorylates the CD3 complex, *Proc. Natl. Acad. Sci. USA*, 86:3277 (1989).

10. C. E. Rudd, P. Anderson, C. Morimoto, M. Streuli, and S. F. Schlossman, Molecular interactions, T-cell subsets and a role of the CD4/CD8:p56^{lck} complex in human T cell activation, *Immunological Reviews* 111:225 (1989).

11. A. Veillette, M. A. Bookman, E. M. Horak, and J. B. Bolen, The CD4 and CD8 T cell surface antigens are associated with the internal membrane tyrosine-protein kinase p56^{lck}, *Cell*, 55:301 (1988).

12. T. R. Hurley, K. Luo, and B. Sefton, Activators of protein kinase C induce dissociation of CD4, but not CD8, from p56^{lck}, *Science*, 245:407 (1989).

13. T. Mustelin, K. M. Coggeshall, and A. Altman, Rapid activation of the T-cell tyrosine protein kinase p56^{lck} by the CD45 phosphotyrosine phosphatase, *Proc. Natl. Acad. Sci. USA*, 86:6302 (1989).

14. H. L. Ostergaard, D. A. Shackelford, T. R. Hurley, P. Johnson, R. Hyman, B. M. Sefton, and I. S. Trowbridge, Expression of CD45 alters phosphorylation of the lck tyrosine protein kinase in murine lymphoma T cell lines, *Proc. Natl. Acad. Sci. USA* 86:8959 (1989).

15. T. Hunter and J. A. Cooper, Protein tyrosine kinases, Ann. Rev. Biochem., 54:876 (1985).

16. A. F. Voronova, and B. M. Sefton, Expression of a new tyrosine protein kinase is stimulated by retrovirus promoter insertion, *Nature (London)*, 319:682 (1986).

17. J. D. Marth, R. Peet, E. G. Krebs, and R. M. Perlmutter, A lymphocyte-specific protein-tyrosine kinase gene is rearranged and overexpressed with murine T cell lymphoma LSTRA, *Cell* 43:393, (1985).

18. J. M. Trevillyan, C. Canna, D. Maley, T. J. Linna, and C. A. Phillips, Identification of the human T-lymphocyte protein-tyrosine kinase by peptide-specific antibodies, *Biochem. Biophys. Res. Commun.*, 140:392 (1986).

19. A. Veillette, I. D. Horak, and J. B. Bolen, Posttranslational alterations of the tyrosine kinase p56^{lck} in response to activators of protein kinase C., *Oncogene Research*, 2:385 (1988).

20. J. D. Marth, D. B. Lewis, C. B. Wilson, M. E. Gearn, E. G. Krebs and R. M. Perlmutter, Regulation of p56^{lck} during T-cell activation: functional implications for the src-like protein kinases, *EMBO J.*, 6:2727 (1987).

21. A. Viellette, M. A. Bookman, E. M. Horak, L. B. Samelson and J. B. Bolen, Signal transduction through the CD4 receptor involves the activation of the internal membrane tyrosine-protein kinase p56^{lck} transduction, *Nature (London)*, 338:257 (1989).

22. A. D. Odysseos, M. E. Drotar, and C. E. Rudd, Regulation of CD4 associated p56^{lck} by antibody binding to CD4, *J. Cell. Bioch.*, Suppl. 14B:279 (1990).

23. A. Veillette, J. C. Zuniga-Pflucker, J. B. Bolen, and A. M. Kruisbeek, Engagement of CD4 and CD8 expressed in immature thymocytes induces activation of intracellular tyrosine phosphorylation pathways, *J. Exp. Med.*, 170:1671 (1989).

24. A. M. Norment, and D. R. Littman, A second subunit of CD8 is expressed in human T cells, *EMBO J.*, 7:3433 (1988).

25. J. P. DiSanta, R. Knowles and N. Flomenberg, The human Lyt-3 molecule requires CD8 for cell expression, *EMBO J.*, 11:3465 (1988).

26. H. Oettgen, W. Kappler, W. J. M. Tax, and C. Terhorst, Characterization of the two heavy chains of the T3 complex on the surface of human T lymphocytes, *J. Biol. Chem.*, 259:12039 (1984).

27. P. Anderson, M. L. Blue, C. O'Brien, and S. F. Schlossman, Monoclonal antibodies reactive with the T cell receptor ζ chain: production and characterization using a new method, *J. Immunol.*, 143:1899 (1989).

28. U. K. Laemmli, Cleavage of structural proteins during the assembly of the head of bacteriophage T4, *Nature (London)*, 227:680 (1970).

29. M. L. Blue, D. Hafler, K. A. Craig, H. Levine, and F. Schlossman, Phosphorylation of CD4 and CD8 molecules following CD3/T cell receptor triggering, *J. Immunol.*, 139:1202 (1987).

30. P. Anderson, M. L. Blue, and S. F. Schlossman, Comodulation of CD3 and CD4: evidence for a specific association between CD4 and lymphocytes, *J. Immunol.*, 140:1732 (1988).

31. K. Saizawa, J. Rojo, and C. A. Janeway, Evidence for a physical association of CD4 and the CD3:α:T-cell receptor, *Nature (London)*, 328:260 (1987).

32. A. S. Shaw, K. E. Amrein, C. Hammond, D. F. Stern, B. M. Sefton, and J. K. Rose, The cytoplasmic domain of CD4 interacts with the tyrosine protein kinase, p56^{lck} through its unique amino-terminal domain, *Cell*, 59:626 (1989).

33. A. Viellette, J. B. Bolen, and M. A. Bookman, Alterations in tyrosine protein phosphorylation induced by antibody-mediated cross linking of the CD4 receptor of T lymphocytes, *Mol. Cell. Biol.* 9:4441 (1989).

34. I. Bank, and L. Chess, Perturbation of the T4 molecule transmits a negative signal to T cells, *J. Exp. Med.*, 162:1294 (1985).

35. A. Weiss, J. Imboden, K. Hardy, B. Manger, C. Terhorst, and J. Stobo, The role of T3/antigen receptor complex in T cell activation, *Ann. Rev. Immunol.*, 4:593 (1986).

CHARACTERIZATION OF TRIPLE NEGATIVE CLONES ISOLATED FROM POST-NATAL

HUMAN THYMUS

Toshiyuki Hori and Hergen Spits

The Department of Human Immunology
DNAX Research Institute of Molecular and Cellular Immunology
Palo Alto, California

ABSTRACT

Human triple negative (CD3⁻ CD4⁻ CD8⁻) thymocytes and double negative (CD4⁻ CD8⁻) thymocytes purified from post-natal thymus were cloned with a feeder cell mixture of irradiated PBL, irradiated JY cells and PHA and expanded with IL-2. The cloning efficiency of triple negative thymocytes was less than 1% and the majority of the clones were triple negative. One out of 11 clones was TCR $\alpha\beta^+$ CD4$^+$. No TCR $\gamma\delta^+$ clones were isolated. On the other hand, the cloning efficiency of double negative thymocytes was about 10% and most of the clones isolated were TCR $\gamma\delta^+$. We could not find any evidence of in vitro differentiation of triple negative thymocytes into TCR $\gamma\delta^+$ cells. Some of the triple negative clones expressed CD16 brightly and were apparently NK cells. All CD16⁻ clones isolated from triple negative thymocytes, however, expressed NKH1, which is also an NK cell marker. Cytoplasmic CD3-δ and CD3-ϵ Ag which have been reported to be expressed in the most immature thymocytes were not detected in any of these clones. Furthermore, the CD16⁻ triple negative clones exhibited significant cytolytic activity against K562. Phenotype of the clones seems to be stable under various conditions in vitro including coculture with human thymic epithelial cells. These data indicate that the CD16⁻ triple negative clones isolated from triple negative thymocytes are similar to a minor subset of NK cells which is CD16⁻ NKH-1$^+$. It is not clear whether they originated from a distinct subset of mature or immature NK cells resident in the thymus tissue or from common precursors of both T and NK lineage.

INTRODUCTION

In vivo reconstitution experiments and in vitro organ cultures of immature thymocytes in murine systems have indicated that the triple negative (CD3⁻ CD4⁻ CD8⁻) thymocyte subset contains precursors for all mature subsets of thymocytes and peripheral T cells.[1-3] In contrast, human T cell precursors have not been well characterized due to methodological limitations. Triple negative thymocytes are also present in the human thymus. These cells express CD45, CD7 and partly CD2 and have been considered to represent the earliest stage of intrathymic T cell development.[4-6] A number of studies have suggested that triple negative thymocytes can differentiate in vitro under certain conditions.[7-10] However, some controversy exists about to which phenotype of cells triple negative thymocytes mature. One group reported that triple negative thymocytes can differentiate in vitro into TCR $\alpha\beta^+$ cells as well as TCR $\gamma\delta^+$ cells.[7,8,10] This in vitro differentiation required IL-2 and thymic stroma cells.[10] Another group presented evidence that triple negative thymocytes can differentiate in vitro into TCR $\gamma\delta^+$ cells but not into TCR$\alpha\beta^+$ cells after culture with anti-CD2 mAb and IL-2.[9] The latter group performed

Mechanisms of Lymphocyte Activation and Immune Regulation III
Edited by S. Gupta *et al.*, Plenum Press, New York, 1991

97

a clonal analysis that supported *in vitro* differentiation of triple negative thymocytes into TCR $\gamma\delta^+$ cells but only a limited characterization of those clones was presented. The claim that triple negative thymocytes can differentiate into TCR $\alpha\beta^+$ cells expressing either CD4 or CD8 has never been substantiated by a clonal analysis and therefore it is difficult to completely exclude the possibility of overgrowth of contaminating more mature cells in bulk culture of triple negative thymocytes.

In the present study, we investigated differentiating potential of highly purified human triple negative thymocytes *in vitro* at the clonal level. These cells were cloned in the presence of a feeder cell mixture and the isolated clones were characterized in detail. It is shown that triple negative thymocytes remained triple negative after cultures with a feeder cell mixture and most of the clones isolated from triple negative thymocytes turned out to be similar to activated NK cells. These results suggest that immature thymocytes include precursors for NK cells.

MATERIALS AND METHODS

Cell Preparations

Normal human thymus tissue was obtained from patients under 2 years of age who underwent median sternotomy and corrective cardiovascular surgery. Fresh thymus fragments were finely minced and pressed through a stainless steel mesh to give rise to single cell suspensions. Thymocyte subsets were separated using magnetic bead columns followed by cell sorting. Thymocytes were incubated with purified mAb Riv-6 (CD4) and WT-82 (CD8) in PBS containing 5 mg/ml BSA 0.2 mg NaN_3 and 1 mM EDTA for 30 min on ice, washed twice and subsequently incubated with biotin-conjugated (Fab')$_2$ fragment of goat anti-mouse IgG (Tago, Burlingame, CA) for 30 min on ice. After two washings, cells were incubated with FITC-conjugated avidin for 5 min, washed twice, incubated with biotin-conjugated magnetic beads for 5 min, and then applied to MACS magnetic columns (Miltenyi-Biotec, Bergisch Gladbach, F.R.G.).[11] Cells which passed through the columns were stained with PE-conjugated Leu4 mAb (CD3), and triple negative cells or double negative cells were sorted using a FACStar-plus (Becton Dickinson, Sunnyvale, CA). Less than 0.5% of purified subsets were of the depleted phenotype.

mAb and Cytofluorometric Analysis

mAb Leu-6 (CD1), Leu-5 (CD2), Leu-9 (CD7), Leu-4 (CD3), Leu-11 (CD16), Leu-19 (NKH-1) were purchased from Becton Dickinson (Sunnyvale, CA). mAb Riv-6 (CD4) and WT-82 (CD8)[12] were the gifts of Drs. Kreeftenberg and W.J.M. Tax, respectively. mAb WT-31[13] detects a common epitope on TCR $\alpha\beta$ heterodimers and mAb TCR $\delta1$[14] (kindly provided by Dr. M. Brenner) reacts with a common framework determinant on the TCE δ chain. Cells were incubated with mAb in PBS containing 5mg/ml BSA and 0.2 mg/ml NaN_3 for 30 min on ice, washed twice and subsequently incubated with FITC-conjugated (Fab')$_2$ fragment of goat anti-mouse IgG (Tago) for 30 min on ice. After two washings, cells were subjected to cytofluorometric analysis using a FACScan (Becton Dickinson).

Cloning of Triple or Double Negative Thymocytes

Yssel's medium[15] supplemented with 1% human serum was used for cell culture throughout this study. Purified triple or double negative thymocytes were seeded at 1 cell/well and 10 cells/well in 96-well round bottom tissue culture plates (Limbro) in a total volume of 100 μl with a feeder cell mixture of 5×10^5 irradiated allogeneic PBL, 5×10^4 irradiated JY cells (an EBV-transformed B cell line) and 50 ng/ml PHA (Wellcome, Beckenham, Kent, UK). After 1 week, 100 μl of fresh medium containing 20 u/ml IL-2 was added to each well. After 2 weeks, growing clones were transferred with feeders into 24-well Limbro plates and further expanded with IL-2.

For comparison, CD16$^-$ as well as CD16$^+$ NK clones were used, which were isolated from NKH-1$^+$ cells freshly prepared from normal PBL using the same cloning procedures as described.

Expression of Cytoplasmic CD3 Ag

mAb SP-64 (CD3-δ chain) and Sp-6 (CD3-ϵ chain) were provided by Dr. C. Terhorst. Cytoplasmic CD3 Ag was detected according to the method described by Dongen et al.[16] Briefly, cytocentrifuge preparations were air-dried, fixed with acid ethanol (ethanol with 5% (volo/vol) acetic acid) for 15 min at -20° C and subsequently incubated with mAb SP-64 CD3-δ chain) or SP-6 (CD3-ϵ chain). FITC-conjugated (Fab')$_2$ fraction of goat anti-mouse IgG (Tago) was used as a second step reagent. Stained preparations were examined using a Zeiss and Leitz fluorescence microscope.

Cytotoxicity Assays

Cytotoxicity assays were performed according to the standard ^{51}Cr release method as described previously.[17] Briefly, 2,000 ^{51}Cr-labeled target cells were mixed with effector cells in Yssel's medium with 1% human serum in round bottom Limbro 96-well plates. The plates were centrifuged for 5 min at 50 x g and incubated for 4 h at 37°C. The samples were harvested by using a Skatron harvester (Lier, Norway) and counted in a gamma counter.

RESULTS

Triple Negative Thymocytes Proliferated in the Presence of Feeder Cells and Remained Triple Negative

Freshly isolated triple negative thymocytes did not proliferate well in response to IL-2 or IL-4 only (data not shown). However, a vigorous proliferation was observed in the cultures with a feeder cell mixture of irradiated PBL, irradiated JY cells and PHA. To examine whether triple negative thymocytes could differentiate under these culture conditions, they were cultured at 10^4 cells/well with feeders in round bottom Limbro 96-well plates and, after 12 days, the cell surface phenotype was determined. The results of a representative experiment are shown in Table 1. Highly purified triple negative thymocytes remained virtually CD3$^-$ CD4$^-$ but some cells weakly expressed CD8. It was noted that about half of the cells were CD16$^+$ and most of the cells were NKH-1$^+$, both of which Ag are known to be expressed on NK cells freshly isolated from PBL.

Clonal Analysis of Triple Negative Thymocytes

Clonal analysis was performed to characterize in more detail the proliferating cells in the culture of triple negative thymocytes with the feeder cell mixture. Double negative and triple negative thymocytes from a single thymus were seeded at 1 cell/well and 10 cells/well in round bottom 96-well plates with feeders. After two weeks, wells with growing cells were counted and clones were transferred to 24-well plates with feeders to be expanded. As shown in Table 2, the cloning efficiency of triple negative thymocytes was less than 1% and the majority of the clones were CD3$^-$. Only one clone was TCR $\alpha\beta$-CD3$^+$ CD4$^+$ CD8$^-$ in this experiment. About two thirds of CD3$^-$ clones expressed CD16 and were apparently NK cells (data not shown). No TCR $\gamma\delta^+$ clones were isolated. On the other hand, the cloning efficiency of double negative (CD4$^-$ CD8$^-$) thymocytes was about 20-fold higher than triple negative cells and most of the clones obtained were TCR $\gamma\delta^+$.

Table 1. Phenotype of Triple Negative Thymocytes Cultured with Feeders for 12 Days

	CD1	CD2	CD3	CD4	CD8	CD16	NKH-1
%positive cells	0	69	0	2	11	45	88

Table 2. Cloning of Triple and Double Negative Thymocytes

Subpopulations	No of cells per well	No of wells seeded	No of wells with growing cell	Cloning efficiency	CD3⁻ clones	TCRαβ⁺ clones	TCRγδ⁺ clones
Triple negatives	1	192	2				
				0.48%	10/11	1[1])/11	0
	10	192	9				
Double negatives	1	192	11				
				9.1%	1/40[2)]	8[1])/40	31/40
	10	192	115				

[1])All TCRαβ clones obtained had CD4⁺ CD8⁻ phenotype.

[2])Phenotype of randomly selected 40 clones was determined.

Thymus-Derived CD3⁻ CD⁻ Clones Expressed NKH-1 But Not Cytoplasmic CD3

A major question was whether CD3⁻ CD16⁻ clones isolated from triple negative thymocytes were immature T cell precursors or cells of other lineages. To answer this question, we characterized 8 representative thymus-derived CD3⁻ CD16⁻ clones in more detail. As shown in Table 3, they were phenotypically all similar; all of them were CD1⁻, CD2⁺, CD3⁻, CD4⁻, CD8⁻ and NKH-1⁺. Phenotype of the clones was stable under various culture conditions including coculture with human thymic epithelial cells (data not shown). Cytoplasmic

Table 3. Phenotypes of Thymus-Derived Triple Negative Clones[1]

Clones	Cell surface marker								Cytoplasmic Ag	
	CD1	CD2	CD3	CD4	CD8	CD7	CD16	NKH-1	CD3-δ	CD3-ε
15-34	-	+	-	-	-	+	±	++	-	-
16-39	-	++	-	-	-	+	-	+	-	-
17-43	-	++	-	-	-	+	-	++	-	-
22-2	-	++	-	-	-	+	-	++	-	-
22-21	-	+	-	-	-	+	-	++	-	-
22-24	-	++	-	-	-	+	-	++	-	-
22-41	-	++	-	-	-	+	-	++	-	-
22-43	-	++	-	-	±	+	-	++	-	-

[1])Thymus-derived CD16⁺ clones which are apparently NK clones are not included.

Table 4. Comparison of Cytotoxicities of Thymus-Derived CD16⁻ Triple Negative Clones and NK Clones Isolated from PBL

| Clones | % specific ^{51}Cr release[1] | | |
| | Target cells | | |
	K562	Daudi	JY
NK clones[2] isolated from PBL			
CD16^{++}.4	5 6	5 2	3
CD16^{++}.10	5 2	2 9	2
CD16^{-}.9	5 1	4 7	8 0
CD16^{-}.13	4 5	3 2	3 6
Thymus-derived CD16^{-} triple negative clones			
15-34	2 5	0	0
16-39	2 6	1 7	5
17-43	2 9	6	2 5
22-2	4 9	4 1	6 9
22-21	3 3	3 1	4 8
22-24	1 2	1 0	2 3
22-41	2 4	1 1	3 3
22-43	8	1	5

[1]Cytotoxicity was determined with a standard ^{51}Cr release assay. The effector to target cell ratio was 1:1.

[2]CD16^{++}.4 and 10 expressed high levels of CD16. CD16^{-}.9 and 13 were negative for CD16.

CD-3δ and CD3-ε Ag which have been reported to be expressed in most immature thymocytes[16] were not detected in any of the clones, suggesting that they were not committed to the T cell lineage.

Thymus-Derived Triple Negative Clones Exhibited Cytolytic Activity Against K562

It has been reported[18] that there is a small subset of NK cells which is CD16⁻ NKH-1⁺ and a number of NK clones of this phenotype as well as CD16⁺ NKH-1⁺ NK clones have recently been isolated from normal PBL (H. Spits et al., in preparation). Since thymus-derived triple negative clones had a similar phenotype to these NK clones, we examined cytolytic activity of the thymus-derived clones in comparison with NK clones. As shown in Table 4, the thymus-derived CD16⁻ clones except one clone exhibited significant cytolytic activity against K562 which is the standard target cell of non-MHC restricted ctyotoxicity. Most of the clones were also cytotoxic for Daudi and JY although the cytolytic activity varied among the clones.

DISCUSSION

In the present study we have shown that highly purified human triple negative thymocytes proliferate in the presence of a feeder cell mixture of irradiated PBL, JY cells and PHA and remain triple negative under these culture conditions. In contrast to our results, several other studies have suggested that human triple negative thymocytes can differentiate into TCR $\gamma\delta^+$ cells[9] or even into both TCR $\alpha\beta^+$ and TCR $\gamma\delta^+$ cells[7,8,10]*in vitro*. The cell separation procedures and the culture conditions used in each of those studies were different from ours. As discussed by others,[2,3] the purity of the thymocyte subsets is crucial for interpretation of differentiation experiments, and use of only complement-mediated cytolysis which some of the above-mentioned studies relied on, has been inadequate in our hands and those of others[9] to obtain acceptably pure subpopulations. In our experience, the combination of magnetic bead columns and cell sorting has been superior compared to complement-mediated cytolysis.

Dennings et al. demonstrated differentiation of triple negative thymocytes into TCR $^+$ cells at the clonal level.[9] The cloning efficiency reported is much higher than that of our experiments, suggesting that we have lost many clones during culture. The difference with our results may be due to the different culture conditions, but not to our inability to grow TCR $\gamma\delta^+$ cells inasmuch as many TCR $\gamma\delta^+$ clones were obtained from double negative thymocytes using the same culture conditions as used for cloning triple negative thymocytes. It is possible that anti-CD2 mAb are crucial for driving triple negative thymocytes into TCR$\gamma\delta^+$ cells at an early phase of culture.

With regard to the origin of the NKH-1$^+$ triple negative clones, the following four possibilities should be taken into consideration. First, triple negative thymocytes contain common precursors for both T cells and NK cells which preferentially differentiate into NK cells under *in vitro* culture conditions used in this study. Second, triple negative thymocytes contain commited precursors for NK cells in addition to those for T cells, both of which are mutually exclusive. Third, the clones represent a distinct subset of NK cells resident in the thymus tissue. Fourth, the clones are derived from contaminating peripheral blood NK cells. The latter is possible but unlikely because purified triple negative thymocytes were virtually CD16$^-$ and less than 2% of them expressed NKH-1 while in PBL more than 95% of NK cells are CD16$^+$ NKH-1$^+$. In addition to this, the cloning efficiency of peripheral blood NKH-1$^+$ NK cells is not so high (in the range of 2-5%) under the same conditions (H. Spits et al., in preparation). It should also be noted that NKH-1$^-$ thymocytes can give rise to NKH-1$^+$ cells with non-MHC restricted cytolytic activity after a short term culture with IL-2,[22,23] indicating that the expression of NKH-1 is inducible on thymocytes *in vitro*.

Recent evidence has suggested that expression of CD3 mRNA and its product is one of the earliest events in T cell differentiation.[24,25] The fact that we could not detect cytoplasmic CD3 Ag expression in any of thymus-derived CD16$^-$ triple negative clones suggests that these clones are not committed to T cell lineage. However, it is still possible that they are derived from the most immature T cell precursors because establishment of murine pro T lymphocyte clones which express neither CD3-δ nor CD3-ϵ mRNA has been described.[22,23] Further studies are needed to elucidate the origin of these clones and the relation between immature thymocytes and NK cells.

ACKNOWLEDGMENT

DNAX Research Institute is supported by Schering-Plough.

REFERENCES

1. B. J. Mathieson and B. J. Fowlkes, Cell surface antigen expression on thymocytes: development and phenotypic differentiation of intrathymic subsets, *Immunol. Rev.* 82:141 (1984).

2. R. Scollay, A. Wilson, A. D'Amico, K. Kelly, M. Egerton, M. Pearse, L. Wu, and K. Shortman, Developmental status and reconstitution potential of subpopulations of murine thymocytes, *Immunol. Rev.* 104:81 (1988).

3. B. J. Fowlkes and D. M. Pardoll, Molecular and cellular events of T cell development, *Adv. Immunol.* 44:207 (1989).

4. E. L. Reinherz and S. F. Schlossman, Discrete stages of human intrathymic

differentiation: analysis of normal thymocytes and leukemia lymphoblasts of T cell lineage, *Cell* 19:821 (1980).

5. D. F. Lobach, L. L. Hensley, W. Ho, and B.F. Haynes, Human T cell antigen expression during the early stages of fetal thymic maturation, *J. Immunol.* 135:1752 (1985).

6. B. F. Haynes, M. E. Martin, H. H. Kay, and J. Kurtzberg, Early events in human T cell ontogeny. Phenotypic characterization and immunohistologic localization of T cell precursors in early human fetal tissues, *J. Exp. Med.* 168:1061 (1988).

7. M. L. Toribio, C. Martínez-A., M. A. R. Marcos, C. Márquez, E. Cabrero, and A. la Hera, Differentiation of mature and functional T-cells from human prothymocytes: towards a role for T3$^+$ 4$^-$ 6$^-$ 8$^-$ transitional thymocytes, *Proc. Natl. Acad. Sci. USA.* 83:6985 (1986).

8. M. L. Toribio, A. de la Hera, J. Borst, M. R. A. Marcos, C. Márquez, J. M. Alonso, A. Barcena, and C. Martínez-A., Involvement of interleukin 2 pathway in the rearrangement and expression of both α/β and γ/δ T cell receptor genes in human T cell precursors, *J. Exp. Med.* 168:2231 (1988).

9. S. M. Denning, J. Kurtzberg, D. S. Leslie, and B. F. Haynes, Human postnatal CD4$^-$ CD8$^-$ CD3$^-$ thymic T cell precursors differentiate *in vitro* into T cell receptor δ-bearing cells, *J. Immunol.* 142:2988 (1989).

10. A. de la Hera, W. Marston, C. Aranda, M. L. Toribio, and C. Martinez-A., Thymic stroma is required for the development of human T cell lineages *in vitro, Internal. Immunol.* 1:471 (1989).

11. H. Abts, M. Emmerich, S. Miltenyi, A. Radbruch, and H. Tesch, CD20 positive human B lymphocytes separated with the magnetic cell sorter (MACS) can be induced to proliferation and antibody secretion *in vitro, J. Immunol. Methods.* 125:19 (1989).

12. W. J. M. Tax, H. F. M. Leeuwenberg, H. M. Willems, P. J. A. Capel and R. A. P. Koene, *in*: "Leukocyte Typing," A. Bernard, ed., Springer, Heidelberg (1984).

13. H. Spits, J. Borst, W. Tax, P. J. A. Capel, C. Terhorst and J. E. deVries, Characteristics of a monoclonal antibody (WT-31) that recognizes a common epitope on the human T cell receptor for antigen, *J. Immunol.* 135:1922 (1985).

14. H. Band, F. Hochstenbach, J. MacLean, S. Hata, M. S. Krangel and M. B. Brenner, Immunochemical proof that a novel rearranging gene encodes the T cell receptor γ subunit. Science 236:682 (1987).

15. H. Yssel, J. E. DeVries, M. Koken, W. van Blitterswijk and H. Spits, A serum free medium for the generation and propagation of functional human cytotoxic and helper T cell clones, *J. Immunol. Methods* 72:219 (1984).

16. J. J. M. van Dongen, G. W. Krissansen, I. L. M. Wolvers-Tettero, W. M. Comans-Bitter, H. J. Adriaansen, H. Hooijkaas, E.R. van Wering and C. Terhorst, Cytoplasmic expression of CD3 antigen as a diagnostic marker for immature T-cell malignancies, *Blood* 71:603 (1988).

17. H. Spits, H. Yssel, A. Voordouw and J. E. de Vries, The role of T8 in the cytotoxic activity of cloned cytotoxic T lymphocytes lines specific for class II and class I major histocompatibility complex antigens, *J. Immunol.* 134:2294 (1985).

18. L. L. Lanier, A. M. Le, C. I. Civin, M. R. Loken, and J.H. Phillips, The relationship of CD16 (Leu-11) and Leu-19 (NKH-1) antigen expression on human peripheral blood NK cells and cytotoxic T lymphocytes, *J. Immunol.* 136:4480 (1986).

19. J. H. Phillips and L. L. Lanier, Acquisition of non-MHC restricted cytotoxic function IL-2 activated thymocytes with an "immature" antigenic phenotype, *J. Immunol.* 139:683 (1987).

20. J. M. Michon, M. A. Caliguiri, S. M. Hazanow, H. Levine, S. F. Schlossman, and J. Ritz, Induction of natural killer effectors from human thymus with recombinant IL-2, *J. Immunol.* 140:3660 (1988).

21. A. J. Furley, S. Mizutani, K. Weilbaecher, H. S. Dhaliwal, A. M. Ford, L. C. Chan, H. V. Molgaard, B. Toyonaga, T. Mak, P. van den Elsen, D. Gold, C. Terhorst, and M. F. Greaves, Developmentally regulated rearrangement and expression of genes encoding the T cell receptor-T3 complex, *Cell* 46:75 (1986).

22. R. Palacios, M. Kiefer, M. Brockhaus, K. Karjalainen, Z. Dembic, P. Kisielow, and H.

von Boehmer, Molecular, cellular, and functional properties of bone marrow T lymphocyte progenitor clones, *J. Exp. Med.* 166:12 (1987).

23.   J. Pelkonen, P. Sideras, H.-G. Rammensee, K. Karjalainen, and R. Palacios, Thymocyte clones from 14-day mouse embryos I. State of T cell receptor genes, surface markers, and growth requirements, *J. Exp. Med.* 166:1245 (1987).

MONOCLONAL ANTIBODIES AGAINST T CELL RECEPTOR/CD3 COMPLEX INDUCE

CELL DEATH OF Th1 CLONES IN THE ABSENCE OF ACCESSORY CELLS

Yang Liu and Charles A. Janeway, Jr.

Section of Immunobiology
Yale University School of Medicine
Howard Hughes Medical Institute
New Haven, Connecticut

The T cells respond to foreign antigens by clonal expansion but respond to self compo-nents by clonal deletion or functional inactivation.[1-3] The specificity of these different responses is dictated by the T cell antigen receptor (TCR) which recognizes peptides associa-ted with MHC antigens and transduces signals. Numerous studies indicate that such recogni-tion can result in a variety of consequences, including cell proliferation, cytokine produc-tion, cell death,[4] and clonal anergy.[5] However, the mechanisms by which the outcomes of T cell recognition is determined are largely unknown.

Previous studies have suggested that one determinant for the outcome of T cell recogni-tion is the developmental stage of T cells. Thus, treatment of fetal thymus culture with anti-CD3 mAbs results in programmed cell death,[4] while similar treatment of spleen cells which contain mature T cells results in T cell proliferation.[6] In addition, dendritic cells, which are the most potent antigen-presenting cells in mixed lymphocyte reactions,[7] are also the most efficient in inducing tolerance of thymus T cells in vitro and in vivo.[8] Further-more, different subpopulations of T cells in fetal thymus, presumably cells at different developmental stages, differ in their susceptibility to death induced by anti-TCR/CD3.[9] As T cells at different stages of differentiation are in very different environments, it is hard to interpret these results without taking into account signals other than TCR litigation.

Recent experiments have shown that mAbs specific for TCR or CD3 can be used as mimic ligands for the T cell receptor.[6,10] Such mAbs offer an opportunity to study the effect of TCR ligation in the absence of any stimulator cell-derived signals. Here we report that anti-CD3 and anti-TCR mAbs induce cell death in Th1 clones. In addition, IFN-γ pro-duced by such clones plays an important role in this process.

Effect of Anti-T Cell mAbs on Th1 Cells in the Presence or Absence of Accessory Cells

Anti-TCR and anti-CD3 mAbs induce proliferation of the murine Th1 clone 5.9 in the presence of accessory cells. As shown in Table 1, optimal proliferation in response to these stimuli depends on the presence of sufficient numbers of accessory cells. Little prolifera-tion was induced by these mAbs in the absence of accessory cells. A control antibody, anti-Thy 1 mAb, failed to induce proliferation of the 5.9 clone. Furthermore, various concen-trations of anti-T cells mAbs coated on tissue culture plates did not induce proliferation of 5.9 clone either. Such proliferation was restored when mitomycin-C-treated, T-depleted spleen cells were added (Table 2). These results demonstrate that costimulatory signals derived from accessory cells are required for the proliferation of 5.9 in addition to the requirement of cross-linking by anti-T cell mAbs.

Mechanisms of Lymphocyte Activation and Immune Regulation III
Edited by S. Gupta *et al.*, Plenum Press, New York, 1991

105

Table 1. Proliferation of Th1 Clone 5.9 to Anti-TCR/CD3 mAbs in the Presence or Absence of Accessory Cells*

mAbs#	Numbers of accessory cells x 10^{-3}			
	100	33	11	3.6
anti-Thy1	188	136	118	662
anti-CD3	38,573	19,364	6,867	3,851
anti-TCR	39,532	31,326	16,998	15,554

* Clone 5.9 was cultured at a density of 2×10^4/well in the presence of soluble mAbs and given numbers of accessory cells for 42 hr, pulsed with ^3H-TdR for additional 6 hr and ^3H-TdR incorporation was determined. Data shown were means of duplicates. T-depleted BALB/c spleen adherent cells were used as accessory cells. Clone 5.9. is derived from BALB/c mouse as described (13). It recognizes ovalbumin in the context of I-A^d.

mAbs specific for Thy1 (Y19), CD3 ε-chain (YCD3-1) were prepared in this laboratory and have been described before (11, 12) and are used at 5 µg/ml.

Table 2. Proliferation of 5.9 to Plate-Coated Anti-T Cell mAbs

mAbs	5.9	AC	concentration of mAbs (ng/ml)			
			330	110	36	12
Anti-CD3	+	+	11,134	13,113	9,020	12,430
Anti-TCR	+	+	7,006	11,570	10,091	8,056
Anti-Thy1	+	+	536	909	977	739
Nil	+	+	525	761	886	809
Anti-CD3	+	−	256	295	486	1,707
Anti-TCR	+	−	319	236	254	448
Anti-CD3	−	+	231	326	481	361
Anti-TCR	−	+	222	365	559	299

* Clone 5.9 was isolated by lymphocyte separation medium 2 weeks after stimulation with antigen. T cells (2×10^4/well) were cultured with plate-coated mAbs in the presence or absence of T-depleted spleen cells (10^5/well). Proliferation was determined by ^3H-TdR incorporation. Data shown were means of duplicates.

Table 3. Death of 5.9 Clone Induced by Monoclonal Antibodies*

mAbs	concentration (μg/ml)	
	5.0	0.05
Anti-Thy1	14.3	20.6
Anti-CD3	60.3	74.6
Anti-TCR	76.0	82.5

* T cells were cultured in 96-well tissue culture plates which were
precoated with anti-T cell mAbs and percentages of specific cell death
were determined by trypan-blue exclusion at 16 h after culture. The
percentages of specific cell death were calculated according to the
following formula: Specific cell death% = (death% in sample-death% in
medium)/(100%-death% in medium).

Table 4.   Inhibition of IL-2 Responsiveness of 5.9 Clone by Anti-TCR/CD3 mAbs*

mAbs	IL-2	Concentration of antibodies (μg/ml)			
		1.0	0.33	0.11	0.36
Anti-Thy1	+	48.8	44.7	40.7	42.9
	−	1.7	0.3	0.6	0.2
Anti-CD3	+	5.2	3.6	5.2	19.6
	−	1.7	3.2	8.0	17.6
Anti-TCR	+	1.0	0.7	1.0	1.8
	−	1.5 .	1.3	1.2	2.3

* T cells were cultured with plate-coated mAbs and recombinant murine IL-
2 (50 U/ml) for 42 h, pulsed with ^{3}H-TdR for 6 h and ^{3}H-TdR incorporation was determined. Data shown (cpm x 10^{-3}) were means of duplicates.

In the absence of accessory cells, anti-CD3 and anti-TCR mAbs induce death of the 5.9 clone as determined by trypan blue exclusion (Table 3), loss of responsiveness to IL-2 (Table 4), and FACS staining with propidium iodide. As shown in Table 3, about 65-70% of specific cell death was recorded by trypan-blue exclusion, and when 5.9 cells were stained with propidium iodide, anti-CD3 or anti-TCR mAb-treated cultures contained a significant proportion of brightly stained cells (about 18%), while anti-Thy 1 or medium treated cells contained less than 3% of brightly stained cells. Note that different amounts of cell death are recorded by FACS staining with propidium iodide than trypan-blue exclusion. This may well reflect differences in the mechanism of staining by these two dyes. The staining of propidium iodide depends on the integrity of the plasma membranes, nuclear membranes and cellular DNA content. Therefore, this method may not detect cells which have damaged plasma membranes but have intact nuclear membranes or which have degraded their DNA. Trypan blue exclusion only reflects membrane integrity, and therefore will record a higher percentage of cell death than does propidium iodide staining. In fact, a significantly higher percentage of cellular debris which is negative for propidium iodide staining is detected in anti-CD3-treated clone 5.9 (data not shown).

In addition, T cells cultured with medium or with plates coated with anti-Thy1 mAb responded to exogenous IL-2, while such proliferative responses were significantly inhibited by anti-CD3 and anti-TCR mAb. A similar level of cell death was observed in the three other Th1 clones tested (Table 5). Thus, the mAbs which activate these cloned Th1 T cells in the presence of accessory cells induce cell death in the T cells. Thus, mAbs which activate Th1 clones in the presence of accessory cells induce death of the same cells in the absence of costimulation.

Inhibition of Cell Death by CsA

CsA significantly inhibits the death of cloned Th1 cells induced by anti-CD3. Table 6 shows the dose-response curve of CsA-mediated inhibition of cell death, while Table 7 shows proliferation of stimulated T cells to exogenous IL-2 in the presence and absence of CsA. It is evident that CsA reverses anti-CD3 induced death of Th1 cells, scored as loss of IL-2 responsiveness and trypan blue exclusion.

Table 5. Effect of mAbs on Cell Death and IL-2 Responsiveness of Th1 Clones*

| Clones | Specific cell death % | | Proliferation to IL-2 cpmx10-3 | | |
	Anti-Thy1	Anti-CD3	Anti-Thy1	Anti-CD3	medium
5.2	3.3	69.5	86.7	3.7	115.4
5.5	-6.0	67.8	ND	7.2	37.3
5.8	4.6	41.7			

* T cell clones were isolated by lymphocyte separation medium 2 weeks after stimulation with antigen and used at a density of 2 x 10⁴/well. T cells were incubated with medium in the presence or absence of anti-T cell mAbs. Specific cell death% was derived from trypan-blue exclusion experiments and anti-cell mAbs were used at concentration of 5 µg/ml. Proliferation data shown were means of duplicates which derived from a titration of anti-cell mAbs, with mAb concentration of 0.1 µg/ml. Similar effects were observed with mAb concentration ranging from 1 to 0.01 µg/ml.

Table 6. Inhibition of Death of 5.9 by Cyclosporin A*

mAbs	Concentration of cyclosporin A (ng/ml)				
	500	170	60	20	0
Anti-Thy1	-13.4	-5.5	-2.8	-12.3	4.5
Anti-CD3	3.3	10.1	7.6	36.2	64.4
Anti-TCR	2.1	15.3	5.1	48.4	62.1

* T cells were cultured with anti-TCR/CD3 mAbs (5 µg/ml) in the presence of various amounts of cyclosporin A. Cell death was determined by trypan-blue exclusion 24 h after culture.

Table 7. Cyclosporin A Reverses the Inhibition of IL-2 Response of Clone 5.9 by Anti-CD3 mAb*

Treatments	IL-2	Concentration of anti-CD3 (µg/ml)			
		1.0	0.2	0.04	0.008
Anti-CD3	+	645	548	983	1,998
	−	578	784	1,409	909
Anti-CD3+CsA	+	7,332	9,645	11,130	29,738
	−	110	78	92	123

* T cells were cultured in the presence or absence of cyclosporin A (100 ng/ml) for 42 h and pulsed with ^{3}H-TdR for additional 6 h. The data shown (cpm) were means of duplicates. Proliferation of 5.9 in the absence of anti-CD3 mAb was 3,680 (without IL-2) and 66,682 (with IL-2).

Table 8. Inhibition of Cell Death by mAbs Against IFN-γ or IL-2[*]

Anti-cytokine mAbs	Specific cell death %
Anti-IL-2 + anti-IFN-γ	-1.3
Anti-IL-2	71.8
Anti-IFN-γ	-3.7
Nil	65.6

* Cloned 5.9 cells were cultured with anti-CD3 (5 μg/ml) for 24 h in the presence of anti-IL-2 and/or anti-IFN-γ mAbs (200 μg/ml) and cell death was evaluated by trypan-blue exclusion.

Table 9. Anti-IFN-γ mAb Rescue IL-2 Responsiveness which was Inhibited by Anti-CD3*

Treatments	Concentration of anti-CD3			
	1.0	0.3	0.1	0.03
Anti-CD3	5,210	3,697	5,256	19,644
Anti-CD3+ anti-IFN-γ	17,639	21,132	23,744	24,817

* T cells were cultured with recombinant murine IL-2 (50 U/ml) in the presence of anti-CD3 mAb and anti-IFN-γ (100 μg/ml) for 42 h and pulsed with ^{3}H-TdR for additional 6 h. Data shown (cpm) were means of duplicates.

Table 10. Effect of IFN-γ on Death of 5.9 Clone*

Treatments	specific cell death %
Anti-Thy1	-6.0
Anti-CD3	63.7
Anti-CD3+CsA	18.1
Anti-CD3+CsA+IFN-γ	69.9

* T cells were cultured with anti-CD3 mAb (5 μg/ml), cyclosporin A (100 ng/ml) and recombinant IFN-γ (1000 U/ml) for 24 h and cell death was determined by trypan-blue exclusion.

IFN-γ Plays an Important Role in Anti-CD3 Induced Cell Death of T Cell Clones

As CsA is a potent inhibitor of cytokine production by T cells, the observation that CsA inhibits cell death induced by anti-CD3 led us to study the possible role of cytokines produced by anti-CD3 stimulation in the death of cloned T cells. The supernatants of anti-CD3-treated 5.9 clone contain IL-2 activity which is equivalent to 100 U/ml of recombinant IL-2. Such activity was neutralized by anti-IL-2 mAb. Anti-CD3 stimulated 5.9 supernatants also contain IFN-γ activity as determined by inhibition of the proliferation of WEHI 279, a B lymphoma cell line, equivalent to 1000 U/ml of recombinant IFN-γ and was neutralized by anti-IFN-γ mAb. No IL-2 or IFN-γ activity was detected in unstimulated 5.9 supernatant.

To address if IL-2 and IFN-γ might contribute to the death of anti-CD3-activated clone 5.9, we tested the ability of anti-IL-2 and anti-IFN-γ mAbs to inhibit cell death. As shown in Table 8, anti-IFN-γ mAb inhibits cell death induced by anti-CD3, while anti-IL-2 mAb has no effect. Similarly, anti-IFN-γ mAb also reversed the inhibition of IL-2 respon-

Table 11. Inhibition of IL-2 Responsiveness by IFN-γ*

IFN-γ (U/ml)	Proliferation to IL-2 (cpm)
1000	8,240
330	7,332
110	9,645
36	11,130
2	29,738

* T cells were cultured with anti-CD3 (0.1 μg/ml), cyclosporin A (100 ng/ml) and various amounts of IFN-γ for 24 h and pulsed with ^{3}H-TdR for additional 6 h. Data shown were means of duplicates.

siveness induced by anti-CD3 mAb (Table 9). The role of IFN-γ was confirmed by reconstitution experiments. IFN-γ stored cell death which had been inhibited by CsA (Table 10) and also reconstituted the inhibition of IL-2 responsiveness (Table 11).

DISCUSSION

The above data demonstrate anti-CD3/TCR mAbs induce death of Th1 clones and that IFN-γ produced by these clones plays an important role in this process. IFN-γ by itself is not sufficient to induce cell death of Th1 clone as IFN-γ does not cause cell death of Th1 clones in the absence of stimulation by the anti-CD3 mAb (data not shown). The inhibitory effect of IFN-γ on T cell activation provides a possible feedback regulation of Th1 cells by their own product. Our results also indicate that costimulatory signals are at least one of the factors that determine the outcome of TCR ligation. This is consistent with the two-signal theory for T cell activation. This theory argues that T cell activation requires TCR ligation as well as the delivery of costimulatory signals. TCR ligation in the absence of costimulation results in inactivation of the T cells. These results show that, in addition to the so-called "T cell clonal anergy" described by Mueller et al,.[3] stimulation via the TCR in the absence of costimulation also causes cell death.

A critical phenomenon in immunology is the elimination of immature T cells by TCR engagement (thymic deletion), while mature T cells can be expanded by an apparently similar engagement. The simplest model to account for these different responses is that in immature T cells, TCR engagement leads to cell death and that this mechanism is altered during development such that different second messenger is utilized in mature T cells. This may correlate with changes in the expression of the TCR-associated proteins ζ and η. Our findings that death can also be induced by CD3/TCR ligation in mature T cells indicates that such a simple model is unlikely to be correct. We would like to propose a unified model which would explain T cell activation, clonal deletion and peripheral tolerance. In this model, the outcome of TCR litigation is also determined by the delivery of costimulatory signals and will therefore be eliminated by TCR ligation, even though the same stimulator cells could be highly immunogenic for mature T cells, as suggested by the work of Matziger and coworkers.[8] Mature T cells express such receptors, and therefore respond to antigens presented by cells with costimulatory activity. Presentation of antigens by cells which do not have costimulatory activity to T cells results in "clonal anergy" or in extreme cases as shown here, clonal deletion, hence generating peripheral tolerance.

This hypothesis has one important caveat. IFN-γ has been shown to be a potent inducer of costimulatory activity in human monocytes[16] and murine B cells.[17] Thus IFN-γ released by Th1-like cells, induced by reacting with self cells, could either result in cell death or in the activation of costimulatory activity leading to T cell proliferation. Resting T cells, which must first be activated to secrete IFN-γ on binding ligand, may neither die nor become activated but rather enter a state of clonal anergy on encountering ligand in tissue cells that lack costimulatory activity. This mechanism could account for the usual tolerance to tissue antigens. The mechanism we have described would thus be a fall-back or fail-safe mechanism to delete those autoreactive T cells that have, perhaps through activation by pathogens, already reached effector states.

SUMMARY

We have used anti-T cell mAbs as mimic ligands to study the effects of TCR/CD3 ligation of Th1 clones in the presence or absence of accessory cells. Our results demonstrated that ligation of TCR/CD3 in the presence of accessory cells induces proliferation of Th1 clone, while the same ligation in the absence of accessory cells results in death. This effect is inhibited by cyclosporin A and by anti-IFN-γ mAbs and is restored by adding exogenous recombinant IFN-γ tb CsA treated cells. We propose a model which could provide a general framework to explain activation, clonal anergy as well as clonal deletion of T lymphocytes during thymic development and in the peripheral.

ACKNOWLEDGMENT

Supported by NIH grant AI-26810 to C. A. Janeway; Y. Liu is a recipient of Irvington post-doctoral fellowshhip.

REFERENCES

1. B. A. Askonas, A. Müllbacher, and R. B. Ashman, Cytotoxic T memory cells in viral infection and specificity of helper T cells, *Immunology* 42:79 (1982).
2. J. W. Kappler, N. Roehm, and P. Marrack, T cell tolerance by clonal elimination in the thymus, *Cell* 59:273 (1987).
3. D. L. Mueller, M. K. Jenkins, and R. H. Schwartz, Clonal expansion versus functional clonal inactivation: a costimulatory signaling pathway determines the outcomes of T cell receptor occupancy, *Annu. Rev. Immunol.* 7:445 (1989).
4. C. A. Smith, G. T. Williams, R. Kingston, E. J. Jenkinson, and J. J. T. Owen, Antibodies to the CD3/T cell receptor complex induce apoptosis (controlled cell death) in immature T cells in thymic culture, *Nature* 337:181 (1989).
5. M. K. Jenkins, D. M. Pardoll, J. Mizuguchi, T. M. Chused, and R. H. Schwartz, Molecular events in the induction nonresponsiveness state in IL-2 producing T lymphocyte clones, *Proc. Natl. Acad. Sci. USA* 84:5409 (1987).
6. O. Leo, M. Foo, D. H. Sachs, L. E. Samelson, and J. A. Bluestone, Identification of mAb specific for a murine T3 polypeptide, *Proc. Natl. Acad. Sci. USA* 84:1374 (1987).
7. K. Inaba, M. D. Witmer-Pack, M. Inaba, S. Muramatsu, and R. M. Steinman, The function of Ia$^+$ dendritic cells and Ia$^-$ dendritic cell precursors in thymocyte mitogenesis to lectin and lectin plus IL-1, *J. Exp. Med.* 167:149 (1988).
8. P. Matzinger and S. Guerder, Does T cell tolerance require a dedicated antigen-presenting cell? *Nature* 338:74 (1989).
9. T. H. Finkel, J. C. Cambier, R. T. Kubo, W. K. Born, P. Marrack, and J. Kappler, The thymus has two functionally distinct population of immature $\alpha\beta^+$ T cells; one population is deleted by ligation of $\alpha\beta$ T cell receptor, *Cell* 58:1041 (1989).
10. T. H. Finkel, P. Marrack, J. W. Kappler, R. T. Kubo, and J. C. Cambier, $\alpha\beta$ cell receptor and CD3 transduce different signals in immature T cells: implications for selection and tolerance, *J. Immunol.* 142:3006 (1989).
11. B. Jones, Functional activities of antibodies against brain associated T cell antigens. II. Stimulation of T cell induced B cell proliferation, *Eur. J. Immunol.* 12:30 (1982).
12. P. Portoles, J. Rojo, A. Golby, M. Bonneville, S. Gromkowski, L. Greenbaum, C. A. Janeway, Jr., D. B. Murphy, and K. Bottomly, Monoclonal antibodies to murine CD3ϵ define distinct epitopes, one of which may interact with CD4 during T cell activation, *J. Immunol.* 142:4169 (1989).
13. P. Conrad and C. A. Janeway, Jr., The expression of I-E^d molecules in F1 hybrid mice detected with antigen-specific, I-E^d-restricted cloned T cell lines, *Immunogenetics* 20:311 (1984).
14. T. R. Mosmann, H. Cherwinski, M. W. Bond, M. A. Giedlin, and R. L. Coffman, Two types of murine T cell clones. I. Definition according to profiles of lymphokine activities and secreted proteins. *J. Immunol.* 136:2348 (1986).
15. F. Finkelman, I. M. Katona, T. R. Mosmann, and R. L. Coffman, Interferon-γ regulates the isotypes of Ig-secreted during *in vivo* humoral immune response, *J. Immunol.* 140:1022 (1988).
16. K. Kawakami, Y. Yamamoto, K. Kashimoto, and K. Onoue, Requirement for delivery of signals by physical interactions and soluble factors from accessory cells in the induction of receptor mediated T cell proliferation, *J. Immunol.* 142:1818 (1989).
17. C. M. Harwryowize and E. R. Unanue, Regulation of antigen-presentation: I. IFN-γ induces antigen-presenting properties of B cells, *J. Immunol.* 141:4083 (1989).

IDENTIFICATION OF A NOVEL THYMOCYTE GROWTH FACTOR DERIVED FROM B

CELL LYMPHOMAS

Takashi Suda, Ian MacNeil, Melissa Fischer, Kevin W. Moore, and Albert Zlotnik

Department of Immunology
DNAX Research Institute of Molecular and Cellular Biology
Palo Alto, California

INTRODUCTION

T cell ontogeny is a rapidly developing area. However, the signals that mediate the differentiation and growth of immature T cells in the thymus through different maturational stages remain largely unknown. While much has been accomplished in understanding the role that molecules of the major histocompatibility complex have in the process of selection of the T cell repertoire,[1,2] little is known about the role that cytokines may play in these processes. Most of these studies have focused on IL-2;[3,4] however, its role in T cell ontogeny is still controversial. More recently, it has been observed that IL-4 can induce thymocyte growth in combination with phorbol ester (PMA).[5,6] Other cytokines that have been shown to have thymocyte growth factor properties include IL-1,[7] IL-6,[8,9] IL-7,[10] TNFα,[11,12] and in fetal thymocytes (day 15 of gestation) P40 (IL-9)[13] and GM-CSF.[14] Many of these cytokines have been shown to be produced by thymocytes and/or thymic stromal cells as well,[13,15] suggesting that they play a role in T cell ontogeny. We have been exploring the response patterns of adult murine thymocyte subsets separated on the basis of their expression of CD3, CD4 and CD8, and in fetal thymocytes at day 15 of gestation. An important conclusion from these studies is that the growth-promoting effects of some of these cytokines are specific for certain thymocyte subsets. For example, IL-1 is specific for the $CD3^+4^-8^-$ subset while IL-6 is specific for the $CD4^+8^-$ and $CD4^-8^+$ subsets.[8,12] These studies have shown that very few cytokine combinations induce the proliferation of the most immature thymocyte subset: $CD3^-4^-8^-$ thymocytes.[12] Therefore, we have endeavored to find other novel cytokines that may activate this subset.

Recently, we have identified a novel cytokine derived from several B cell lymphomas with unique thymocyte growth cofactor properties.[14] We found this activity while studying the ability of B cell lymphomas expressing allogeneic class II MHC antigens to stimulate thymocyte subsets. Some subclones derived from the CH12 B cell lymphoma[16] induced strong thymocyte proliferation in the presence of IL-2 and IL-4. However, these responses were not MHC-restricted, and control experiments indicated that the activity was mediated by a soluble molecule produced by these B cell lymphomas. Thus, we named this activity B cell-derived T cell growth factor (B-TCGF). Here, we describe some characteristics of this molecule and its effects on thymocyte proliferation. In addition, we have recently observed that this molecule is similar to a factor produced by "type 2" helper T cell (Th2) clones which inhibits cytokine production in "type 1" helper T cell (Th1) clones (Cytokine Synthesis Inhibitory Factor; CSIF).[17] The cDNA encoding this novel molecule was recently isolated at our institution[18] and the recombinant material has all the characteristic activity of B-TCGF. Given its expanding range of activities, we have now termed this activity Interleukin 10 (IL-10).

Mechanisms of Lymphocyte Activation and Immune Regulation III
Edited by S. Gupta *et al.*, Plenum Press, New York, 1991

115

MATERIALS AND METHODS

Preparation of Adult and Fetal Thymocytes

3-7 week old female BALB/c mice (Simonsen Laboratories, Gilroy, CA) were used as source of adult thymocytes. Day 15 FT were obtained from timed pregnant BALB/c mice (Simonsen Labs). The day of gestation was calculated by plug date and verified by fetal morphology. Fetal thymic lobes were obtained using microdissection forceps under a dissecting microscope. Single cell suspensions were prepared by gentle teasing between two glass slides.

B Cell Lymphoma Cell Lines

Various subclones derived from CH12 B cell lymphomas were kind gifts of Dr. G. Haughton (University of North Carolina).

Cytokines and Antibodies Against Cytokines

The cytokines and the neutralizing anti-cytokine antibodies used in this study have been described previously.[12,14]

Separation of Thymocyte Subsets

Adult thymocyte subsets were sorted on a Becton Dickinson FACS IV as described.[19] Briefly, thymocytes were incubated with either monoclonal anti-CD4 antibodies (RL172) and/or monoclonal anti-CD8 (Lyt-2.2) antibodies (AD4, purchased from Cedarlane Laboratories Limited, Ontario, Canada) as needed for 45 min on ice, followed by treatment with low toxicity rabbit complement (Cedarlane) for 30 min at 37°C. Cells were then stained with phycoerythrin-conjugated monoclonal anti-CD4 antibodies (GK 1.5) and fluorescein-conjugated monoclonal anti-CD8 antibodies (53-6.7, Becton Dickinson, Mountain View, CA). $CD4^-8^-$ double negative $CD4^-8^-$ (DN) and $CD4^+$ single positive (SP) cells were routinely greater than 99% pure, while the $CD8^+$ SP were greater than 98% and $CD4^+8^+$ double positives (DP) were greater than 97% pure.

Thymocyte Proliferation Assay

10^5 cells were cultured with supernatants (SN) from B lymphomas or various recombinant cytokines in the presence of IL-2 (500 units/ml) and/or IL-4 (250 units/ml) in 300 μl of RPMI medium supplemented with L-glutamine (200 mM), MEM-amino acids, MEM-vitamins, sodium bicarbonate, penicillin/streptomycin (Sigma), 5×10^{-5} M 2-mercaptoethanol and 10% fetal calf serum (JR Scientific) in a flat bottom 96 well culture plate for 4 days. The wells were then pulsed with 1 μCi/well [^{3}H]-thymidine and harvested 18 hrs later.

RESULTS

Supernatants from the B Cell Lymphoma CH12LX.4866 Produce B-TCGF

We first detected this activity while studying the response of thymocytes to B cell lymphomas plus IL-2 and IL-4. Adult unseparated thymocytes proliferated with several lines derived from the CH12 B cell lymphoma. Control experiments showed that the activity was found in SN of many of these lymphomas, with one of the highest titers found in SN from CH12.LX 4866. We therefore selected this lymphoma in order to characterize and isolate the molecule responsible for this activity.

B-TCGF is Not Mediated by a Known Cytokine

We have previously studied the ability of several cytokines to induce proliferation of day 15 fetal thymocytes (day 15 FT)[13] or various subsets of adult thymocytes.[12] We found two types of thymocyte growth factors: primary growth factors that induce substantial proliferation of thymocytes when used alone, and secondary growth factors that enhance prolifer-

116

Table 1. Summary of Growth Factor Properties of Various Known Cytokines on Thymocyte Subsets

Cytokine	Primary or Secondary	Target Thymocyte Subsets
IL-1	Secondary (IL2, IL7)[1]	Adult CD3+4-8-
IL-2	Primary	All tested[2]
IL-4	Primary	Adult CD4+8, CD4-8+
IL-6	Secondary (IL2, IL4, IL7)[1]	Adult CD4+8, CD4-8+
IL-7	Primary	All tested[2]
IL-9 (P40)	Secondary (IL2)[1]	Day 15 FT
TNFα	Secondary (IL2, IL7[3])[1]	All tested[2]
GM-CSF	Secondary (IL2, IL7)[1]	Day 15 FT

1: When used with these cytokines as primary growth factors.
2. Except CD4+8+ thymocytes.
3: TNFa enhanced IL-7 response only when adult thymocytes were used

ation when added to a primary growth factor, although they do not induce proliferation by themselves. These findings are summarized in Table 1. When compared with these patterns, B-TCGF induced a novel pattern of responses in various thymocyte subsets used with IL-2 and/or IL-4.[14] These results are summarized in Table 2. A comparison of Table 1 and Table 2 indicates that no single known cytokine could account for the pattern of responses observed with B-TCGF.

We had selected the B cell lymphoma CH12.LX 4866 as our source of B-TCGF. We knew that this lymphoma produces the following cytokines as detected by bioactivity: IL-3, IL-6, TNFα, and GM-CSF. It does not produce IL-1, IL-2, IL-5, IL-7, P40 or IFNγ.[19] IL-6 and TNFα were ruled out as possible sources of the activity since neutralizing monoclonal antibodies directed against these cytokines failed to inhibit the B-TCGF activity.[14]

Table 2. B-TCGF Enhances Proliferative Responses of Thymocyte Subsets to IL-2 or IL-4

Thymocyte Subset	B-TCGF enhances the response to	
	IL-2	IL-4
Day 15 FT	Yes	No
Adult CD4-8-3-	Yes	N.D.
Adult CD4-8-3+	Yes	N.D.
Adult CD4+8+	No	No
Adult CD4+8-	Yes	Yes
Adult CD4-8+	Yes	Yes

N.D.. Not Done

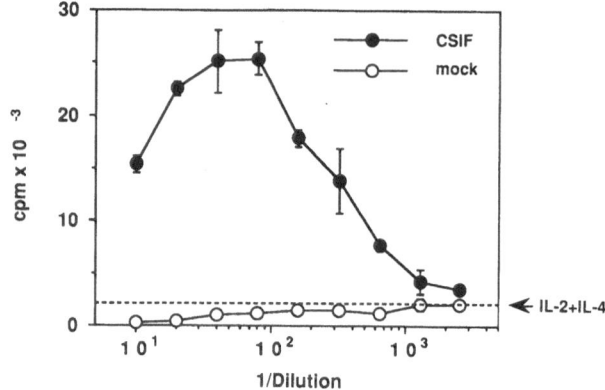

Figure 1. IL-10 enhances IL-2/IL-4 driven proliferation of adult thymocytes. Unseparated adult thymocytes were incubated at 5×10^5 cells/ml with 500 units/ml IL-2 and 250 units/ml IL-4. Supernatants from IL-10 transfected COS7 cells (solid circles) or mock transfected COS7 cells (open circles) were added at various dilutions. Wells were pulsed at day 4 with tritiated thymidine.

B-TCGF activity eluted as a single peak with an apparent molecular weight of 27-50 Kd (data not shown). Given these observations, we concluded that the B-TCGF activity was mediated by a novel cytokine.

B-TCGF is CSIF

During the isolation and characterization of B-TCGF, another cytokine activity, designated Cytokine Synthesis Inhibitory Factor (CSIF), was described by Fiorentino et al.[17] A cDNA encoding this novel molecule has recently been isolated at our institution and has been designated IL-10.[18] We tested recombinant IL-10 in the B-TCGF assay. Supernatants derived from COS 7 cells transfected with the IL-10 cDNA were added to fetal or adult thymocyte populations. Both adult (Figure 1) and fetal (data not shown) thymocytes proliferated strongly in response to recombinant IL-10 in a dose-dependent fashion. IL-10 by itself had no effect. Recombinant IL-10 was able to stimulate various subsets of adult thymocytes separated on the basis of their CD4 and CD8 surface expression. Moreover, the pattern of proliferative responses of thymocyte subsets to IL-10 was exactly the same as the response pattern to B-TCGF (summarized in Table 2). In addition to its ability to stimulate murine thymocytes, the CH12.LX 4866 B cell lymphoma expresses IL-10 mRNA.[18,19] Thus, we conclude that IL-10 is very likely to mediate the B-TCGF activity observed in supernatants of CH12.LX 4866.

DISCUSSION

We have demonstrated thymocyte proliferation induced by a novel cytokine that we initially designated B-TCGF.[14] We conclude that this activity is probably mediated by the same molecule which is produced by Th2 clones and inhibits cytokine synthesis (Cytokine Synthesis Inhibitory Factor, CSIF) in Th1 clones.[17] This conclusion is based on the following observations: 1) the B cell lymphoma CH12.LX 4866 expresses CSIF mRNA and 2) recombinant CSIF has B-TCGF activity. We have also recently observed that anti-CSIF neutralizing antibodies inhibit the activity of natural B-TCGF.[20] Given its unique nucleotide sequence[18] and its various biological properties, this novel cytokine has been designated IL-10.

As we reported here, one of the main characteristics of this molecule is its ability to act as a thymocyte growth cofactor in the presence of IL-2 and/or IL-4. The requirement for

IL-4 is not absolute and varies depending on the target thymocyte population used in the assay. Fetal thymocytes, for example, appear to require only IL-2 for optimum proliferation with IL-10. Finally, we have observed IL-10 production by several stimulated adult thymocyte subsets as well as by day 15 FT,[20] suggesting that IL-10 plays a role in normal thymocyte development. Thus, IL-10 can be added to an expanding list of cytokines that have demonstrated effects on thymocyte growth and differentiation. The availability of recombinant IL-10 should facilitate studies aimed at establishing its role in T cell ontogeny.

ACKNOWLEDGEMENTS

The authors wish to thank James Cupp, Josephine Polakoff, Anne O'Garra, and Tim Mosmann for their help during these studies, as well as Allan Waitz for reading the manuscript. DNAX Research Institute is supported by Schering-Plough Corporation.

REFERENCES

1. J. Kappler, N. Roehm, and P. Marrack, T cell tolerance by cloncal elimination in the thymus, *Cell* 49:273 (1987).
2. P. Kisielow, H. Teh, H. Bluthmann, and H. von Boehmer, Positive selection of antigen-specific T cells in the thymus by restricting MHC molecules, *Nature* 332:730 (1988).
3. R. Palacios and H. von Boehmer, Requirements for growth of immature thymocytes from fetal and adult mice *in vitro, Eur. J. Immunol.* 16:12 (1986).
4. R. Raulet, Expression and function of interleukin-2 receptors on immature thymocytes, *Nature* 314:101 (1985).
5. A. Zlotnik, J. Ransom, F. Frank, M. Fischer, and M. Howard, Interleukin 4 is a growth factor for activated thymocytes: possible role in T cell ontogeny, *Proc. Natl. Acad. Sci. USA* 84:3856 (1987).
6. R. Palacios, P. Sideras, and H. von Boehmer, Recombinant interleukin 4 promotes growth and differentiation of intrathymic T cell precursors from fetal mice *in vitro, EMBO J.* 6:91 (1987).
7. P. Conlon, C. Henney, and S. Gillis, Cytokine-dependent thymocyte responsess: Characterization of IL-1 and IL-2 target subpopulations and mechanism of action, *J. Immunol.* 128:797 (1982).
8. P. Hodgkin, J. Cupp, A. Zlotnik, and M. Howard, IL-2, IL-6 and IFNγ have distinct effects on the IL-4 plus PMA induced proliferation of thymocyte subpopulations, *Cell Immunol.* 126:5 (1990).
9. W. F. Chen, M. Fischer, G. Frank, and A. Zlotnik, Distinct patterns of lymphokine requirement for the proliferation of various subpopulations of activated thymocytes in a single cell assay, *J. Immunol.* 143:1598 (1989).
10. R. Murray, T. Suda, N. Wrighton, F. Lee, and A. Zlotnik, IL-7 is a growth and maintenance factor for mature and immature thymocyte subsets, *Intl. Immunol.* 1:526 (1989).
11. G. E. Ranges, A. Zlotnik, T. Espevik, C. Dinarello, A. Cerami, and M. Palladino, Tumor necrosis factor α/cachectin is a growth factor for thymocytes, *J. Exp. Med.* 167:1472 (1988).
12. T. Suda, R. Murray, C. Guidos, and A. Zlotnik, Growth promoting activity of IL-1α, IL-6, and TNFα in combination with IL-2, IL-4 or IL-7 on murine thymocytes: Differential effects on CD4/CD8 subsets and on $CD3^+$/$CD3^-$ double negative thymocytes, *J. Immunol* 144:3039 (1990).
13. T. Suda, R. Murray, M. Fisher, T. Yokota, and A. Zlotnik, Tumor necrosis factor α and P40 induce day 15 murine fetal thymocyte proliferation in combination with Interleukin 2, *J. Immunol.* 144:1783 (1990).
14. T. Suda, A. O'Garra, I. MacNeil, M. Fischer, M. Bond, and A. Zlotnik, Identification of a novel thymocyte growth promoting factor derived from B cell lymphomas *Cell Immunol.* 129:228 (1990).
15. J. Ransom, M. Fischer, T. Mosmann, T. Yokota, D. DeLuca, J. Schumacher, and A. Zlotnik, Interferon-γ is produced by activated immature thymocytes: Effects on Ia induction on thymic epithelial cells and on interleukin-4-mediated immature thymocyte proliferation, *J. Immunol.* 139:4102 (1987).

16. P. Willoughby, J. Jenette, and G. Haughton, Analysis of a murine B cell lymphoma, CH44, with an associated non-neoplastic T cell population. I. Proliferation of normal T lymphocytes is induced by a secreted product of the malignant B cells, *J. Pathol.* 133:507 (1988).

17. D. Fiorentino, M. Bond, and T. Mosmann, Two types of mouse helper T cell. IV. Th2 clones secrete a factor that inhibits cytokine production by Th1 clones, *J. Exp. Med.* 170:2081 (1989).

18. K. Moore, P. Vieira, D. Fiorentino, M. Trounstine, T. Khan, and T. Mosmann, Homology of cytokine synthesis inhibitory factor (IL-10) to the Epstein-Barr virus gene BCRF1, Science 248:1230 (1990).

19. A. O'Garra, G. Stapleton, V. Dhar, M. Pearce, J. Schumacher, H. Rugo, D. Barbis, A. Stall, J. Cupp, K. Moore, P. Vieira, T. Mossman, A. Whitmore, L. Arnold, G. Haughton, and M. Howard, Production of known and novel cytokines by mouse B lymphomas and normal Ly$^-$1$^+$ B cells *Internat. Immunol.* (in press).

20. I. MacNeil, T. Suda, K. Moore, T. Mossman, and A. Zlotnik, Interleukin 10: A novel cytokine growth cofactor for mature and immature murine lymphocytes (submitted for publication).

EXPRESSION OF RECEPTORS FOR INTERLEUKIN 4 AND INTERLEUKIN 7 ON HUMAN T CELLS

Richard J. Armitage, Steven F. Ziegler, M. Patricia Beckmann, Rejean L. Idzerda, Linda S. Park, and William C. Fanslow

Immunex Corporation, Seattle, WA, USA

SUMMARY

Human recombinant interleukin 4 (IL-4) and interleukin 7 (IL-7) have been modified with biotin-N-hydroxysuccinimide and used to examine the expression of human IL-4 and IL-7 receptors (R) on activated peripheral blood T cells by flow cytometry. Freshly isolated T cells expressed only a low level of IL-4R which remained unchanged when cells were cultured in the absence of stimuli. In the presence of IL-4, IL-7, phytohemagglutinin A (PHA) or immobilized CD3 monoclonal antibody the intensity of biotinylated IL-4 staining increased approximately twofold on the majority of cells. A combination of mitogen with either IL-4 or IL-7 caused a considerable increase in IL-4 receptor expression over that seen in the presence of mitogen alone. IL-2 alone failed to induce IL-4R although it was able to cause a significant increase in receptor expression on T cells co-cultured with PHA or CD3.

Freshly isolated T cells expressed high levels of IL-7R, as determined by biotinylated IL-7 binding and flow cytometry, which did not change significantly with culture in medium alone. Stimulation with PHA, Concanavalin A (Con A) or CD3 had little effect on the intensity of staining. In contrast, activation with phorbol ester resulted in a decrease in IL-7R expression. Similarly, in the presence of IL-4 or IL-7, but not IL-2, the intensity of staining with biotinylated IL-7 was lowered. Analysis of purified T-cell populations showed that IL-7R were present, and IL-4R could be induced, on both CD4[+] and CD8[+] populations.

Analysis of IL-4 receptor expression by this flow cytometric technique was supported by results from ^{125}I-labeled IL-4 binding and by Northern blot analysis of mRNA levels. Taken together, the results of these studies show that the use of biotinylated cytokines and flow cytometry provides a very sensitive method with which to study the expression and regulation of cytokine receptors.

INTRODUCTION

Interleukin 4 was originally described as a cytokine which induced proliferation of B cells pre-activated with anti-IgM[1] and was subsequently found to stimulate normal T cells[2,3] and T cell lines.[4] On B cells, IL-4 has been shown to induce increased expression of class II MHC,[5,6] CD23[78] and surface IgM[9] as well as augmenting secretion of IgE and IgG1 antibody.[10,11] IL-4R have been detected on a wide range of murine and human hematopoietic and non-hematopoietic cell lineages[12-14] suggesting that IL-4 might be involved in growth regulation of a wider variety of cells and tissue.

Mechanisms of Lymphocyte Activation and Immune Regulation III
Edited by S. Gupta *et al.*, Plenum Press, New York, 1991

121

Interleukin 7, originally described as having the ability to support growth of early B cells,[15] has also been shown to induce IL-2 receptor expression and to be a potent costimulus for mitogen-activated murine and human mature T cells.[16,17]

Using a biotinylation procedure to chemically modify recombinant human IL-4 and IL-7, we have examined the binding of these cytokines to peripheral blood T cells in various states of activation by flow cytometric analysis. Data presented here demonstrate that binding of biotinylated IL-4 and IL-7 to receptor-expressing cells is an extremely sensitive means by which the regulation of expression of receptors for these cytokines can be studied. Using this technique, it was found that IL-4 and IL-7 each enhanced expression of IL-4R on T cells, as did PHA and immobilized CD3 mAb either alone, or in concert with IL-2, IL-4 or IL-7. These findings suggest that the capacity to enhance expression of IL-4R on T cells may be a common feature shared by various lymphokines and mitogens. In contrast, IL-7R were found to be expressed at high levels on freshly isolated T cells and this expression either remained unchanged or was reduced upon exposure of cells to a range of stimuli. Results from flow cytometry using biotinylated IL-4 were substantiated by ^{125}I-labeled IL-4 binding studies and were found to be proportional to levels of IL-4R gene expression seen following T-cell activation.

MATERIALS AND METHODS

Cell Separation

Human peripheral blood mononuclear cells were isolated by centrifugation over Histopaque (Sigma Chemical Co., St. Louis MO). T cells were further isolated by rosetting with AET-treated sheep RBC. Adherent cells were removed from the T cell-enriched population by passage over Sephadex G10. Resultant T cells were $\geq 98\%$ pure with $< 1\%$ of cells being of the reciprocal T-cell subset as determined by flow cytometry.

Culture Conditions

All cultures were carried out in RPM-I + 10% heat inactivated FCS at 37°C in a 10% CO_2 atmosphere. For measurement of IL-4R and IL-7R expression, cells were cultured with the appropriate mitogen or lymphokine for 48 hrs prior to analysis by FACScan (Becton Dickinson, Mountain View, CA). Cultures containing IL-4 or IL-7 were washed twice with RPMI + 10% FCS after 24 hours and recultured for a further 24 hours in medium alone to prevent subsequent inhibition of biotinylated IL-4 or IL-7 binding by the relevant cytokine present in the culture medium. Activation with immobilized OKT3 was performed by coating wells with 5 μg/ml purified mAb in PBS for 1 hr at 37°C. Wells were washed 3X with PBS prior to the addition of cells.

Biotinylation of Cytokines

Biotin-N-hydroxysuccinimide (ICN Immunobiologics) and purified human IL-4 were mixed at a molar ratio of 2:1. 100μg rIL-4 (8 mg/ml) in 0.1 M $NaHCO_3$pH 8.5 was mixed with 1.7 μl biotin-NHS (10 mg/ml) in DMSO) for 30 min at room temperature with mixing every 10 min. At the end of the incubation period the reaction mixture was microfuged through a 1 ml Sephadex G-25 (Pharmacia) desalting column and the eluate adjusted to 100 μg/ml in PBS + 0.02% NaN_3 assuming 100% recovery of IL-4.

IL-7 was coupled by the same procedure except using a ratio of 6:1 biotin-NHS to IL-7 (100 μg IL-7 (5 mg/ml) + 0.85μl biotin-NHS). These conditions for the biotinylation of IL-4 and IL-7 were previously determined to be optimal for each cytokine.

Both IL-4 and IL-7 have previously been shown to provide a proliferative signal to human T cells activated with PHA of CD3 mAb.[17] Following biotinylation, labeled IL-4 and IL-7 were found to have comparable biological activity to the unmodified forms of both cytokines in a costimulation assay using PHA-activated PB T cells.

Radioreceptor Binding Assays

Radiolabeling of IL-4 was performed by the enzymobead method as described.[21] Specific activities of 0.5 - 1.0 x 10^{16} cpm/mmol were routinely achieved. Saturation binding assays were performed using the phthalate oil separation method.[22] Briefly, 50 μl of PBT cells (4 x 10^7 cells/ml) treated with various cytokines were incubated with a range of concentrations of ^{125}I-IL-4 in binding medium (RPMI medium containing 2.5% BSA, 0.2% sodium azide and 0.02 M HEPES, pH 7.4) for one hour at 37°C. Replicate aliquots of cell suspension were then transferred to pre-cooled polyethylene centrifuge tubes containing a phthalate oil mixture. Cells binding ^{125}I-IL-4 were separated from unbound cytokine by centrifugation and quantitation was performed on a gamma counter. Binding data analyses were done on RS/1 (Bolt, Beranek and Newman, Boston, MA), a commercially available data processing package.

Northern Blot Analysis

Peripheral blood T cells from three donors were isolated and cultured as for flow cytometry and then pooled just prior to RNA isolation. Total RNA was isolated using guanidine hydrochloride, essentially as previously described,[23] and 9 μg samples were electrophoresed through formaldehyde/agarose.[24] 28S and 18S ribosomal RNA bands were visualized by staining with methylene blue, verifying that each land had equivalent amounts of RNA. The RNA was transferred to a Hybond-N filter (Amersham Corporation, Chicago IL) and hybridized to a ^{32}P-labeled human IL-4 receptor riboprobe.[25]

RESULTS

Enhancement of huIL-4R Expression

Purified T cells were cultured for 48 hr with IL-4 or IL-7 in the presence or absence of PHA. Binding of biotinylated IL-4 to the cultured cells was assessed by flow

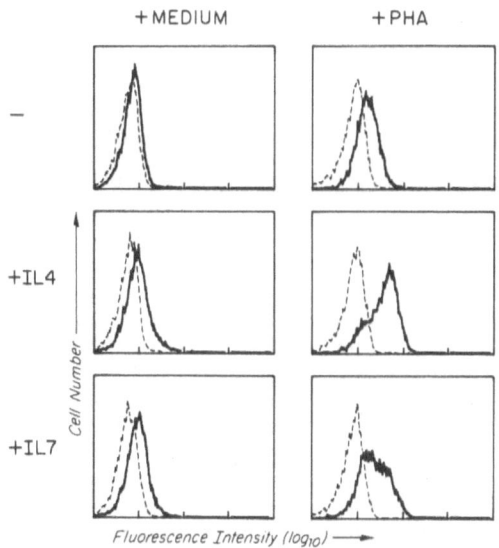

Figure 1. Flow cytometric analysis of biotinyl-
lated IL-4 binding in the presence (----) or
absence (___) of excess non-biotinylated
IL-4. PBT cells were cultured 48 hr with
medium or PHA (0.25%) in the presence or
absence of IL-4 (10 ng/ml) or IL-7 (10
ng/ml).

Table 1. IL-4 Receptor Expression on Unfractionated, CD4[+] and CD8[+] T Cells

Stimulus[a]	E+		CD4+		CD8+	
	% Positive[b]	MFI[c]	% Positive	MFI	% Positive	MFI
Medium	2	8	10	10	2	6
IL-2	2	7	8	9	3	6
IL-4	19	17	21	18	14	14
IL-7	22	19	18	16	22	18
PHA	22	20	20	16	22	19
PHA + IL-4	52	31	54	30	45	25
PHA + IL-7	47	29	49	27	49	26

a Unfractionated (E+), CD4+ or CD8+ PB T cells were cultured for 48 hrs with medium or PHA
   (0.25%) in the presence or absence of IL2, IL4 or IL7 (each 10ng/ml).

b Flow cytometric determination of percent cells staining with biotinylated IL-4 and strep-
   tavidinphycoerythrin with percent background fluorescence subtracted.

c Mean fluorescence intensity determined on a log scale.

cytometry (Figure 1). Incubation with IL-4 or IL-7 caused a significant increase in IL-4
expression over that seen on cells cultured in medium alone which showed the same low
level of expression as freshly isolated T cells (data not shown). Culture in the
presence of PHA induced a similar level of IL-4R to that seen in the presence of IL-4 or
IL-7 alone. The highest levels of IL-4R expression were observed on T cells co-activated
with IL-4 or IL-7 in the presence of PHA, which showed an increase in mean fluorescence
intensity (MFI) over cells cultured in medium alone of 31 and 25 respectively.
Activation of T cells with immobilized OKT3 resulted in enhancement of IL-4R to a level
comparable with that seen with PHA. This expression was further enhanced by co-culturing
with IL-4 or IL-7 (Table 1). IL-4R expression was not increased when T cells were
cultured with IL-2 alone, although IL-2 in combination with immobilized OKT3 or PHA
induced a considerably higher level of IL-4 receptor expression than seen with either
mitogen alone (Table 1).

Both CD4[+] and CD8[+] Subpopulations Display Enhanced IL-4R

Highly purified T cells were further fractionated into CD4[+] and CD8[+] subsets as
described in Materials and Methods. Cells were cultured with IL-4 or IL-7 in the
presence or absence of PHA for 48 hrs and analyzed for biotinylated IL-4 staining by flow
cytometry (Table 1). Cells of both subpopulations cultured with medium alone showed
relatively low levels of staining, although the percent positivity was consistently
higher on the CD4[+] cells than on the CD8[+] subset. A significant increase over this
level of staining was seen when either subpopulation was cultured with IL-4, IL-7, or
PHA. As was seen with staining of unfractionated (E[+]) T cells (Table 1), the highest
levels of IL-4R expression were induced when CD4[+] or CD8[+] T cells were stimulated
with IL-4 or IL-7 in the presence of PHA. In three experiments performed no clear
differences were noted in the level of enhanced IL-4R expression on CD4[+] versus CD8[+]
T cells after culture. Purified, unfractionated T cells cultured with PHA + IL-4 were
double stained with biotinylated IL-4 and CD4 or CD8 mAb confirming the presence of
enhanced IL-4R expression on both subpopulations (Figure 2A & B). Double staining of the
same cells with biotinylated IL-4 and the anti-IL-2 receptor mAb, 2A3, showed that
virtually all the IL-4R+ cells also expressed IL-2R, while a significant proportion of T
cells expressed IL-2 receptors but bound little or no biotinylated IL-4 (Figure 2C).

Radiolabeled IL-4 Binding to T. Cells

IL-4R expression was assessed by binding of ^{125}I-IL-4 to unfractionated T cells

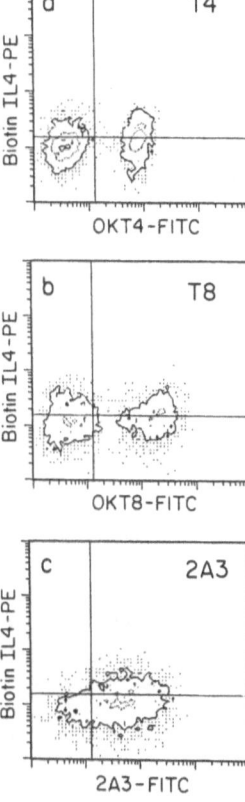

Figure 2. PBT cells were cultured 48 hr with
PHA + IL-7 and double stained with biotinylated IL-4 and a) Leu3a (CD4) b) Leu2a
(CD8) c) 2A3 (CD25) mAb as described in
the Materials and Methods.

after 48 hr culture with IL-4, IL-7, PHA or PHA + IL-7 (Table 2). An increase was
observed in the mean number of receptors per cell determined by ^{125}I-IL-4 binding, on
cells cultured with IL-4, IL-7 or PHA compared to cells cultured in medium or those
examined immediately following isolation. Activation with PHA + IL-7 together resulted
in an approximately three-fold increase in mean receptor number per cell.

Table 2. Binding of Radiolabeled-Human IL-4 to Activated T Cells

Stimulus	Receptor number	Ka
-	140	2.43 $\times 10^9$
IL-4	200	9.78 $\times 10^8$
IL-7	170	4.22 $\times 10^9$
PHA	240	2.53 $\times 10^9$
PHA + IL-7	420	3.67 $\times 10^9$

PB T cells were cultured for 48 hrs in the presence of IL-4
(10 ng/ml), IL-7 (10 ng/ml) or PHA (0.25%) prior to determination of ^{125}I-IL4 binding as described in Materials and
Methods.

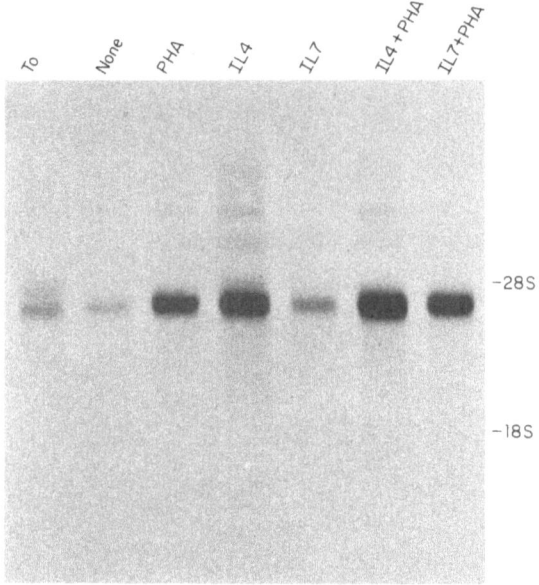

Figure 3. Induction of IL-4R mRNA. RNA gel blot containing total RNA from PB T cells freshly isolated (Lane To), and after culture for 45 hr with the following additions as labeled: None, PHA (0.25%), IL-4 (10 ng/ml), IL-7 (10 ng/ml), PHA + IL-4, PHA + IL-7. Positions of 28S and 18S ribosomal RNAs are shown.

Induction of Human IL-4R mRNA in Peripheral Blood T Lymphocytes

Highly purified PB T cells were cultured for 45 hr with PHA, IL-4, IL-7 or a combination of PHA and IL-4 or IL-7. RNA was extracted and analyzed by Northern blot using a human IL-4R probe. Low levels of IL-4R mRNA were detected from freshly isolated T cells or those cultured in the absence of stimuli (Figure 3). In contrast, cells activated with PHA or IL-4 alone contained elevated levels of IL-4R mRNA. Addition of

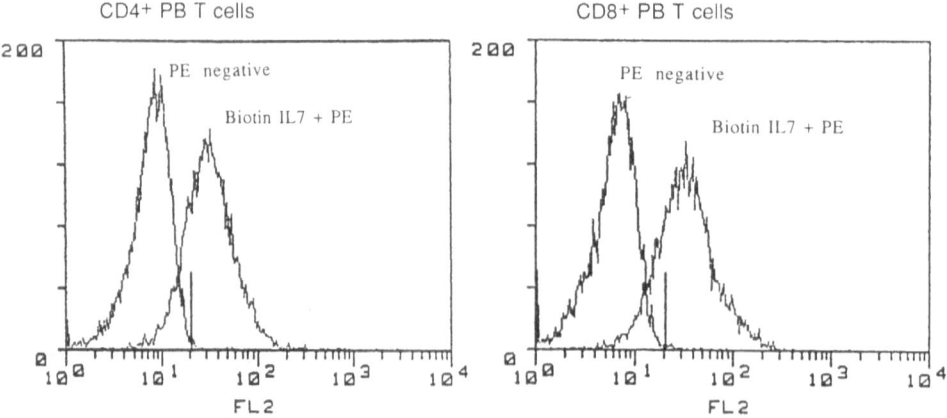

Figure 4. Flow cytometric analysis of freshly isolated CD4$^+$ and CD8$^+$ PB T cells stained with biotinylated IL-7 + streptavidin-phycoerythrin (PE) or PE alone as a negative control.

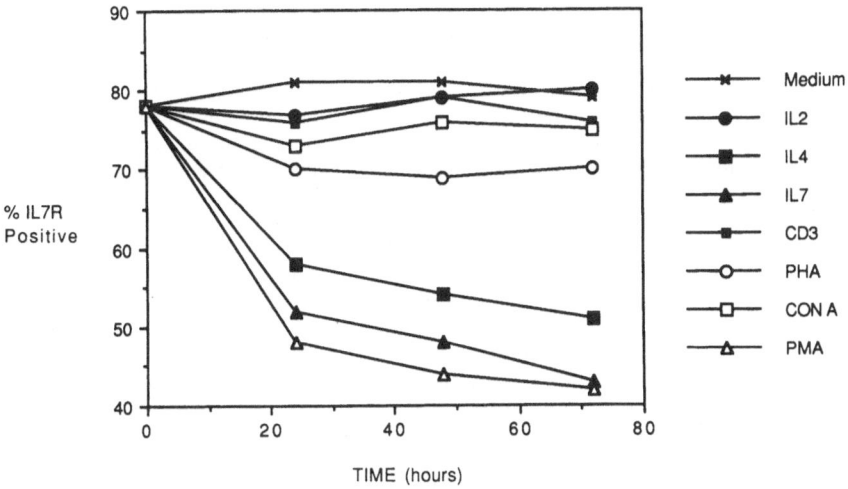

Figure 5. IL-7 receptor expression on cultured T cells. PB T cells were stained
with biotinylated IL-7 following culture for up to 72 hrs in the presence of
IL-2, IL-4, IL-7, PMA (all at 10 ng/ml), immobilized CD3 mAb OKT3,
PHA (0.25%) or Con A (2.5 μg/ml).

PHA and IL-4 together showed no increase over that seen with IL-4 alone. The presence of
IL-7 had little effect on its own or in combination with PHA.

After only 16 hr culture no significant increase in IL-4R mRNA levels was seen in the
presence of any stimuli used. This finding was supported by the failure to detect a
significant increase in surface expression of IL-4R by flow cytometry at this time point
(data not shown).

IL-7 Receptor Expression

In contrast to the low level of IL4R detected on freshly isolated T cells, IL7R, as
determined by biotinylated IL7 staining and flow cytometry, were expressed at high levels
on both CD4$^+$ and CD8$^+$ populations (Figure 4). Culture of T cells with a range of
stimuli failed to significantly increase the intensity of staining with biotinylated
IL7. In fact, exposure of T cells to IL4, IL7 or PMA for up to 72 hours resulted in
decreased expression of IL7R compared to that seen on freshly isolated cells (Figure 5).
Culture in the presence of IL2, CD3 mab, Con A or medium alone had no significant effect
on the level of IL7R detectable up to 7 hours, while the addition of PHA resulted in a
small, but consistently observed, decrease in receptor level with time.

DISCUSSION

In this study, using biotinylated human IL-4 and flow cytometric analysis IL-4R
expression on highly purified human peripheral blood T cells was found to be enhanced by
various stimuli. IL-4 enhanced expression of its own receptor on virtually all
peripheral blood T cells as indicated by the complete peak shift seen by flow cytometry.
A similar increase in IL-4R expression was seen when T cells were activated with either
PHA or immobilized OKT3. A combination of either one of these stimuli and IL-4 resulted
in a significantly higher level of IL-4R expression. Addition of IL-7 resulted in
enhancement of IL-4R similar to that seen in the presence of IL-4 either alone or
together with PHA or OKT3. Previous reports showing IL-2 alone to be ineffective at
inducing IL-4R upregulation on T cells[26] were confirmed, although we found that in
the presence of PHA or OKT3 IL-2 could induce a similar level of IL-4R expression to that
seen when IL-4 or IL-7 were added with these stimuli.

In all the experiments the shape of the histograms obtained by flow cytometry suggested that the great majority of T cells expressed more IL-4R after culture in IL-4, IL-7 or PHA than cells cultured in medium alone. These observations are in agreement with results of previous studies using biotinylated murine IL-4,[27] which found that activation with IL-4 induced an increase in IL-4R expression on virtually all murine T cells isolated from mesenteric lymph nodes. Although our data shows a similar change in MFI on human T cells cultured with IL-4 we see only a small increase--140 to 200--in the number of receptors per cell determined by ^{125}I-IL-4 binding compared to a much larger increase observed in the murine system.[27] This difference in enhancement of IL-4R on mouse and human T cells may indicate a real biological difference in the capacity to express IL-4R on T cells of these two species. However, it should be noted that the T cells used in these two studies were obtained from different tissue, murine lymph node and human peripheral blood. This could account for the differences observed in IL-4R expression on murine and human T cells. The fact that we consistently see a substantial increase in the binding of biotinylated IL-4 by flow cytometry on T cells cultured in the presence of IL-4, and yet only a relatively low number of receptors by ^{125}I-IL4 binding, suggests that in our hands the biotinylated human IL-4 has a sufficient amount of streptavadin-accessible biotin to provide an extremely sensitive method for the detection of changes in the level of IL-4R on human cells. The observation that IL-4, IL-7 and PHA enhance IL-4R on human T cells is supported by Northern blot analysis of IL-4R mRNA. It is of interest that after 45 hours of culture IL-4 alone induced a considerably higher level of IL-4R mRNA than IL-7 or PHA alone while the presence of these three stimuli resulted in a similar level of IL-4R surface expression determined by flow cytometry and ^{125}I-IL-4 binding. Higher expression of surface IL-4R was seen after stimulation with IL-4 or IL-7 in the presence of PHA, although the addition of PHA had no effect on IL-4- or IL-7-induced IL-4R mRNA levels. Activation of T cells with IL-4 alone may have resulted in post-transcriptional processing leading to a lower level of surface IL-4R expression than suggested by the level of mRNA present. Alternatively, binding of IL-4 to its receptor in the absence of the mitogenic signal provided by PHA could cause rapid internalization of IL-4R, thus lowering the level of surface receptor detectable. One further possibility is that maximal IL-4R gene expression is induced by IL-7 or PHA alone at a time other than those examined here. However, under no condition of stimulation was increased IL-4R expression detected by flow cytometry earlier than 36 hrs after initiation of culture (data not shown). These observations are therefore incompatible with high IL-7 or PHA-induced IL-4R gene expression occurring at a time earlier than 45 hours. This apparent discrepancy between surface receptor expression and mRNA levels is worthy of further study.

The results presented here suggest that regulation of IL-4R and IL-7R occurs through distinct mechanisms. In contrast to the low levels of IL-4R detected on unstimulated T cells, IL-7R are present at high levels. While T-cell expression of IL-4R can be increased upon activation, under no culture conditions was enhancement of IL-7R observed. Stimulation of T cells with IL-4, IL-7 or PMA resulted in increased IL-4R expression, while the level of IL-7R on these cells was reduced. Other stimuli which have been shown to elevate IL-4R levels, PHA, CD3 mAb and con A, had little or no effect on IL-7R expression. Although the majority of freshly isolated human T cells express high levels of IL-7R, only a minority of these cells are induced by IL-7 to express IL-2R[17] or are driven into cell cycle by IL-7, either alone or in the presence of costimulus (personal observation). Why such a high proportion of T cells should express IL-7R and yet not respond to IL-7 is not clear. IL-7 receptors are expressed on a wide range of hemopoietic cell types and have been shown to bind IL-7 with both high and low affinity.[28] Future studies may determine whether a relationship exists between the stimulatory response of a particular cell type to IL-7 and the affinity with which receptors on that cell binds IL-7.

In combination with cell surface phenotyping and cell cycle analysis, staining with biotinylated cytokines such as IL-4 and IL-7 should prove to be a useful tool in the dissection of functionally and phenotypically distinct subpopulations of cell lineages on which receptors for these cytokines are expressed.

ACKNOWLEDGMENTS

We wish to thank Ky Clifford, Tim Vanden Bos, Wenie Din and Brian Macduff for excellent technical assistance, and Alan Alpert for assistance with flow cytometric analysis.

REFERENCES

1. M. Howard, J. Farrar, M. Hilfiker, B. Johnson, K. Takatsu, T. Hamaoka, and W. E. Paul, Identification of a T cell-derived B cell growth factor distinct from interleukin 2, *J. Exp. Med.* 155:914 (1982).

2. K. H. Grabstein, L. S. Park, P. J. Morrissey, H. Sassenfeld, V. Price, D. L. Urdal, and M. B. Widmer, Regulation of murine T cell proliferation by B cell stimulatory factor, *J. Immunol.* 139:1148 (1987).

3. J. Hu-Li, E. M. Shevach, J. Mizuguchi, J. Ohara, T. Mosmann, and W. E. Paul, B cell stimulatory factor 1 (interleukin 4) is a potent costimulant for normal resting T lymphocytes, *J. Exp. Med.* 165:157 (1987).

4. T. R. Mosmann, M. W. Bond, R. L. Coffman, J. Ohara, and W. E. Paul, T cell and mast cell lines respond to B cell stimulatory factor-1, *Proc. Natl. Acad. Sci. USA* 83:5654 (1986).

5. R. Noelle, P. Krammer, J. Ohara, J. W. Uhr, and E. S. Vitetta, Increased expression Ia antigens on resting B cells: an additional role for B-cell growth factor, *Proc. Natl. Acad. Sci. USA* 81:6149 (1984).

6. N. W. Roehm, J. Liebson, A. Zlotnik, J. Kappler, P. Marrack, and J. C. Cambier, Interleukin induced increase in Ia expression by normal mouse B cells, *J. Exp. Med.* 160:679 (1984).

7. T. Defrance, J. P. Aubry, F. Rousset, B. Vanbervliet, J. Y. Bonnefoy, N. Arai, Y. Takebe, T. Yokata, F. Lee, K. Arai, J. deVries, and J. Banchereau, Human interleukin IL-4 induces Fcε receptors (CD23) on normal human B lymphocytes, *J. Exp. Med.* 165:1459 (1987).

8. S. A. Hudak, S. O. Gollinick, D. H. Conrad, and M. R. Kehry, Murine B cell stimulatory factor 1 (interleukin 4) increases expression of the Fc receptor for IgE on mouse B cells, *Proc. Natl. Acad. Sci. USA* 84:4666 (1987).

9. J. G. Shields, R. J. Armitage, B. N. Jamieson, P. C. L. Beverley, and R. E. Callard, Increased expression of surface IgM but not IgD or IgG on human B cells in response to IL-4, *Immunol.* 66:224 (1989).

10. E. S. Vitetta, J. Ohara, C. Myers, J. Layton, P. H. Krammer, and W. E. Paul, Serological, biochemical and functional identity of B cell stimulatory factor-1 and B cell differentiation factor for IgG1, *J. Exp. Med.* 162:1726 (1985).

11. R. L. Coffman, J. Ohara, M. W. Bond, J. Carty, A. Zlotnik, and W. E. Paul, B cell stimulatory factor-1 enhances the IgE response of lipopolysaccharide-activated B cells, *J. Immunol.* 136:4538 (1988).

12. L. S. Park, D. Friend, K. Grabstein, and D. L. Urdal, Characterization of the high affinity cell-surface receptor for murine B cell stimulating factor 1, *Proc. Natl. Acad. Sci. USA* 84:1669 (1987).

13. J. Ohara and W. E. Paul, Receptors for B-cell stimulatory factor 1 expressed on cells of haematopoietic lineage, *Nature (London)* 325:537 (1987).

14. J. W. Lowenthal, B. E. Castle, J. Christiansen, J. Schreurs, D. Rennick, N. Arai, P. Hoy, Y. Takebe, and M. Howard, Expression of high affinity receptors for murine interleukin 4 (BSF-1) on hemopoietic and non-hemopoietic cells, *J. Immunol.* 140:456 (1988).

15. A. E. Namen, A. E. Schmierer, C. J. March, R. W. Overell, L. S. Park, D. L. Urdal, and D. Y. Mochizuki, B cell precursor growth-promoting activity. Purification and characterization of a growth factor active on lymphocyte precursors, *J. Exp. Med.* 167:988 (1988).

16. P. J. Morrissey, R. G. Goodwin, R. P. Nordan, D. Anderson, K. H. Grabstein, D. Cosman, J. Sims, S. Lupton, B. Acres, S. G. Reed, D. Mochizuki, J. Eisenman, P. J. Conlon, and A. E. Namen, Recombinant interleukin 7 pre-B growth factor has costimulatory activity on purified mature T cells, *J. Exp. Med.* 169:707 (1989).

17. R. J. Armitage, A. E. Namen, H. M. Sassenfeld, and K. H. Grabstein, Regulation of human T-cell proliferation by interleukin 7, *J. Immunol.* 144:938 (1990).

18. D. L. Urdal, D. Mochizuki, P. J. Conlon, C. J. March, M. L. Remerowski, J. Eisenman, C. Ramthun, and S. Gillis, Lymphokine purification by reversed phase high performance liquid chromatography, *J. Chromatogr.* 296:171 (1984).

19. M. B. Widmer, R. B. Acres, H. M. Sassenfeld, and K. H. Grabstein, Regulation of cytolytic cell populations from human peripheral blood by B cell stimulatory factor 1 (interleukin 4), *J. Exp. Med.* 166:1447 (1987).

20. R. G. Goodwin, S. Lupton, A. Schmierer, K. J. Hjerrild, R. Jerzy, W. Clevenger, S. Gillis, D. Cosman, and A. E. Namen, Human interleukin 7: molecular cloning and growth factor activity on human and murine B-lineage cells, *Proc. Natl. Acad. Sci. USA* 84:302 (1989).

21. L. S. Park, D. Friend, H. M. Sassenfeld, and D. L. Urdal. Characterization of the human B cell stimulatory factor 1 receptor, *J. Exp. Med.* 166:476 (1987).

22. S. K. Dower, K. Ozato, and D. M. Segal, The interactions of monoclonal antibodies with MHC class I antigens on mouse spleen cells. I. Analysis of the mechanism of binding, *J. Immunol.* 120:2027 (1984).

23. C. J. March, B. Mosley, A. Larsen, D. P. Cerretti, G. Braedt, V. Price, S. Gillis, C. S. Henney, S. R. Kronheim, K. Grabstein, P. J. Conlon, T. P. Hopp, and D. Cosman, Cloning, sequence and expresson of two distinct human interleukin-1 complementary DNAs, *Nature* 315:641 (1985).

24. L. M. Hoffman, M. K. Fritsch, and J. Gorski, Probable nuclear precursors of preprolactin mRNA in rat pituitary cells, *J. Biol. Chem.* 256:2597 (1981).

25. R. L. Idzerda, C. J. March, B. Mosley, S. D. Lyman, T. Vanden Bos, S. D. Gimpel, W. S. Din, K. H. Grabstein, M. B. Widmer, L. S. Park, D. Cosman, and M. P. Beckmann, Human interleukin-4 receptor confers biological responsiveness and defines a novel receptor superfamily, *J. Exp. Med.* 171:861 (1990).

26. D. L. Jankovic, M. Gibert, D. Baran, J. Ohara, W. E. Paul, and J. Theze. Activation by IL-2, but not IL-4, up-regulates the expression of the p55 subunit of the IL-2 receptor on IL-2- and IL-4-dependent T cell lines, *J. Immunol.* 142:3113 (1989).

27. J. Ohara and W. E. Paul, Up-regulation of interleukin 4/B-cell stimulatory factor 1 expression, *Proc. Natl. Acad. Sci. USA* 85:8221 (1988).

28. L. S. Park, D.J. Friend, A. E. Schmierer, S. K. Dower, and A. E. Namen, Murine interleukin-7 receptor: characterization on an IL-7-dependent cell line, *J. Exp. Med.* (in press).

EFFECTS OF LEUKEMIA INHIBITORY FACTOR (LIF) ON GENE TRANSFER

EFFICIENCY INTO MURINE HEMATOLYMPHOID PROGENITORS

Frederick A. Fletcher,* Kateri A. Moore,* Douglas E. Williams,^ Dirk Anderson,^
Charles Maliszewski,^ and John W. Belmont#*

#Howard Hughes Medical Institute
*Institute for Molecular Genetics
 Baylor College of Medicine
 Houston, Texas
^Immunex Research and Development Corporation
 Seattle, Washington

ABSTRACT

We have investigated the effects of the cytokine leukemia inhibitory factor (LIF) on recovery and retroviral infection of murine hematopoietic stem cells maintained in short-term culture. Up to a two-fold increase in $CFU-S_{13}$ recovery was observed, from 9.7×10^{-5} cells in untreated controls to 17.6×10^{-5} cells when 10U/ml LIF is added to the culture medium. Intermediate concentrations of LIF (.1U/ml and 1U/ml) were not significantly different from the control. Histological analysis of spleen colonies harvested thirteen days posttransplant demonstrated that LIF does not cause a detectable alteration in the differentiative potential of $CFU-S_{13}$. The efficiency of retroviral-vector infection in $CFU-S_{13}$ is also improved, from 15% (24/158) in untreated controls to 91% (116/127) at a LIF concentration of 10U/ml. LIF concentrations of .1U/ml and 1U/ml increased infection efficiency to 35% (14/40) and 71% (37/51), respectively. Analysis of proviral insertion sites in spleen colonies indicated that some $CFU-S_{13}$ precursors were infected in the LIF-treated marrows, but no identical pairs were detected in the controls. Finally, long-term expression of provirally-encoded human adenosine deaminase (hADA) was measured in hematopoietic tissues of bone marrow transplant recipients six months posttransplant. In all tissues analyzed (spleen, thymus, bone marrow, splenic B cells, peritoneal macrophages, and blood) differentiated progeny of LIF-treated marrows had higher levels of hADA than untreated controls. Tenfold increases in levels of hADA are detected in some tissues, but levels were variable. These experiments demonstrate that LIF directly or indirectly enhances retroviral infection efficiency of hematopoietic stem cells, and might be used to improved existing gene transfer protocols.

INTRODUCTION

The hematopoietic system is continuously replenished throughout life by a limited set of pluripotent stem cells. As a transplantable, self-renewing entity, the hematopoietic stem cell presents an attractive target for correcting genetic disease by *in vitro* gene transfer methodologies. We are studying retroviral vector-mediated delivery of a human adenosine deaminase (hADA) cDNA to cultured murine bone marrow, followed by transplantation into irradiated syngeneic recipients, as a model system for gene therapy of inherited metabolic disease. Retroviral vector infection efficiency of hematopoietic stem cells in short term culture has previously been low (5-10%),[1,2] and expression of proviral sequences

Mechanisms of Lymphocyte Activation and Immune Regulation III
Edited by S. Gupta *et al.*, Plenum Press, New York, 1991

131

variable. The hematopoietic stem cell is thought to be predominantly quiescent,[3] with a cell cycle transit time of approximately one week. Most of this time is spent in the G_0 phase of the cell cycle. It is a widely accepted hypothesis that target cell replication is required for productive retroviral infection to occur, and low replication rate of stem cells has been cited as an explanation for low infection efficiency. Normal replication stimuli for hematopoietic stem cells are unknown, but might include the effects of soluble growth factors. Stimulation of bone marrow cultures with the hematopoietic growth factors IL-6 and IL-3 can improve retroviral vector infection efficiency of stem cells.[4]

The novel cytokine Leukemia Inhibitory Factor (LIF) was identified by its ability to induce differentiation of the murine myeloid leukemia cell line M1,[5] a property shared with the hematopoietic growth factors IL-6 and G-CSF.[6] In addition to this differentiative effect, LIF is known to support proliferation of the murine IL3-dependent cell line, DA-1a.[7] The apparent functional similarity of LIF with known hematopoietic growth factors, and recent identification of unique LIF receptors on normal murine monocytes and macrophages,[10] has suggested a role for LIF in normal hematopoiesis. Functional identity of LIF with the totipotent embryonic stem (ES) cell Differentiation Inhibitory Factor (DIA) has also been demonstrated,[9,10] and has suggested a similar effect of LIF on the pluripotent hematopoietic stem cell.

We have examined the effects of recombinant LIF on recovery and retroviral vector infection efficiency of cultured, myeloid-restricted hematopoietic stem cells (CFU-S$_{13}$); and long-term (> six months) expression of vector-encoded hADA in differentiated progeny of pluripotent hematopoietic stem cells. Frequency of CFU-S$_{13}$ was measured in groups of murine bone marrow cells cultured in the presence or absence of murine LIF. LIF-supplemented cultures allowed greater recovery of CFU-S$_{13}$, when compared to untreated

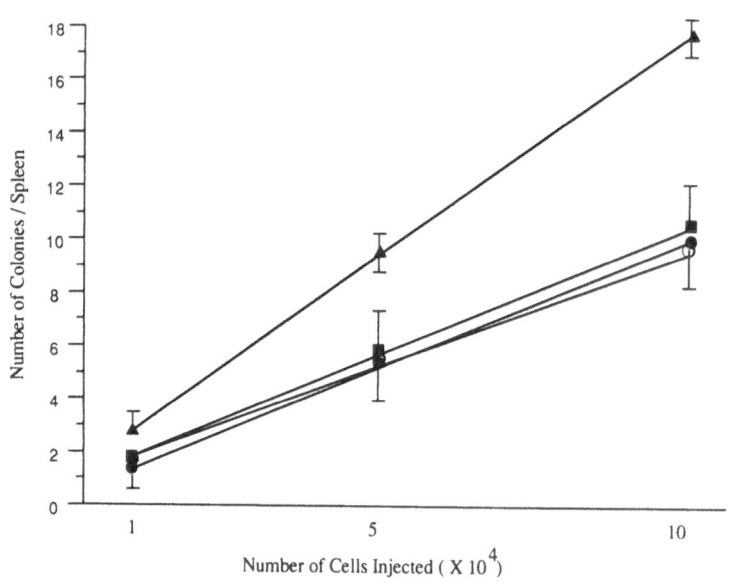

Figure 1. The effects of LIF on recovery of CFU-S$_{13}$ from suspension culture. Bone marrow was harvested from normal 10-12 week old FVB/N female mice and plated onto irradiated (2500 R) monolayers of retrovirus-vector-producing fibroblasts. Irradiated (1000 R) mice were injected IV with the indicated number of bone marrow cells co-cultured as described for 72 h in the presence of .1U/ml (O, n=9), 1U/ml (■, n=13), 10U/ml (▲, n=23) LIF, or no added growth factor (●, n=30). The points represent the average number of macroscopic spleen colonies (± SEM) on the indicated number (n) of spleens on day thirteen post-transplant.

(noGF) cultures. Histological analysis of spleen colonies derived from LIF-treated bone marrow provides evidence that no detectable alteration in differentiative potential is occuring in the CFU-S_{13} population in response to LIF treatment. Analysis of infection efficiency of CFU-S_{13} and determination of clonal relationships between the resulting spleen colonies provides further evidence for effects of LIF on these populations. Finally, quantitative analysis of hADA levels in hematopoietic tissues from transplanted animals demonstrates that *in vitro* LIF-stimulation of bone marrow during co-culture improves long-term expression in progeny of infected stem cells.

RESULTS

CFU-S Frequency Determination

A LIF concentration of 1000U/ml has been reported to allow maintenance of pluripotentiality of ES cells and to induce differentiation of M1 cells[10]. Half-maximal effect is seen on these cells at 50-100U/ml LIF. We have tested the effect of murine LIF on CFU-S_{13} in short term culture at concentrations up to 10U/ml. The LIF is derived from transfected cos cell supernatants; therefore, appropriate volumes of untransfected cos cells supernatants were included in separate control cultures. Cos cell supernatants do not contain endogenous LIF activity (data not shown).[9] The effect of LIF or control supernants on *in vitro* survival and proliferation of CFU-S_{13} was measured as a function of recovery of CFU-S_{13} from an additional control cultured without any growth factor (Figure 1). Supplementation of cultures with cos cell supernatants did not yield a significantly higher CFU-S_{13} recovery (11.7 x 10^{-5}, p > 0.01) when compared to controls without growth factor (9.5 x 10^{-5}). Cultures supplemented with .1U/ml and 1U/ml LIF also yielded CFU-S_{13} recoveries similar to the noGF control (8.7 x 10^{5} and 9.7 x 10^{-5}, respectively). A LIF concentration of 10U/ml increased CFU-S_{13} recovery to 17.8 x 10^{-5} cells injected (p < 0.001).

Histological Analysis of Spleen Colonies

Histological sections were analyzed from spleen in the noGF and 10U/ml LIF experimental groups to determine whether LIF affects the differentiative program of the CFU-S_{13}.

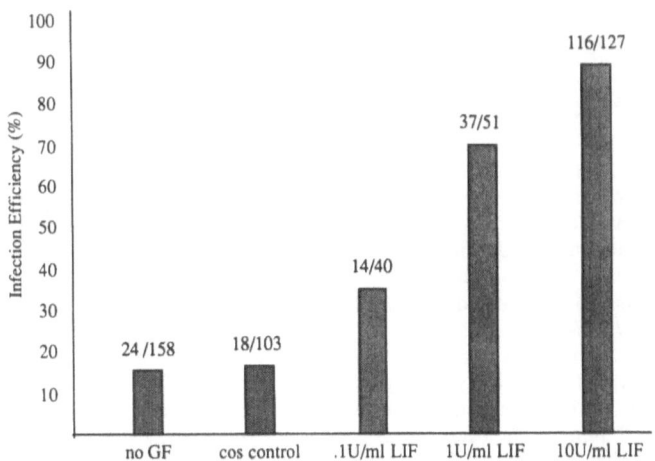

Figure 2. The effects of LIF on *in vitro* retroviral-vector infection efficiency of CFU-S_{13}. Genomic DNA (500 mg) from individual spleen colonies was analyzed by PCR amplification, as described,[11] for the presence of integrated provirus. Infection efficiency for each experimental group is reported graphically and numerically, as number of infected colonies divided by total number of colonies analyzed.

Individual, non-overlapping, colonies in each section were microscopically analyzed and scored as mixed or mono/biopotent. Of the colonies examined in each group, virtually all (> 90%) contained differentiated progeny of three or four lineages. No alteration in the differentiative potential of CFU-S_{13} was noted in either group.

CFU-S Infection Efficiency

We employed the polymerase chain reaction (PCR) to identify infected spleen colonies. Primers were synthesized to amplify the hADA cDNA encoded by the provirus without also amplifying the endogenous mouse locus. The sensitivity of the PCR was reduced to allow discrimination between bona fide infection events in single colonies (provirus single copy or greater) and small numbers of provirus-bearing cells mixed with an uninfected colony. Infection efficiencies were calculated for each experimental group (Figure 2). The noGF control provides a baseline infection efficiency of 15% (24/158); and the low volume cos supernatant control is not significantly different at 17% (18/103) infection efficiency. LIF increases infection efficiency at all concentrations tested, .1U/ml increases efficiency to 35% (14/40), 1U/ml increases infection to 73% (37/51), and a LIF concentration of 10U/ml increases infection efficiency to 91% (116/127) of colonies analyzed.

Clonal Relationship of Spleen Colonies

Retroviral infection marks individual cells, and all their progeny, in a unique fashion. By characterizing the retroviral insertion site(s) of infected colonies, it is possible to compare clonal relationships of individual CFU-S_{13} progenitors. To characterize these sites, we digested genomic DNA from each retrovirally-marked colony with a restriction enzyme that does not cut within the integrated provirus. Because the enzyme cuts only in the flanking DNA, a unique-length restriction fragment specific for each insertion site can be detected. Pairs of colonies bearing the same restriction pattern were scored as being derived from a common precursor. No identical pairs were detected in the noGF or the cos supernatant control group, out of nine animals studied. Five pairs were detected in seven animals from the 10U/ml LIF group (Figure 3).

Expression of Vector-Encoded Adenosine Deaminase

Expression of hADA was examined in hematopoietic tissues of recipient animals six months post transplant. Protein extracts were prepared from the spleen, thymus, bone marrow, blood, splenic B cells, and peritoneal macrophages. Western blots of these extracts were

Figure 3. The effects of LIF on retroviral-vector infection of CFU-S_{13} precursors. An example of two individual spleen colonies from a single animal bearing identical retroviral junction patterns. DNAs (10 μg) were digested with HindIII, an enzyme that does not cut within the integrated provirus, yielding a single, unique-length, restriction fragment for each insertion site. Digested DNAs were electrophoresed in 0.8% agarose, transferred to a charged nylon membrane by the method of Southern[12] and probed with a ^{32}P-labelled fragment of the hADA cDNA.[13]

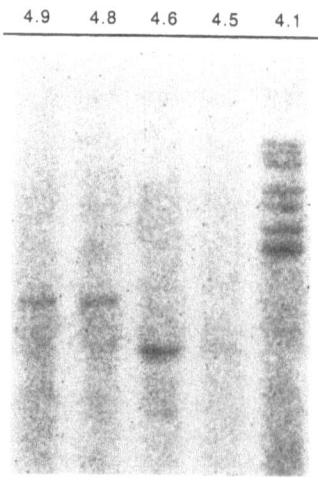

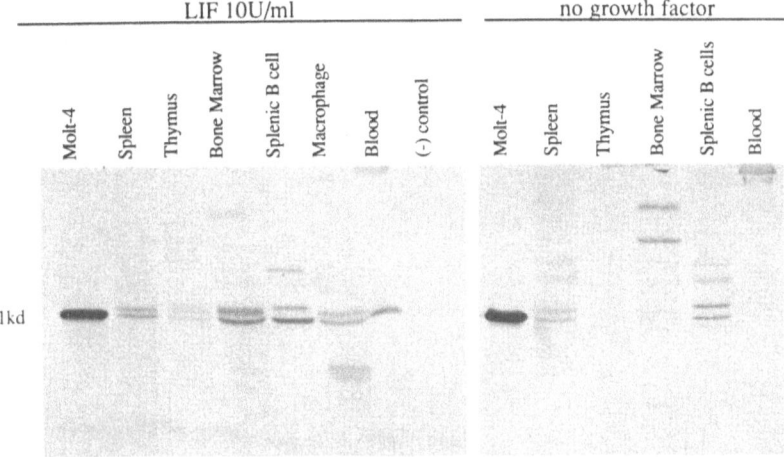

Figure 4. Western blots of hematopoietic tissues from two long-term transplants. Protein extracts were separated by 10% SDS-PAGE and transferred to nitrocellulose. The 41 kd hADA band was detected by a monospecific hADA antibody that does not cross react with mouse ADA. Molt-4 is a human leukemic T-cell line, the negative (-) control is derived from a mouse transplanted with cultured but non-retrovirus infected marrow.

developed with hADA-specific antibody, revealing increased levels of expression in animals receiving LIF-stimulated marrow, compared to controls (Figure 4). Quantitative analysis hADA levels in these animals was performed by video densitometry of bands on Western blots, normalized to a positive control, and compared to standardized activity/density curves developed with purified human ADA (Figure 5). All tissues from the LIF animals contained higher mean levels of hADA than tissues derived from control (noGF) animals.

DISCUSSION

Similarity in function with known hematopoietic growth factors on differentiation of myeloid leukemic cell lines, and developmental regulation of LIF receptor number in some committed hematopoietic cells, suggests that LIF has a role in normal hematopoiesis. LIF alone does not affect *in vitro* survival or differentiation of normal monocyte or macrophage progenitors, even though their progeny have a high number of LIF receptors.[14] More primitive progenitors bear few, if any, LIF receptors and have been considered to be unaffected by LIF. The effect of LIF that we have noted is not necessarily direct, as we have cultured whole bone marrow aspirates in the presence of virus-producing fibroblasts. It is possible that LIF is playing an indirect role in stimulating marrow accessory cells or the virus-producing fibroblasts to exert other proliferative signals. Another possibility is that the presence of a stroma (virus-producing fibroblasts) enhances GF presentation to sensitive cells. Studies are in progress to test the effect of LIF on purified CFU-S$_{13}$ in the absence of other cell types.

In addition to an increase in CFU-S$_{13}$ number, we have detected a large effect on retroviral infection efficiency of the CFU-S$_{13}$ population. This could be the result of induction of replication in CFU-S$_{13}$ or the result of an undefined enhancement of retroviral infection unrelated to replication status. It has been widely assumed that target cell replication is required for productive retroviral infection, and this hypothesis has gained experimental support.[15-18] Recently, ^{3}H-thymidine suicide assays have been used to demonstrate a correlation between replication and retroviral infection of CFC-preB and CFU-GM *in vitro*.[17] A reported inability to infect T lymphocytes arrested in G$_0$[15] suggests that the replication requirement is more critical in cells that spend a significant proportion

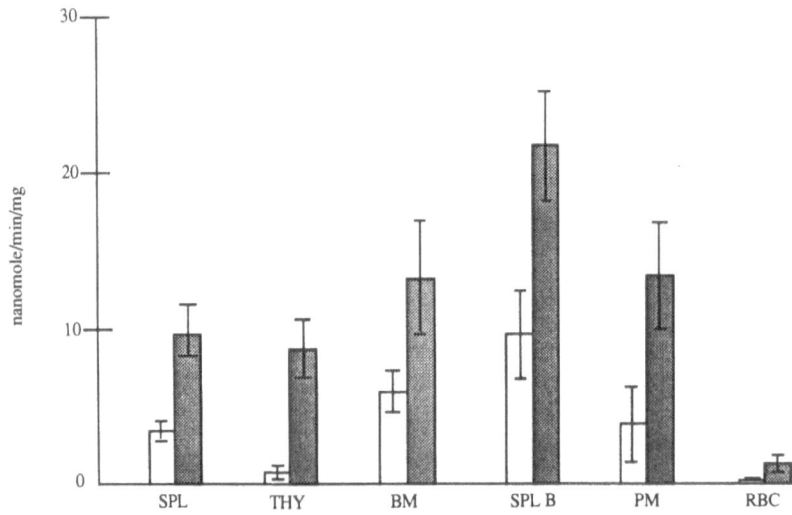

Figure 5. Quantitative comparison of hADA expression. Tissues were analyzed by Western blotting and hADA detected by a monospecific hADA
antibody. The blots were scanned by a digital immage analyzer and
normalized to a positive control present on each blot. The level of
hADA was extrapolated from a standard curve developed from purified
recombinant hADA. Open bars represent tissues obtained from mice
transplanted with bone marrow co-cultivated without added growth
factors (n=7). Hatched bars represent tissues repopulated by LIF
(10U/ml) treated marrow (n=10). SPL, spleen; THY, thymus; BM, bone
marrow; SPL B, splenic B-cells; PM, peritoneal macrophages; RBC, red
blood cells.

of the cell cycle in G_0, such as CFU-S.[3] $CFU-S_{13}$ infection efficiency in our
no GF control is 15%, a result consistent with previously reported estimates of *in vitro*
CFU-S replication.[19,20] Leary et al.[21] have recently reported that LIF can act in synergy
with IL-3 to reduce the emergence time of blast *in vitro*, another LIF function in common
with IL-6 and G-CSF. The hypothesized mechanism of the effect is inducement of the G_0-G_1
phase transition in cytokine-sensitive cells.[22] The present data are consistent with a
hypothesis that nearly all $CFU-S_{13}$ present in culture replicated once in response to LIF
stimulation.

The data also suggest that LIF acts on precursor of $CFU-S_{13}$, as we have observed
multiple pairs of spleen colonies with identical proviral insertion patterns derived from
LIF-treated marrow. We postulate that, in these instances, a primitive progenitor infected
in vitro was transplanted and divided *in vivo*, yielding identical daughters that seeded the
irradiated spleen. The result is a pair of spleen colonies with identical proviral insertion
patterns.

Improved long-term expression of vector-encoded hADA in hematopoietic tissues of
transplanted animals is interpreted as increased infection efficiency of hematopoietic stem
cells under LIF stimulation. Analysis of proviral copy number in these tissues is proceeding
to support this conclusion.

These experiments demonstrate that LIF directly or indirectly enhances retroviral infection efficiency of hematopoietic stem cells, and might be used to improve existing gene
transfer protocols.

136

REFERENCES

1. J. W. Belmont, G. R. MacGregor, K. Wager-Smith, F. A. Fletcher, K. A. Moore, D. Hawkins, D. Villalon, S. M.-W. Chang, and C. T. Caskey, Expression of human adenosine deaminase in murine hematopoietic cells, *Mol. Cell Biol.* 8:5116 (1988).

2. K. A. Moore, F. A. Fletcher, D. K. Villalon, A. E. Utter, and J. W. Belmont, Human adenosine deaminase expression in mice, *Blood* 75:2085 (1990).

3. L. G. Lajtha, Stem cell concepts, *Differentiation* 14:23 (1979).

4. D. M. Bodine, S. Karlsson, and A. W. Nienhuis, Combination of interleukins 3 and 6 preserves stem cell function in culture and enhances retrovirus-mediated gene transfer into hematopoietic stem cells, *Proc. Natl. Acad. Sci. USA* 86:8897 (1989).

5. D. J. Hilton, N. A. Nicola, N. M. Gough, and D. Metcalf, Resolution and purification of three distinct factors produced by Krebs ascites cells which have differentiation-including activity on murine myeloid leukemic cell lines, *J. Biol. Chem* 263:9238 (1988).

6. D. Metcalf, Actions and interactions of G-CSF, LIF, and IL-6 on normal and leukemic murine cells, *Leukemia* 3:349 (1989).

7. A. Godard, H. Gascan, J. Naulet, M.-A. Peyrat, Y. Jacques, J.-P. Soulillou, and J.-F. Moreau, Biochemical characterization and purification of HILDA, a human lymphokine active on eosinophils and bone marrow cells, *Blood* 71:1618 (1988).

8. D. J. Hilton, N. A. Nicola, and D. Metcalf, Specific binding of murine leukemia inhibitory factor to normal and leukemic monocytic cells, *Proc. Natl. Acad. Sci. USA* 85:5971 (1988).

9. A. G. Smith, J. K. Heath, D. D. Donaldson, G. G. Wong, J. Moreau, M. Stahl, and D. Rogers, Inhibition of pluripotential embryonic stem cell differentiation by purified polypeptides, *Nature* 336:688 (1988).

10. R. L. Williams, D. J. Hilton, S. Pease, T. A. Willson, C. L. Stewart, D. P. Gearing, E. F. Wagner, D. Metcalf, N. A. Nicola, and N. M. Gough, Myeloid leukemia inhibitory factor maintains the developmental potential of embryonic stem cells, *Nature* 336:684 (1988).

11. S. C. Kogan, M. Doherty, and J. Gitschier, An improved method for prenatal diagnosis of genetic diseases by amplified DNA sequences, *N. Engl. J. Med* 317:985 (1987).

12. E. M. Southern, Detection of specific sequences among DNA fragments separated by gel electrophoresis, *J. Mol. Biol.* 98:503 (1975).

13. G. S. Adrian, D. A. Wiginton, and J. J. Hutton, Structure of adenosine deaminase mRNAs from normal and adenosine deaminase deficient human cell lines, *Mol. Cell. Biol.* 4:1712 (1984).

14. D. Metcalf, T. Maekawa, D. Hilton, N. Nicola, D. Gearing, and N. Gough, Interactions of leukemia inhibitory factor (LIF) IL-6 and colony-stimulating factors on murine and human leukemic cells. *Exp. Hematol.* 17:483 (1989).

15. J. Harel, E. Rassart, and P. Jolicoeur, Cell cycle dependence of synthesis of unintegrated viral DNA in mouse cells newly infected with muine leukemia virus, *Virology* 110:202 (1981).

16. I. S. Y. Chen, and H. Temin, Establishment of infection by spleen necrosis virus: inhibition in stationary cells and the role of secondary infection, *J. Virol.* 41:183 (1982).

17. D. E. Williams, A. E. Namen, D. Y. Mochizuki, and R. W. Overell, Clonal growth of murine pre-B colony forming cells and their targeted infection by a retroviral vector: dependence on interleukin-7, *Blood* 75:1132 (1990).

18. G. M. Springett, R. C. Moen, S. Anderson, R. M. Blaese, and W. F. Anderson, Infection efficiency of T lymphocytes with amphotropic retroviral vectors is cell cycle dependent, *J. Virol* 63:3865 (1989).

19. A. J. Becker, E. A. McCulloch, L. Siminovitch, and J. E. Till, The effect of different demands for blood cell production on DNA synthesis by haemopoietic colony forming cells of mice, *Blood* 26:296 (1965).

20. B. I. Lord, L. G. Lajtha, and J. Gidali, Measurement of the kinetic status of bone marrow precursor cells: three cautionary tales, *Cell and Tissue Kin* 7:507 (1974).
21. A. G. Leary, G. C. Wong, S. C. Clark, and M. Ogawa, Leukemia inhibitory factor (LIF)/differentiation-inducing activity (DIA) is a synergistic factor for human hemopoietic stem cells, *Exp. Hematol.* 17:524 (1989).
22. J. Ikebuchi, G. G. Wong, S. C. Clark, J. N. Ihle, Y. Hirai, and M. Ogawa, Interleukin 6 enhancement of interleukin 3-dependent proliferation of multipotential hemopoietic progenitors, *Proc. Natl. Acad. Sci. USA* 84:9035 (1987).

THYMIC MECHANISMS FOR INDUCING TOLERANCE TO Mls

Fred Ramsdell, Tracy Lantz, Frances Hausman, and B. J. Fowlkes

Laboratory of Cellular and Molecular Immunology
National Institute of Allergy and Infectious Diseases
National Institutes of Health
Bethesda, Maryland

INTRODUCTION

As T cells develop in the thymus, they are subjected to a number of selective events which determine the antigen recognition repertoire of mature T cells. Also, as a result of selection, T cells emerge that are able to distinguish self from non-self. A number of mechanisms have been proposed for establishing self tolerance.[1] In this report, it is demonstrated that T cells developing in the thymus become tolerant of the minor lymphocyte stimulatory antigens, Mls,[2-4] by at least two distinct mechanisms. Through the use of $V\beta$-specific antibodies that identify receptors with Mls specificity,[5-7] it is possible to show that tolerance to this self antigen can be acquired by clonal deletion or functional inactivation (clonal anergy). Whether tolerance to Mls is achieved and the type of tolerance induced appears to be critically dependent upon a number of factors, most notably the MHC haplotype, the tissue involved in antigen presentation, and the type of Mls antigen (Mls-1 or Mls-2). Here, chimeric animals are used in order to manipulate the site of tissue expression of MHC and Mls antigens. The results obtained reveal that clonal deletion is dependent on the type of MHC expressed by the bone marrow-derived elements and not on the tissue source of the Mls antigen. Furthermore, in the absence of clonal deletion, tolerance to Mls can be achieved by clonal anergy. This latter form of tolerance induction appears to be the result of interactions with radiation-resistant host elements, most likely the thymic epithelium.

RESULTS AND DISCUSSION

Recent studies using antibodies to TCR that are specific for certain self antigens[5-10] and transgenic mice in which the majority of T cells bear a single receptor specific for a self antigen[11-13] clearly show that clonal deletion of self-reactive cells is a major mechanism for attaining tolerance to self. This mechanism can occur at a $CD4^+8^+$ precursor stage of development[12,14,15] and appears to involve an interaction with bone marrow-derived cells.[16] It is somewhat controversial as to whether T cells develop functional tolerance to thymic epithelial antigens.[17] In the system described, we investigate and discuss the intrathymic mechanism for achieving tolerance to Mls and the role of thymic stromal cells in inducing these events.

Like allogeneic MHC antigens, Mls antigens induce high primary *in vitro* proliferative responses.[2,3] In addition to the ability of the Mls-1^a and Mls-2^a antigens to stimulate a large fraction of T cells *in vitro*, mice bearing the Mls-1^a antigens show *in vivo* deletion of

Mechanisms of Lymphocyte Activation and Immune Regulation III
Edited by S. Gupta *et al.,* Plenum Press, New York, 1991

139

Table 1.　Genes Affecting Selection of Vβ6 Cells in Various Inbred Mouse Strains

| Strain[a] | Mls-1 | MHC Class II | |
		I-A	I-E
SJL/JCr	b	s	—
A.SW/J	b	s	—
B10.S	b	s	—
AKR/NCr	a	k	k
CBA/CaHNCr	b	k	k
C3H/HeJNCr	b	k	k
DBA1/J	a	q	—
B10.Q	b	q	—

(a) Strain designations J or Cr indicate that mouse strains were obtained from Jackson Laboratories (Bar Harbor, ME) or the National Cancer Institute (Frederick, MD), respectively. B10.S and B10.Q were a kind gift of Dr. Ronald Schwartz and were produced on NIAID Contract by Bioqual, Inc., Rockville, MD.

Vβ8.1[+] and Vβ6[+] T cells,[5,6] whereas those bearing Mls-2[a] antigens delete Vβ3[+] T cells.[7,18,19] The Mls-1[b] and Mls-2[b] gene products are monstimulatory and fail to promote clonal deletion. Mls appears to be dependent on class II MHC (especially I-E) for presentation to T cells, and antibodies to class II antigens block proliferative responses to Mls.[20,21,22] Mls does not show conventional MHC restriction, however, and thus T cells recognize Mls in the context of many types of MHC molecules. In addition, there appears to be a hierarchy in efficiency of Mls presentation, such that H-2[k], H-2[d] > H-2[b] > H-2[s] > H-2[q].[7] This presentation capacity by various MHC is reflected in both proliferative responses and clonal deletion.

Antigens like Staphylococcal Enterotoxin B (SEB) share certain properites with Mls in that they appear to be mitogens for T cells bearing specific Vβ chains (regardless of the Vα) and promote deletion of such cells when administered in vivo during development. Thus, Mls and SEB have been referred to as "superantigens".[24] The fact that SEB cannot be presented following cleavage into peptides[25] supports the concept that such antigens may bind MHC molecules as unprocessed antigens outside of the conventional peptide binding groove.[26] Therefore, it is reasonable to propose that "presentation" of Mls may be very different from that of conventional antigens.

In the present study, chimeras were constructed in order to manipulate the site of Mls-1[a] and MHC antigen expression. To determine the fate of developing Vβ6[+] thymocytes encountering Mls-1[a] on either radiation-resistant or -sensitive cellular elements, chimeras were made by injection of T-depleted bone marrow stem cells into irradiated hosts. The Mls and MHC type for the strains used and discussed in this study are presented in Table 1.

In previous studies using H-2[k] donors (CBA/Ca or C3H) to reconstitute (B10.S x AKR) F$_1$ recipients, deletion of Vβ6[+] thymocytes was quite good, even though the donor-derived cells were unable to produce the Mls-1[a] antigen.[23] Since the irradiated host provided the antigen in this case, deletion appeared to be mediated by transfer of the Mls antigen from the host to be presented by the donor. In contrast, if the donor-derived cells were of the H-2[s] type (SJL or A.SW), which is nonpermissive or poor for presentation of Mls, little or no deletion of Vβ6[+] thymocytes was observed. Since deletion was a function of the donor MHC type and not of the host providing the relevant antigen (the host was the same in both cases), it appears that clonal deletion of the Vβ6[+] cells occurred by transfer of Mls from the host

Table 2. Clonal Deletion in Chimeras Made from Strains Which Do Not Mediate Deletion of Vβ6$^+$ Thymocytes[a]

Chimeras/Strains	% Vβ6 Thymocytes[b]		
	Total	CD4$^+$8$^-$	CD4$^-$8$^+$
1. (CBA/Ca x B10.Q)F$_1$ → DBA/1	1.9 ± 0.5	1.5 ± 0.4	3.3 ± 1.3
2. DBA/1 → (CBA/Ca x B10.Q)F$_1$	3.9 ± 0.7	3.7 ± 0.7	4.4 ± 0.6
(CBA/Ca x B10.Q)F$_1$	10.5 ± 0.3	9.0 ± 1.5	17.5 ± 2.6
DBA/1 → DBA/1	5.2 ± 0.1	4.3 ± 0.0	8.3 ± 0.3
DBA/1	5.1 ± 0.4	4.1 ± 0.6	7.9 ± 0.5

	Donor	Host
1.	Mls-1^b x Mls-1^b →	Mls-1^a
2.	Mls-1^a →	Mls-1^b x Mls-1^b

(a) Bone marrow chimeras were constructed by 1000R γ irradiation (Cs source) followed by intravenous injection of 1 x 10^7 or 2 x 10^7 T-depleted bone marrow cells 12-24 hr later. Chimeras were maintained in sterile cages and on antibiotic water (Biosol, Upjohn) throughout the study. T cell depletions were done using a mixture of mAb to Thy-1.2, Jlj[36], and CD5.2, C3PO[40] or CD5.1 (anti-Ly-1.1, New England Nuclear) and complement. F$_1$ animals were bred in our facility. Animals were sacrificed 35-80 days after reconstitution, when animals > 97% donor type as determined by expression of CD5 allelic markers. Thymocytes were treated with J11d antibody and analyzed by FC analysis as described in Figures 1 and 2. Arithmetic mean and S.E.M. are given for 3-5 individual animals. All values have been normalized to include only CD3^{Hi} cells.

(b) Values represent total Vβ6 cells divided by total CD4$^+$8$^-$ + CD4$^-$8$^+$ (x100), Vβ6 CD4$^+$8$^-$ cells divided by total CD4$^+$8$^-$ (x100), and Vβ6 CD4$^-$8$^+$ cells divided by total CD4$^-$8$^+$ (x100).

to be presented by the donor whenever the donor expressed a permissive MHC for Mls presentation. Furthermore, the fact that the host expressed Mls and also a permissive MHC type (H-2^{sxk}) and yet no deletion occurred with H-2^s donors, indicated that the thymic epithelium is poor at promoting Mls-mediated deletion. These findings have been confirmed using other chimeras.[27,28,29]

Since transfer of Mls has not been easily demonstrated *in vitro*, a further test for transfer of Mls antigen was made using another set of chimeras in which neither strain alone is able to mediate deletion of Vβ6$^+$ T cells. The DBA/1 strain encodes Mls-1^a antigen but is unable to mediate deletion of Vβ8.1 and Vβ6-bearing T cells, presumably because H-2^q is nonpermissive for Mls presentation. While DBA/1 cells cannot stimulate Mls-1^a proliferative responses, F$_1$ progeny of DBA/1 and MHC-permissive strains (H-2^k or H-2^d) mediate deletion and yield Mls-stimulatory cells.[30] Thus, by creating chimeras using a 1) (CBA/Ca x B10.Q)F$_1$ host which possesses a permissive H-2^{kxq} MHC, but is unable to delete Vβ6$^+$ T cells because it does not make Mls-1^a antigen (Tables 1 and 2), and 2) a DBA/1 bone marrow donor, the issue of Mls antigen transfer could be more rigorously tested.

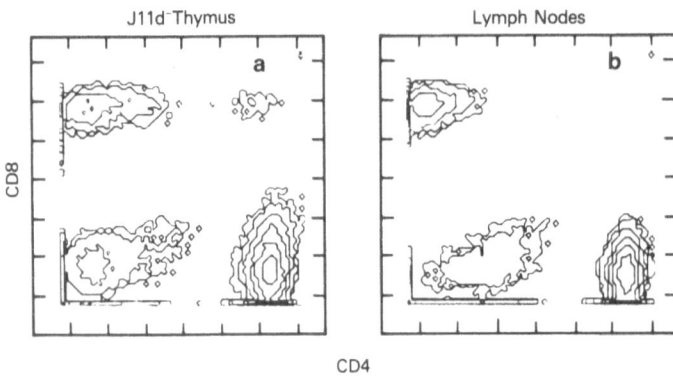

Figure 1. Two-color flow cytometric analysis (FC) of CD4 and CD8 on
mature thymocytes (a) or lymph node cells (b) from (CBA/Ca
x B10.Q)F$_1$ mice (made as described in Table 2). To enrich
for mature T cells, thymocytes were treated with J11d mAb[36]
plus complement (Low-Tox-M, Cedar Lane). Cells were stained
with mAb to CD4, H129.19[37] and second antibody, fluorescein-
isothiocyanate (FITC)- labeled goat anti-rat IgG (Kierkegaard
and Perry), followed by biotinylated mAb to CD8, 53-6.7
(Becton-Dickinson)[38] and allophycocyanin (APC)-labeled avidin)
(Caltag). 100,000 cells were analyzed using a FACS 440 (Becton-
Dickinson).

In order to analyze deletion in such chimeras, two-color-flow cytometry was performed
to assay for Vβ6 usage. To facilitate analysis of thymocytes, mature T cells were enriched
using the J11d monoclonal antibody (mAb) and complement. As shown in Figure 1, this
treatment severely depletes CD4$^+$8$^+$ thymocytes, leaving a population of thymocytes that is
> 90% CD3^{Hi} (Figure 2). Such treatment greatly increases the reliability of the analysis,
since mature T cells (CD3^{Hi}) are < 15% of total thymocytes, and Vβ6$^+$ T cells comprise
< 15% of this subset.

The Vβ6 data obtained from analysis performed as shown in Figure 2 (a and b) were
normalized for the number of CD3^{Hi}, CD4$^+$8$^-$ or CD4$^-$8 cells (Figure 2, c and d).

Two types of chimeras were examined using these analyses: 1) (CBA/Ca x B10.Q)F$_1$)
→ DBA/1, and 2) the reciprocal chimera, DBA/1 → (CBA/Ca x B10.Q)F$_1$. Vβ6 usage
by J11d$^-$ thymocytes from these chimeras is shown in Table 2.

While normal (CBA/Ca x B10.Q)F$_1$ animals express no Mls antigen and therefore contain approximately 10% Vβ6$^+$ cells, the F$_1$ → P chimera is deleted by > 80%. Thus, a chimera
made from two non-deleting strains is able to delete Vβ6$^+$ T cells by transfer of Mls from
the host to donor, when the donor expresses a permissive MHC haplotype for Mls presentation. The fact that > 75% of the Vβ6$^+$ T cells are retained in the reciprocal chimera,
DBA/1 → (CBA/Ca x B10.Q)F$_1$ (when comparing this chimera to the DBA/1 → pseudo-
chimeric control) suggests that either 1) Mls cannot be transferred in the direction of donor →
host for presentation by the thymic epithelium, or 2) the thymic epithelium of the permissive
F$_1$, H-2^{kxq}, presents Mls-1^a but is in some way defective in its ability to induce
clonal deletion. Since transfer of Mls can occur from host to donor to mediate deletion (as
demonstrated by the first chimera, Table 2), it appears likely that the opposite direction of
transfer should also occur. A more tenable hypothesis to explain the failure to delete is
that clonal deletion involves some accessory molecule or second signal supplied by the bone
marrow-derived cells (most likely the macrophages or dendritic cells) which is not supplied by
the thymic epithelium.

Table 3 summarizes the results obtained using various chimeras, highlighting the impor-

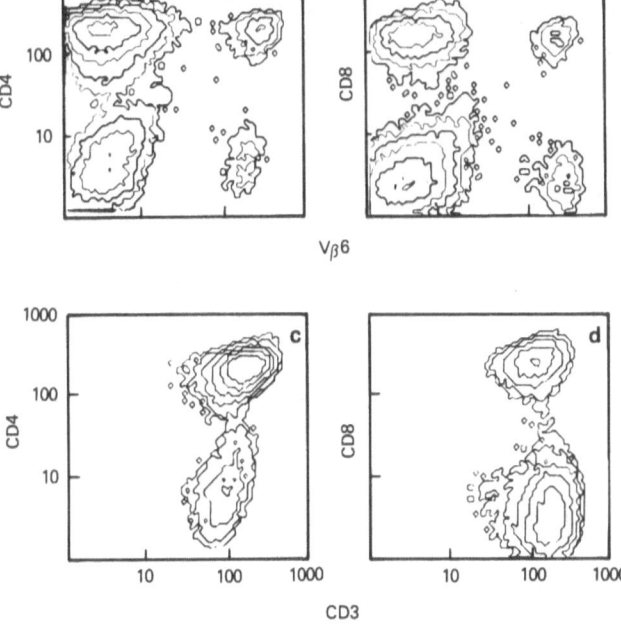

Figure 2. Two-color FC analysis of Vβ6 (RR4-7) versus
(a) CD4 or (b) CD8 and CD3 versus (c) CD4 or (d) CD8
on mature thymocytes from (CBA/Ca x B10.Q)F$_1$ mice.
J11d mAb and complement were used to enrich for mature
thymocytes (described in Figure 1). Cells were stained
with a mAb to Vβ6 (RR4-7)[9] detected with
FITC-labeled goat anti-rat IgG or mAb 500A2[39]
detected with FITC-labeled goat anti-hamster IgG
(Caltag), followed by biotinylated mAb to CD4 (H129.19)
or CD8 (53-6.7) and APC-avidin. 100,000 cells were
analyzed.

tance of the MHC type of the donor, the inability of the radiation-resistant host to induce clonal deletion, and the host-to-donor transfer of Mls in achieving clonal deletion.

Since the majority of Vβ6$^+$ T cells were not eliminated in certain chimeras (Table 3), it was of interest to determine if the T cells from these chimeric mice were tolerant of the epithelial-type antigens to which they were exposed during development. As reported previously[23,31] and as illustrated in Figure 3, T cells from these P $\rightarrow$ F$_1$ chimeras are tolerant of the epithelial-type antigens when they are presented by T-depleted spleen cells from normal F$_1$ mice in a mixed lymphocyte culture (MLR).

T cells from SJL $\rightarrow$ (B10.S x AKR)F$_1$ chimeras show a reduced ability to respond to F$_1$ stimulators when compared to normal SJL (Figures 3a and c) or pseudo-chimeric controls (SJL $\rightarrow$ SJL; data not shown). This failure to respond is specific, since T cells from the chimeras respond well to third-party BALB/c stimulators (Figures 3b and d). Thus, T cells developing in such chimeric animals have been tolerized to the host antigens.

The fact that Vβ6$^+$ T cells in such P $\rightarrow$ F$_1$ were not deleted in the chimeric animals but were tolerant to Mls as assayed by MLR suggested that the Vβ6$^+$ T cells were in some way functionally inactivated. As shown previously[23,30] and here in Figure 4b, Vβ6-specific, plate-bound antibodies stimulate Vβ^+ mature thymocytes from the pseudo-chimeric SJL $\rightarrow$ SJL controls; however, they fail to stimulate Vβ6$^+$ T cells from the SJL $\rightarrow$ (B10.S x AKR)F$_1$ chimeras. This nonresponsiveness is specific, since such chimeras are able to respond to anti-TCR$\alpha\beta$ stimulation (Figure 4a).

Table 3. Clonal Deletion of Vβ6⁺, CD4⁺ Thymocytes in Chimeras in which Mls-1ᵃ is Contributed by Either the Donor or the Host is a Function of the MHC Haplotype of the Donor

Chimera	Mls-1 of Donor and Host	MHC Donor	MHC Host	Deletion[a]
SJL → (B10.S x AKR)F_1	b → b x a	NP	P	—
A.SW → (B10.S x AKR)F_1	b → b x a	NP	P	—
CBA/J → (B10.S x AKR)F_1	a → b x a	P	P	+
CBA/Ca → (B10.S x AKR)F_1	b → b x a	P	P	+
C3H → (B10.S x AKR)F_1	b → b x a	P	P	+
(CBA/Ca x B10.Q)F_1 → DBA/1	b x b → a	P	NP	+
DBA/1 → (CBA/Ca x B10.Q)F_1	a → b x b	NP	P	—

(a) Deletion is designated as negative if the majority of Vβ6-bearing mature thymocytes are not deleted. Table 3 is derived from data presented in Table 2 and previously published.[23]

P, permissive; NP, nonpermissive

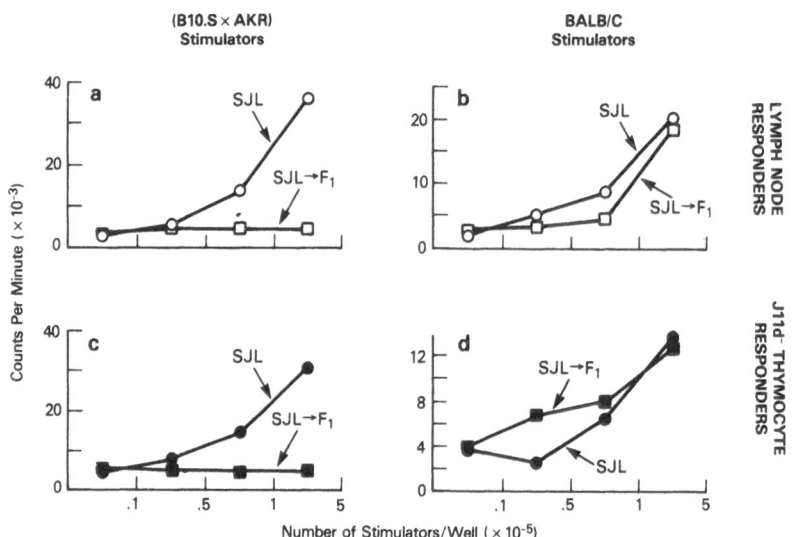

Figure 3. Tolerance of SJL → (B10.S x AKR) F_1 chimeras to F_1 spleen cells in an MLR. Responder cells were obtained from normal SJL (circles) or SJL → F_1 (boxes). Chimera cells were > 95% donor type. Lymph node (a and b) or J11d⁻ thymocytes (c and d) responders were plated at 2 x 10⁵ cells per well in 96-well, round-bottom plates with varying doses of irradiated (3000R), T-depleted (B10.S x AKR)F_1 (a and c) or BALB/c (b and d) stimulator cells. Stimulators were added, the plates were incubated at 37°C for 4 days in complete medium (RPMI 1640 plus fetal calf serum, 10%), and the cells were pulsed with 1 μCi or [³H]thymidine for an additional 12-18 hours. Samples were processed for standard scintillation counting.

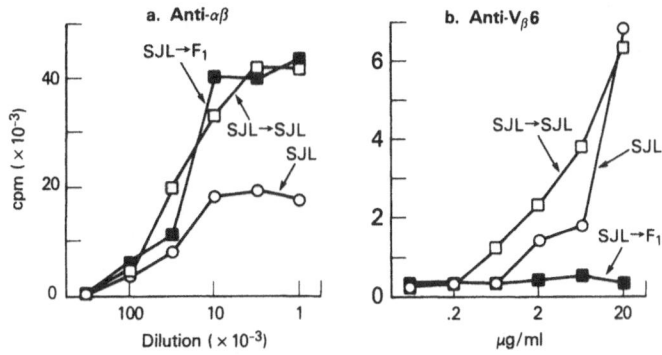

Figure 4. P → F$_1$ chimeric T cells show poor responses to specific anti-Vβ stimulation. J11d⁻ thymocyte responders (prepared as described in Figure 1) from SJL (open circles), SJL → (B10.S x AKR)F$_1$ chimeras (closed boxes) were plated at 1 x 10⁵ cells per well on 96-well, round-bottom microtiter wells previously coated with the indicated concentrations of (a) a diluted anti-TCRαβ (H57-597 mAb)[41] ascites or (b) purified anti-Vβ6 (RR4-7). Cells were incubated for 72 hr, pulsed, and harvested as in Figure 3. Responder cells from chimeras were > 95% donor type.

Similarly, in another chimera, B6 → (B6 x DBA/2) F$_1$ (where the Mls-1ᵃ antigen is contributed by the host and the donor is a poor presenter for Mls) Vβ6 is also not deleted. Similar to the above results, T cells from this chimera respond well to stimulation with plate-bound anti-Vβ8 as a control, but not to anti-Vβ6 (data not shown). These results, taken together, demonstrate the clonal specificity of the antibody stimulated responses. Most importantly, these studies indicate that the thymus is able to induce tolerance not only by clonal deletion but also by clonal anergy.

It is noteworthy that this *in vivo* chimeric model, in a number of aspects, resembles an *in vitro* system using T helper type 1 clones which can be anergized.[32,33] The cells from the chimeras do make a partial response, in that they express interleukin-2 (IL-2) receptors in response to Mls or anti-Vβ6,[23,21] respond to exogenously-added IL-2 (although not to the level of the controls),[23] and appear not to be suppressed by mixing experiments.[17] Thus, similar to the *in vitro* model, the anergized cells from the chimeras appear to be defective in the production of IL-2. The inability to fully reconstitute the response with exogenous IL-2 suggests that additional functional defects must also exist.

The anergy in the mature T cell clones is attributed to TCR engagement in the absence of an obligatory second signal provided by bone marrow-derived dendritic cells or macrophages. Invoking this concept for the chimera system, one could argue that clonal deletion is mediated by bone marrow-derived cells, while clonal anergy is induced by the radiation-resistant thymic epithelium. In support of this notion, epithelial cells can present antigen to T cell hybridomas but not to T cell clones,[35,39] implicating, perhaps, the lack of a second signal. An alternative possibility is that deletion versus anergy is not a result of interaction with different inducing cells which deliver qualitatively different signals but is, instead, attained as a result of quantitative differences in signaling. That is, H-2ˢ + Mls on bone marrow-derived cells provides a weaker signal, leading to clonal anergy. It should be mentioned that, in a chimeric animal in which only the bone marrow donor provided Mls-2ᵃ, and only the radiation-resistant host expressed a permissive MHC, neither clonal deletion nor clonal anergy was obtained.[17] The failure to anergize may be due to the direction of transfer of Mls (donor → host) or the type of Mls (Mls-2ᵃ as opposed to Mls-1ᵃ). Mls-2ᵃ is considered to be a weak stimulatory agent *in vitro*. In any case, appropriate radiation bone marrow and thymus-grafted chimeras can be made to test these two opposing hypotheses for thymic-induced anergy.

ACKNOWLEDGMENTS

The authors gratefully acknowledge Dr. Ronald H. Schwartz for continued advice and support and Brenda Rae Marshall for expert editorial assistance.

REFERENCES

1. R. H. Schwartz, Acquisition of immunologic self-tolerance, *Cell* 57:1073 (1989).
2. R. Abe and R. J. Hodes, T-cell recognition of minor lymphocyte stimulating (Mls) gene products, *Ann. Rev. Immunol.* 7:683 (1989).
3. H. Festenstein, S. Kumura, and G. Biasi, Mls and tolerance, *Immunol. Rev.* 107:29 (1989).
4. C. A. Janeway, Jr. and M. E. Katz, The immunobiology of the T cell response to Mls-locus-disparate stimulator cells. I. Unidirectionality, new strain combinations and the role of Ia antigens, *J. Immunol.* 134:2057 (1985).
5. J. W. Kappler, U. Staerz, J. White, and P. C. Marrack, Self-tolerance eliminates T cells specific for Mls-modified products of the major histocompatibility complex, *Nature* 332:35 (1988).
6. H. R. MacDonald, R. Schneider, R. K. Lees, R. C. Howe, H. Acha-Orbea, H. Festenstein, R. M. Zinkernagel, and H. Hengartner, T-cell receptor Vβ use predicts predicts reactivity and tolerance to Mls[a]-encoded antigens, *Nature* 332:40 (1988).
7. A. M. Pullen, P. Marrack, and J. W. Kappler, The T-cell repertoire is heavily influenced by tolerance to polymorphic self-antigens, *Nature* 335:796 (1988).
8. J. Bill, O. Kanagawa, D. O. Woodland, and E. Palmer, The MHC molecule I-E is necessary but not sufficient for the clonal deletion of Vβ11-bearing T cells, *J. Exp. Med.* 169:1405 (1989).
9. O. Kanagawa, E. Palmer, and J. Bill, A T cell receptor Vβ6 domain that imparts reactivity to the Mls[a] antigen, *Cell. Immunol.* 119:412 (1989).
10. J. W. Kappler, N. Roehm, and P. Marrack, T cell tolerance by clonal elimination in the thymus, *Cell* 49:273 (1987).
11. L. J. Berg, B. F. de St. Groth, A. M. Pullen, and M. M. Davis, Phenotypic differences between $\alpha\beta$ versus β T-cell receptor transgenic mice undergoing negative selection, *Nature* 340:559 (1989).
12. P. Kisielow, H. Blüthmann, U. D. Staertz, M. Steinmetz, and H. von Boehmer, Tolerance in T-cell-receptor transgenic mice involves deletion of nonmature C4[+]8[+] thymocytes, *Nature* 333:742 (1988).
13. W. C. Sha, C. A. Nelson, R. D. Newberry, D. M. Kranz, J. H. Russell, and D. Y. Loh, Positive and negative selection of an antigen receptor on T cells in transgenic mice, *Nature* 336:73 (1988).
14. B. J. Fowlkes, R. H. Schwartz, and D. M. Pardoll, Deletion of self-reactive thymocytes occurs at a CD4[+]CD8[+] precursor stage, *Nature* 335:830 (1988).
15. H. R. MacDonald, H. Hengartner, and T. Pedrassini, Intrathymic deletion of self-reactive cells prevented by neonatal anti-CD4 antibody treatment, *Nature* 335:174 (1988).
16. P. Marrack, D. Lo, R. Brinster, R. Palmiter, L. Burkly, R. H. Flavell, and J. Kappler, The effect of thymus environment on T cell development and tolerance, *Cell* 53:627 (1988).
17. F. Ramsdell and B. J. Fowlkes, Clonal deletion vs. clonal anergy: the role of the thymus in inducing self tolerance, *Science* 248:1342 (1990).
18. R. Abe, M. S. Vacchio, B. Fox, and R. J. Hodes, Preferential expression of the T-cell receptor Vβ3 gene by Mls[c] reactive T cells, *Nature* 335:827 (1988).
19. A. M. Fry and L. A. Matis, Self-tolerance alters T-cell receptor expression in an antigen-specific MHC restricted immune response, *Nature* 335:830 (1988).
20. C. A. Janeway, Jr., E. A. Lerner, J. M. Jason, and B. Jones, T lymphocytes responding to Mls-locus antigens are Lyt-1[+]2[+] and I-A restricted, *Immunogenetics* 10:481 (1980).

21. C. A. Janeway, Jr., and M. E. Katz, The immunobiology of the T cell response to Mls-locus-disparate stimulator cells. I. Unidirectionality, new strain combinations and the role of Ia antigens, *J. Immunol.* 134:2057 (1985).

22. S. MacPhail and O. Stutman, Independent inhibition of IL2 synthesis and cell proliferation by anti-Ia antibodies in mixed lymphocyte responses to Mls, *Eur. J. Immunol.* 14:318 (1984).

23. F. Ramsdell, T. Lantz, and B. J. Fowlkes, A nondeletional mechanism of thymic self tolerance, *Science* 246:1038 (1989).

24. J. White, A. Herman, A. M. Pullen, R. Kubo, J. W. Kappler, and P. Marrack, The Vβ-specific superantigen staphylococcal enterotoxin B: Stimulation of mature T cells and clonal deletion in neonatal mice, *Cell* 56:27 (1989).

25. J. D. Fraser, High-affinity binding of staphylococcal enterotoxins A and B to HLA-DR, *Nature* 339:221 (1989).

26. C. A. Janeway, Jr., J. Yagi, P. J. Conrad, M. E. Katz, B. Jones, S. Vroegop, and S. Buxser, T-cell responses to Mls and to bacterial proteins that mimic its behavior, *Immunol. Rev.* 107:61 (1989).

27. D. E. Speiser, R. K. Lees, H. Hengartner, R. M. Zinkernagel, and H. R. MacDonald, Positive and negative selection of an antigen receptor on T cells in transgenic mice, *Nature* 336:73 (1988).

28. D. E. Speiser, R. Schneider, H. Hengartner, H. R. MacDonald, and R. M. Zinkernagel, Clonal deletion of self-reactive T cells in irradiation bone marrow chimeras and neonatally tolerant mice: evidence for intercellular transfer of Mls$_a$, *J. Exp. Med.* 170:595 (1989).

29. Y. Yoshikai, M. Ogimoto, K. Matsumoto, M. Sakumoto, G. Matsuzaki, and K. Nomoto, Deletion of Mls-reactive T cells in H-2-compatible but Mls-incompatible bone marrow chimeras, *Eur. J. Immunol.* 19:1009 (1989).

30. R. Abe and R. J. Hodes, Properties of the Mls system: a revised formulation of Mls genetics and an analysis of T-cell recognition of Mls determinants, *Immunol. Rev.* 107:5 (1989).

31. J. L. Roberts, S. O. Sharrow, and A. Singer, Clonal deletion and clonal anergy in the thymus induced by cellular elements with different radiation sensitivities, *J. Exp. Med.* 171:935 (1990).

32. M. K. Jenkins, D. M. Pardoll, J. Mizuguchi, T. M. Chused, and R. H. Schwartz, Molecular events in the induction of a nonresponsive state in interleukin 2-producing helper T-lymphocyte clones, *Proc. Natl. Acad. Sci. U.S.A.* 84:5409 (1987).

33. D. L. Mueller, M. K. Jenkins, and R. H. Schwartz, Clonal expansion versus functional clonal inactivation: a costimulatory signalling pathway determines the outcome of T cell antigen receptor occupancy, *Ann. Rev. Immunol.* 7:445 (1989).

34. R. G. Lorenz and P. M. Allen, Thymic cortical epithelial cells can present self-antigens *in vivo, Nature* 337:560 (1989).

35. R. G. Lorenz and P. M. Allen, Thymic cortical epithelial cells lack full capacity for antigen presentation, *Nature* 340:557 (1989).

36. J. Bruce, F. W. Symington, T. J. McKean, and J. Sprent, A monoclonal antibody discriminating between subsets of T and B cells, *J. Immunol.* 127:2486 (1981).

37. A. Pierres, P. Naquet, A. van Agthoven, F. Bekkhoucha, F. Denizot, Z. Mishal, A. Schmitt-Verhulst, and M. Pierres, A rat anti-mouse T4 monoclonal antibody (H129.19) inhibits the proliferation of Ia-reactive T cell clones and delineates two phenotypically distinct (T4$^+$, Ly-2,3$^-$, and T4$^+$, Ly-2,3$^+$) subsets among anti-Ia cytolytic T cell clones, *J. Immunol.* 132:2775 (1984).

38. J. A. Ledbetter and L. A. Herzenberg, Xenogeneic monoclonal antibodies to mouse lymphoid differentiation antigens, *Immunol. Rev.* 47:63 (1979).

39. W. L. Havran, M. Poenie, J. Kimura, R. Tsien, A. Weiss, and J. P. Allison, Expression and function of the CD3-antigen receptor on murine CD4$^+$8$^+$ thymocytes, *Nature* 300:170 (1987).

40. C. Mark, F. Figueroa, Z. A. Nagy, and J. Klein, Cytotoxic monoclonal antibody specific for Lyt1.2 antigen, *Immunogenetics* 16:95 (1982).

41. R. T. Kubo, W. Born, J. W. Kappler, P. Marrack, and M. Pigeon, Characterization of a monoclonal antibody which detects all murine $\alpha\beta$ T cell receptors, *J. Immunol.* 142:2736 (1989).

CELLS INDUCING TOLERANCE TO Mls AND H-2 ANTIGENS

Susan R. Webb and Jonathan Sprent

Department of Immunology
Research Institute of Scripps Clinic
La Jolla, California

INTRODUCTION

The thymus shapes the T cell repertoire by selecting T cells which recognize foreign antigens in association with self major histocompatibility complex (MHC) molecules and by tolerizing those cells which have the potential to react to self antigens.[1,2] There is now convincing evidence that negative selection (tolerance induction) of potentially self-reactive cells can occur by a process of clonal deletion.[3-5] Direct support for this viewpoint has come from studies with monoclonal antibodies (mAB) specific for particular T cell receptor (TCR) $V\beta$ gene products. Thus, it has been found that mature T cells bearing $V\beta11$, $V\beta17$, and $V\beta5$-positive TCRs are selectively deleted in I-E-positive mouse strains, presumably because these TCRs have high reactivity for I-E molecules (perhaps complexed with self peptides).[6-8] Similar results are found for Mls antigens, a set of poorly characterized self antigens which have the unusual property of stimulating very high primary mixed lymphocyte responses (MLR).[9,10] Thus, Mls[a]-positive mouse strains delete $V\beta6$ and $V\beta8.1$-positive T cells and Mls[c]-positive strains delete T cells bearing $V\beta3$.[11-14]

Recently, a second mechanism for the maintenance of T cell tolerance has received considerable attention. Reports show that, both *in vitro* and *in vivo*, T cells can display functional inactivation (anergy) as a consequence of interaction with inappropriate antigen pre senting cells (APC).[15-21] In this case, the potentially-reactive T cells are present but TCR triggering fails to induce proliferation or IL 2 production.

The relative contribution of clonal deletion and anergy to the induction and maintenance of tolerance remains to be clarified. Likewise the factors which favor the induction of one form of tolerance over the other are unclear. One possibility is that the different forms of tolerance reflect contact with different types of APC. Recent studies on T cell tolerance developing in bone marrow (BM) chimeras[20,21] are relevant to this question. These studies suggest that, when T cells confront antigen expressed not on typical APC but on other cell types, anergy rather than clonal deletion predominates.

To shed light on the factors controlling tolerance induction we have undertaken a detailed study to identify the cell types inducing tolerance to Mls[a] antigens. One of the main aims here is to determine whether there is any correlation between the cell type inducing tolerance and the mechanism of tolerance, either anergy or deletion.

Mechanisms of Lymphocyte Activation and Immune Regulation III
Edited by S. Gupta *et al.*, Plenum Press, New York, 1991

CELL TYPES WHICH EXPRESS Mls^a DETERMINANTS

Since Mls determinants are not yet serologically defined, we and others have relied on the ability to stimulate an anti-Mls^a proliferative response of mature T cells as an indication of cellular expression. Using this criteria, Mls^a expression seems to be restricted to B cells;[22-26] other typical APC such as dendritic cells and macrophages fail in our hands to stimulate either primary anti-Mls^a MLR or IL-2 production by Mls^a-specific T cell hybridomas.[25,26] Since typical APC, especially DC, are potent stimulators for allo-H-2 reactive T cells,[27] the failure of Mϕ and DC to stimulate Mls^a-responses most likely reflects lack of expression of Mls antigens by these cells. The inability of other cell types, e.g. T cells, to stimulate anti-Mls^a responses may be a reflection, not of lack of expression of Mls antigen, but rather a failure to express class II antigens or important costimulatory signals. In this respect it should be mentioned that there is general agreement that anti-Mls^a responses depend upon joint recognition of class II (Ia) molecules.[10,28] Whether Mls^a antigens and class II molecules are recognized as a complex or as separate structures is disputed.

DO B CELLS CONTROL TOLERANCE TO Mls ANTIGENS?

To examine whether B cells are the only cells which induce Mls-specific tolerance, we prepared B-cell-depleted mice of Mls^a genotype by daily injection from birth of anti-μ antibodies.[29] Despite the fact that these mice had no detectable Ig$^+$ B cells, DBA/2 (H-2^d, Mls^a) and CBA/J (H-2^k, Mls^a) μ-suppressed mice (μsm) were functionally tolerant to Mls^a antigens in MLR. Furthermore, adult μsm showed partial deletion of Vβ6$^+$ cells.

TOLEROGENICITY OF THYMIC EPITHELIUM

The observation that the deletion of Vβ6$^+$ cells in Mls^a mice is evident at the level of mature thymocytes[12] favors the idea that self tolerance induction to Mls^a antigens is induced intrathymically. Which cell types in the thymus induce Mls^a tolerance, however, is unclear. Although B cells are obvious candidates, the possibility that thymic epithelium plays a role in Mls^a tolerance requires serious consideration for two reasons. First, thymic epithelium shows strong expression of class II molecules. Second, recent studies on bone marrow chimeras have led to the conclusion that Mls^a antigens are expressed on a non-marrow-derived component of the thymus (presumably epithelial cells) and that T cell contact with this component leads to strong tolerance of mature Vβ6$^+$ thymocytes.[20,21] Interestingly, the tolerance seen in this situation appeared to reflect anergy rather than deletion. The critical question here is whether such tolerance reflects contact with thymic epithelial cells *per se* or with some other type of cell in the thymus. The most popular approach for assessing the tolerogenicity of thymic epithelium is to treat fetal thymuses with deoxyguanosine (DG) *in vitro*; this treatment destroys marrow-derived cells but leaves epithelial cells apparently intact.[30] In the case of H-2 tolerance, studies with thymic organ cultures and thymus-grafted athymic nude mice have shown that DG-treated thymuses induce only minimal tolerance induction to the H-2 antigens of the graft. This evidence rests largely on experiments with class I-reactive CD8$^+$ cytotoxic T lymphocyte (CTL) precursors and MLR of unseparated T cells.[31-32] Whether DG-thymus induces tolerance of H-2 class II-reactive CD4$^+$ cells has not received close attention. Likewise, it has yet to be established whether DG-thymus induces Mls^a tolerance.

To compare the tolerogenicity of DG-thymus for allo-H-2 reactive T cells and Mls-reactive T cells, we studied tolerance induction at the level of isolated CD4$^+$ and CD8$^+$ cells using athymic nu/nu mice grafted with normal and DG-treated thymuses. These thymuses were taken from mice expressing both an Mls (Mls^a) difference as well as an allo-H-2 (H-2^k) difference.

C57BL/6 (H-2^b, Mls^b) nu/nu mice were grafted under the kidney capsule with day 14 fetal thymus lobes taken from Mls^a-positive (B6 X CBA/J)F$_1$ (H-2^b, Mls^b x H-2^k, Mls^a) mice. When normal (non-DG-treated) thymuses were grafted, the T cells developing in these thymuses showed complete tolerance to both graft-type H-2^k antigens and to Mls^a antigen. The Mls^a-tolerance observed in these mice was accompanied by deletion of

$V\beta6^+$ T cells. We conclude, therefore, that the cells inducing Mls-specific tolerance in the thymus are already present at day 14 of fetal life. To examine whether epithelium contributed to tolerance in this situation, we prepared mice in which only the epithelial component of the thymus was of F_1 origin. C57BL/6 nu/nu mice were grafted with day 14 fetal thymus lobes which had been cultured for 5 days with DG to remove the BM-derived cells. Since DG treatment might leave behind a few BM-derived cells, these mice were left for a period of two months to allow any residual cells to be replaced with host B6 BM-derived cells. The mice were irradiated at this point (800 rad whole body irradiation) and reconstituted with Thy 1-marked (B6.PL-Thy 1^a/cy) BM cells; the T cells subsequently developing in the grafted thymuses were tested in MLR for tolerance to Mls^a and H-2^k antigens. In the case of lymph node (LN) T cells the CD4$^+$ subset of cells was significantly, though not completely, tolerant to thymic H-2^k class II antigens; in six experiments the response to H-2^k was reduced by 50-85% when compared to normal B6 CD4$^+$ cells. When assayed for CTL activity, LN CD8$^+$ cells from the DG-thymus-grafted mice showed little evidence of tolerance to thymic H-2^k class I antigens. Thus H-2 tolerance induced by these epithelial grafts appeared to be largely restricted to the CD4$^+$ subset. In marked contrast to H-2 antigens, however, we saw no evidence for tolerance to Mls^a antigens. Thus, T cells developing in DG-thymuses of Mls^a genotype responded normal to Mls^a determinants in MLR and expressed normal percentages of $V\beta6^+$ T cells.

The above findings lead to two broad conclusions. First, the data indicate that the epithelial component of the thymus plays no detectable role in tolerance induction to Mls^a antigens. The simplest explanation for this finding is that Mls^a antigens are expressed solely on the BM-derived component of the thymus and are not found on epithelial cells. Which particular intrathymic subset of BM-derived cells express Mls^a antigens in tolerogenic form is still unclear. Although intrathymic B cells might play an important role, the possibility that other cell types could be involved is not excluded. In a later section we will consider the role of T cells in causing intrathymic Mls^a tolerance.

The second conclusion from the above studies is that thymic epithelial cells do induce significant tolerance of H-2 reactive T cells. Such tolerance is especially pronounced at the level of class II-reactive CD4$^+$ cells. A key issue is whether tolerance induced by epithelial cells reflects clonal deletion or anergy. At face value, the data would seem to support anergy as the major mechanism of tolerance induction. Thus, despite the strong functional tolerance of CD4$^+$ cells in the DG-thymus-grafted mice, we observed only a small reduction in $V\beta11^+$ cells (the grafted thymuses being I-E$^+$). This finding must be interpreted with caution, however, because the affinity of $V\beta11^+$ T cells for I-E antigens might be quite low. In this respect it is notable that, despite the strong deletion of $V\beta11^+$ T cells in I-E$^+$ mice, $V\beta11^+$ cells from normal I-E-negative mice show conspicuously weak alloreactivity for I-E antigens *in vitro*; thus $V\beta11^+$T cells are not significantly enriched in anti-I-E MLR[1], nor do $V\beta11^+$ T hybridomas make IL-2 in response to I-E$^+$ stimulators.[7] Under *in vivo* conditions, however, $V\beta11$ T cells do show I-E reactivity,[33] perhaps because the *in vivo* environment favors responses of low affinity T cells. Similar reservations apply to monitoring $V\beta17a$ T cells. Thus, although $V\beta17a^+$ T cells do show I-E alloreactivity at the level of T hybridomas,[6] typical primary MLR of I-E$^-$ T cells to I-E$^+$ APC show little or no enrichment for $V\beta17a$ T cells.[33] In view of these findings we suggest that the failure of thymic epithelium to cause more than moderate clonal deletion of $V\beta17^+$ and $V\beta17a^+$ T cells does not exclude the possibility that thymic epithelium is fully capable of causing strong deletion of high-affinity CD4$^+$ cells. We argue that most T cells detected by anti-$V\beta11$ and anti-$V\beta17a$ antibodies do not have high affinity for I-E antigens. Hence even minor deletion of $V\beta11^+$ or $V\beta17a^+$ T cells might be highly significant.

With regard to anergy, we have yet to obtain definitive evidence on whether the nondeleted $V\beta11^+$ T cells from our DG-thymus-grafted mice are refractory to stimulation with anti-$V\beta11$ mAb (the currently accepted approach for assessing anergy). The problem we have encountered is that, at the level of purified CD4$^+$ cells, even $V\beta11^+$ T cells from normal I-E$^-$ mice fail to respond detectably to anti-$V\beta11$ mAb in our hands. This finding is in accord with the report that stimulation of CD4$^+$ T cells with anti-TCR/CD3 Ab (in the absence of IL-2) is not a property of all CD4$^+$ cells but is limited to the subset of "memory" (CD45R^{lo}) cells.[34] For these reasons it is arguable whether stimulation with anti-$V\beta$ mAb is a valid approach for testing anergy at the level of virgin (non memory) CD4$^+$ cells.

Table 1. CD8[+] T Cells are Potent Tolerogens for Mls-Reactive T Cells[1].

CD4[+] responders from B10.BR mice injected with:	% of $V_\beta 6$	Stimulators (cpm x 10^3)[2]		
		B10.BR (syngeneic)	AKR/J (Mls^a)	B10.P (allo-H-2^P)
None (Normal B10.BR)	9.1	1.4	169.8	42.2
Mls^a CD4[+] T (2 x 10^6)[3]	2.5	1.1	25.2	25.4
(2 x 10^5)	9.3	1.0	161.4	30.5
(2 x 10^4)	10.7	0.9	214.8	41.3
Mls^a CD8[+] T (2 x 10^6)	1.3	7.2	12.3	30.4
(2 x 10^5)	2.1	3.0	13.0	25.6
(2 x 10^4)	2.6	1.1	28.8	33.6
Mls^a B cells (2 x 10^6)	2.4	2.0	8.4	37.9
(2 x 10^5)	6.5	1.5	207.5	37.0
(2 x 10^4)	10.4	1.7	150.3	53.0

[1]B10.BR neonates were injected i.v. with highly purified T cells or B cells from Mls^a-positive (B10.BR x CBA/J)F$_1$ mice (see ref. 38 for purification procedure). 6 wk later CD4[+] T cells were isolated from the LN of these mice and tested for their response in MLR to either Mls^a or to irrelevant, third party allo-H-2^P. $V_\beta 6$ expression was assessed by FACS analysis following staining with RR47 ascites fluid and Fitc-mouse anti-rat Ig (Pel Freez).
[2]Response of 1 x 10^5 CD4[+] responders to 5 x 10^5 mitomycin C-treated T-depleted spleen stimulators.
[3]Number of cells injected into neonates.

Despite this reservation we are impressed by the finding that the unresponsiveness of CD4[+] cells from our DG-thymus-grafted mice appears to be almost complete at the level of mature thymocytes. Since the unresponsiveness is less marked at the level of LN CD4[+] cells, it is possible that thymic epithelium does indeed induce anergy but that anergy wanes when the cells reach the extrathymic environment.

THE ROLE OF T CELLS IN THE INDUCTION OF TOLERANCE TO Mls ANTIGENS

Two lines of evidence suggest that, under defined conditions, tolerance of class I-reactive T cells can reflect contact with antigen expressed on T cells. First, intrathymic transfer of immature (CD4$^-$8$^-$) thymocytes across class I barriers induces tolerance to self (as well as allo) class I molecules.[35] Second, immunization with minor histocompatibility (H) antigens expressed on T cells tolerizes ("vetoes") the host response to the donor minor H antigens.[36]

Since Mls recognition requires co-recognition of class II antigens, one might presume at T cells would be unlikely candidates for cells inducing Mls tolerance. However, it has en reported that Mls^a tolerance occurs when Mls^b mice are injected as neonates with cells om a T cell line derived from an Mls^a-positive strain.[37] This finding raised the possibil- y that T cells might play a role in tolerance to Mls^a antigens.

We have examined this issue in detail by injecting neonatal Mls^b mice with highly irified subpopulations of cells from Mls^a-positive donors.[38] The T cells from these mice ere then evaluated 4-7 weeks later for tolerance to Mls^a in MLR and expression of $V\beta6^+$ cells. The data in Table 1 are representative of many such experiments (summarized in ıble 2). In this particular experiment, BIO.BR ($H-2^k$, Mls^b) neonates were injected ith $CD4^+$ T cells, $CD8^+$ T cells or B cells from (BIO.BR x CBA/J)F_1 ($H-2^k$, Mls^b x $H-2^k$, [ls^{a,c}) mice. The number of cells injected varied from 2 x 10^4 to 2 x 10^6. Compared the response of normal (uninjected) BIO.BR mice, mice injected with 2 x 10^6 cells from l three cell types showed a reduced response to the Mls^a antigens expressed on AKR/J T- ıpleted spleen stimulator cells. Lower doses of $CD4^+$ cells or B cells failed to induce detect)le tolerance. Of particular interest was the finding that mice injected with low doses of D8$^+$ T cells (as few as 2 x 10^4) were still unresponsive to Mls^a antigens. The injections ıd no significant affect on the ability of the T cells from these mice to respond to irrelevant, ird party allo-H-2 antigens. The tolerance seen in MLR was paralleled by the deletion of $\beta6^+$ T cells, indicating that the tolerance induced by all three cell types reflected clonal ıletion. We also examined tolerance to Mls^c antigens in this experiment; for the sake of evity the data are not shown, however the results were identical to those seen for Mls^a. hese data show that not only do T cells express both Mls^a and Mls^c self antigens, but that lerance to these antigens is controlled predominantly by $CD8^+$ T cells. To approach the ıestion of whether the tolerogenic cells have to be functional, we have injected $CD8^+$ or -depleted spleen cells exposed to irradiation (1000 rad) prior to injection. In no case did we e any evidence of functional tolerance in MLR or deletion of $V\beta6^+$ T cells when the injec- d cells were irradiated.

Analysis of $CD4^+$ single-positive mature thymocytes have shown that tolerance in the ɔonatally injected mice is evident in the thymus itself. Thus, thymocytes from mice injec- d with Mls^a $CD8^+$ T cells are functionally tolerant in anti-Mls^a MLR and show a signifi- ınt reduction (70-80%) in the number of $V\beta6^+$ cells (data not shown).

We have examined the extent of chimerism of the donor-derived T cells in the neonatal- injected mice (using antibody to Thy 1.1, Lyt 1.1, or Lyt 2.1, depending on the strain ɔmbination). In nearly all mice examined, we found a variable but significant (5-30%)

Table 2. Tolerance to Mls Antigens in Neonatally Tolerized Mice

Cells injected	No. cells injected:			
	20 x 10^6	2 x 10^6	2 x 10^5	2 x 10^4
Mls^+ B cells	+[1]	+/-[2]	-[3]	-
Mls^+ $CD4^+$ T cells	+	+/-	-	-
Mls^+ $CD8^+$ T cells	+	+	+	+

Summary of data on tolerance to Mls^a (and Mls^c) antigens in neonatally tolerized mice. Mice were injected i.v. within 1 day of birth with various numbers of purified lymphoid subsets expressing Mls^a or $Mls^{a,c}$ antigens. Lymph node cells were removed at 2 months post-injection and tested for $V_\beta6$ expression and responsiveness to Mls^a and Mls^c antigens in vitro. Mls^+ refers to expression of either Mls^a or $Mls^{a,c}$ antigens.
[1]Strong tolerance: marked deletion of $V_\beta6^+$ cells and near complete unresponsiveness to Mls^a (or Mls^c) antigens in primary MLR.
[2]Variable tolerance.
[3]No tolerance.

number of donor-derived T cells in the LN of these mice. (It should be noted that donor T cells were removed with antibody and C' prior to the analysis of tolerance above). Given that tolerance is long lasting (at least 7 weeks) and is evident within the thymus itself, it would appear that the injected T cells must be rapidly entering the thymus and remaining there for prolonged periods as a constant source of tolerogen. Preliminary experiments indicate that, even as late as 7 weeks post-injection, we can detect a very small (0.4%) but significant percentage of donor derived cells in the thymus of animals injected with either CD4$^+$ or CD8$^+$ T cells.

TOLERANCE INDUCED IN MATURE T CELLS

In the mid-1970's, Lillihook and his collaborators reported that the injection of adult Mls^b mice with spleen cells from Mls^a-positive strains led to a pronounced tolerance of the host Mls^a-specific T cells.[39] This finding was interpreted in terms of a suppressor cell network. More recently, Ramensee and his colleagues assessed Vβ6 expression under similar conditions and concluded that this form of unresponsiveness reflected anergy of the Mls^a-reactive T cells.[19] Using the same highly purified T cell subsets and B cells described in the preceding section, we have investigated the cell types responsible for this type of tolerance. One modification we made was to thymectomize the adult recipient mice to avoid confusion with the continuing flow of recent thymus emigrants. A brief summary of the data is given below.

In general terms the results are very similar to what we see in the neonatal tolerance system. Thus, injection of CD8$^+$ T cells and T-depleted spleen from Mls^a spleen from Mls^a-donors into adult Mls^b mice induces profound and long-lasting tolerance to Mls^a antigens. (Tolerance is less marked with injection of CD4 cells). Tolerance in this model is specific for the Mls-type of the injected cells and (except with injection of very high doses of cells) responses to third party allo-H-2 antigens are not significantly affected. The functional tolerance seen in MLR is almost always associated with deletion of Vβ6$^+$ T cells. Thus, when the host and the donor cells are of the H-2^k haplotype, the deletion of Vβ6$^+$ CD4$^+$ T cells is generally in the order of 60-80%. Two points should be made with regard to the deletion seen in this adult model. First, in approximately 20-30% of the experiments using H-2^k haplotype mice, significantly less deletion was seen for inexplicable reasons; in a very small number of experiments there was no deletion at all. Second, when the donor and host are of H-2^b origin, the host T cells reproducibly show little deletion of Vβ6$^+$ T cells. Nevertheless, both for H-2^k and H-2^b combinations the T cells from the injected mice still manifest marked tolerance in MLR. With regard to H-2^b mice it is of interest that Mls^a H-2^b mice show incomplete deletion of Vβ6$^+$ T cells, both in normal mice and in μsm.[12,29] These findings suggest that I-A^b may function less well in deleting Vβ6$^+$ T cells than I-E or I-A^k. The possibility arises therefore that inefficient interactions between T cells and APC tend to favor the induction of anergy rather than clonal deletion.

CONCLUDING REMARKS

The main conclusion from these studies on Mls^a antigens is that immunogenicity and tolerogenicity may be controlled by different cell types. In the case of immunogenicity, we have been able to demonstrate Mls expression only on a single class of cells, i.e., B cells. At the level of T hybridomas we have failed to demonstrate immunogenic expression of Mls^a antigens on various types of non-B Ia$^+$ cells, including Mϕ, DC and thymic epithelium. T cells also fail to stimulate anti-Mls^a responses. This applies even when T cells (including T blasts treated with neuraminidase) are supplemented with APC, i.e. are cultured under conditions designed to allow transfer of Mls antigens to class II$^+$ cells (see below). In terms of tolerance induction, however, the data presented here suggest that Mls^a tolerance is controlled largely by a subset of T cells, i.e., by CD8$^+$ cells. These cells show extraordinary potency. Thus in the neonatal tolerance model, injection of as few as 2 x 10^4 CD8$^+$ cells induce strong functional Mls^a tolerance and near complete deletion of Vβ6$^+$ cells. Since, cell-for-cell, CD8$^+$ cells are 50-100 fold more potent than other cells (including B cells) in this model, the possibility arises that tolerance induction to Mls^a antigens is controlled solely by CD8$^+$ cells. Given that CD8$^+$ are Ia$^-$ and that Mls^a responses are Ia-restricted, how do CD8$^+$ cells induce Mls^a tolerance? The answer to this question is not immediately apparent. There are several possibilities:

CD8$^+$ Cells Do Express Class II Molecules. Although murine T cells are thought to be incapable of class II expression, it is possible that T cells do express very low (but functionally significant) levels of class II molecules[40-42] or unusual forms of these molecules. Such expression might be limited to a small subset of CD8$^+$ T cells. It is also conceivable that CD8$^+$ cells are capable of absorbing class II molecules from other cells. In either situation the joint expression of Ia and Mls^a antigens on CD8$^+$ cells might be directly tolerogenic.

Mls Antigens May Be Transferred to Typical Ia$^+$ APC. Several studies using BM chimeras have claimed to prove that Mls antigens can be transferred from one cell to another.[20,43,44] And yet, experiments that we have carried out to directly test this have thus far failed to reveal evidence for transfer. For example, we constructed double BM chimeras where a mixture of H-2^k, Mls^b + H-2^b, Mls^a BM cells was used to reconstitute an irradiated (H-2^k x H-2^b, Mls^b)F$_1$ mouse.[25] Although the H-2^k-marked Mls^b cells (T cells, B cells and DC) developed alongside the H-2^b-different Mls^a cells, we could find no indication that the Mls^b cells picked up the Mls^a antigen; thus, the H-2^k, Mls^b cells purified from the chimeras failed to become stimulatory for anti-Mls^a responses. Similarly, as discussed earlier, *in vitro* mixing experiments have been conspicuously unsuccessful in converting Mls^b nonstimulatory APC into Mls^a stimulators. For example, we have never succeeded in generating Mls^a stimulators by mixing Mls^a-negative APC with fixed or killed Mls^a B cells (unpublished data). In view of these findings we suggest that the evidence for transfer of Mls^a antigens reported for BM chimeras is far from conclusive. The data are equally compatible with a model in which tolerance reflects transfer of donor derived class II molecules onto the radioresistant CD8$^+$ cells of the Mls^a host. As discussed below, tolerance could also reflect a three cell interaction.

Tolerance Reflects Joint Contact With Mls^a Antigens on CD8$^+$ Cells and Ia Molecules on APC. This possibility might seen unlikely because most groups argue that Mls antigens and Ia molecules are recognized as a complex. However, direct evidence on this issue is lacking. Based on studies showing segregation of Mls recognition from alloantigen recognition in T hybridomas,[45,46] we suggested several years ago that Mls antigens might be accessory molecules interacting with complementary accessory molecules (Mls^a receptors) on T cells. We and others[46,47] suggested that interaction of these Mls-anti-Mls accessory molecules stabilizes T-APC binding and facilitates TCR contact with Ia molecules (engagement of the TCR being essential for cell triggering). A feature of this model is that Mls^a antigens and Ia molecules are not necessarily recognized as a complex. But what would happen if these two ligands were recognized on different cells? Perhaps these conditions might favor tolerance induction rather than immunity. A three cell interaction might also be envisaged in a model in which Mls antigens are recognized by the TCR itself. If the Mls-binding site on the TCR were outside of the peptide binding site,[48] the TCR β chain might have sufficient affinity to bind to Mls antigen in the absence of class II.

At present we obviously cannot distinguish between the above models. We are skeptical of the idea that tolerance reflects movement of Mls^a antigens from one cell to another. Nevertheless the alternatives discussed above have yet to be supported with direct evidence. Establishing the precise mechanism of tolerance induced by Mls^a-bearing CD8$^+$ cells will thus require further experimentation.

REFERENCES

1. G. Moller, ed., Acquisition of the T cell repertoire, *Immunol. Rev.*, 42:3 (1978)
2. J. Sprent and S. Webb, Function and specificity of T cell subsets in the mouse, *Adv. Immunol.* 41:39 (1987).
3. P. Kisielow and H. von Boehmer, Negative and positive selection of immature thymocytes: timing and the role of the ligand for $\alpha\beta$ T cell receptor, *Seminars in Immunol.* (1990, in press).
4. W. C. Sha, C. A. Nelson, R. A. Newberry, D. M. Kranz, J. H. Russel, and D. Y. Loh, Positive and negative selection of an antigen receptor on T cells in transgenic mice, *Nature* 336:73 (1988).

5. P. Marrack and J. Kappler, T cell tolerance, *Seminars in Immunol.* 2:45 (1990).
6. J. Kappler, T. Wade, J. White, E. Kushnir, M. Blackman, J. Bill, R. Roehm, and P. Marrack, A T cell receptor Vβ segment that imparts reactivity to a class II major histocompatibility complex product, *Cell* 49:263 (1987).
7. J. Bill, O. Kanagawa, D. W. Woodland, and E. Palmer, The MHC molecule I-E is necessary but not sufficient for the clonal deletion of Vβ11-bearing T cells, *J. Exp. Med.* 169:1405 (1989).
8. E. Palmer, D. E. Woodland, M. P. Happ, J. Bill, and O. Kanagawa, A third set of genes regulates thymic selection, *Cold Spring Haebor Symp. Quant. Biol.* LIV:135 (1990).
9. H. Festenstein, The Mls system, *Transplant. Proc.* 8:339 (1976).
10. G. Moller, ed., T cell tolerance, Mls and MHC antigen, *Immunol Rev.* 107:1 (1989).
11. J. W. Kappler, U. Staerz, J. White, and P. C. Marrack, Self-tolerance eliminates T cells specific for Mls-modified products of the major histocompatibility complex, *Nature* 332:35 (1988).
12. H. R. MacDonald, R. Schnieder, R. K. Lees, R. C. Howe, H. Acha-Orbea, J. Festenstein, R. M. Zinkernagel, and H. Hengartner, T-cell receptor Vβ use predicts reactivity and tolerance to Mls^a-encoded antigens, *Nature* 332:40 (1988).
13. R. Abe, M. S. Vacchio, B. Fox, and R. Hodes, Preferential expression of the T-cell receptor Vβ gene by Mls^c reactive T cells, *Nature* 335:827 (1988).
14. A. Fry and L. Matis, Self-tolerance alters T-cell receptor expression in an antigen-specific MHC restricted immune response, *Nature* 335:830 (1988).
15. M. K. Jenkins and R. H. Schwartz, Antigen presentation by chemically modified splenocytes induces antigen-specific T cell unresponsiveness *in vitro* and *in vivo*, *J. Exp. Med.* 165:302 (1987).
16. H. Quill and R. H. Schwartz, Stimulation of normal inducer T cell clones with antigen presented by purified Ia molecules in planar membranes: Specific induction of a long lived state of proliferative unresponsiveness, *J. Immunol.* 138:3704 (1987).
17. D. Lo, L. Burkly, R. Flavell, R. Palmiter, and R. J. Brinster, Tolerance in transgenic mice expressing class II major histocompatibility complex on pancreatic acinar cells, *J. Exp. Med.* 170:87 (1989).
18. S. Qin, S. Cobbold, R. Benjamin, and H. J. Waldmann, Induction of classical transplantation tolerance in the adult, *J. Exp. Med.* 169:779 (1989).
19. H. G. Rammensee, R. Kroschewski, and B. Frangoulis, Clonal anergy induced in mature Vβ6$^+$ T lymphocytes on immunizing Mls-1^b mice with Mls-1^a expressing cells, *Nature* 339:541 (1989).
20. F. Ramsdell, T. Lantz, and B. J. Fowlkes, A nondeletional mechanism of thymic self tolerance, *Science* 246:1038 (1989).
21. J. L. Roberts, S. O. Sharrow, and A. Singer, Clonal deletion and clonal anergy in the thymus induced by cellular elements with different radiation sensitivities, *J. Exp. Med.* 171:935 (1990).
22. H. von Boehmer and J. Sprent, Expression of M locus differences by B cells but not T cells, *Nature* 249:363 (1974).
23. A. Ahmed, I. Scher, A. H. Smith, and K. W. Sell, Studies on non H-2 linked lymphocyte activating determinants. I. Description of the cell type bearing the Mls product, *J. Immunogenet.* 4:201 (1977).
24. S. R. Webb, J. Hu Li, D. B. Wilson, and J. Sprent, Capacity of small B cell enriched populations to stimulate mixed lymphocyte reactions: marked differences between irradiated vs. mitomycin C-treated stimulators, *Eur. J. Immunol* 115:92 (1985).
25. S. R. Webb, A. Okamoro, Y. Ron, and J. Sprent, Restricted tissue distribution of Mls^a determinants: Stimulation of Mls^a-reactive T cells by B cells but not by dendritic cells or macrophages, *J. Exp. Med.* 169:1 (1989).
26. I. J. Molina, N. A. Cannon, R. Hyman, and B. R. Huber, Macrophages and T cells do not express Mls^a determinants, *J. Immunol* 143:39 (1989).
27. R. M. Steinman, K. Inaba, G. Shuler, and M. Witmer, Stimulation of the immune response: contribution of dendritic cells, *in*: "Mechanisms of Host Resistance to Infectious Agents, Tumors and Allografts", R. M. North and R. M. Steinman, eds., The Rockefeller University Press, New York (1986).
28. C. A. Janeway, Jr. and M. E. Katz, The immunology of the T cell response to Mls-locus-disparate stimulator cells. I. Unidirectionality, new strain combinations, and the role of Ia antigens, *J. Immunol.* 134:2057 (1985).
29. S. R. Webb and J. Sprent, T cell responses and tolerance to Mls^a determinants, *Immunol. Rev.* 107:141 (1989).

30. E. J. Jenkinson, L. L. Franchi, R. Kingstrom, and J. J. T. Owen, Effect of deoxyguanosine on lymphopoiesis in the developing thymus rudiment *in vitro*: application in the production of chimeric thymus rudiments, *Eur. J. Immunol.* 12:583 (1982).

31. H. von Boehmer and K. Schubiger, Thymocytes appear to ignore class I major histocompatibility complex antigens expressed on thymus epithelial cells, *Eur. J. Immunol.* 14:1048 (1984).

32. E. J. Jenkinson, P. Jhittay, R. Kingston, and J. J. T. Owen, Studies of the role of the thymic environment in the induction of tolerance to MHC antigens, *Transplant* 39:331 (1985).

33. E.-K. Gao, O. Kanagawa, and J. Sprent, Capacity of unprimed CD$^+$ and CD8$^+$ T cells expressing Vβ11 receptors to respond to I-E alloantigens *in vivo*, *J. Exp. Med.* 170:1947 (1989).

34. J. A. Byrne, J. L. Butler, and M. D. Cooper, Differential activation requirements for virgin and memory T cells, *J. Immunol* 141:3249 (1988).

35. R. P. Shimokevitz and M. J. Bevan, Split tolerance induced by the intrathymic adoptive transfer of thymocyte stem cells, *J. Exp. Med.* 168:143 (1988).

36. P. J. Fink, R. P. Shimonkevitz, and M. J. Bevan, Veto cells, *Ann. Rev. Immunol.* 6:115 (1988).

37. D. Waite and G. Sunshine, Neonatal tolerance induction to Mls[a]. II. T cells induce tolerance, *Cell. Immunol.* 117:78 (1988).

38. S. Webb and J. Sprent, Induction of neonatal tolerance to Mls[a] antigens by CD$^+$ T cells, *Science* 248:1643 (1990).

39. B. Lillihook, H. Jacobson, and H. Blomgren, Specifically decreased MLC response of lymphocytes from CBA mice injected with cells from the H-2 compatible, M-antigen-incompatible strain C3H, *Scand. J. Immunol.* 4:209 (1975).

40. L. Hudson and J. Sprent, Specific adsorption of IgM antibody onto H-2 activated mouse T lymphocytes, *J. Exp. Med.* 143:444 (1976).

41. E. Nagy, M Nabholz, P. H. Krammer, and B. Pernis, Specific binding of alloantigens to T cells activated in the mixed lymphcyte reaction. *J. Exp. Med.* 143:648 (1976).

42. I. Melchers and H. O. McDevitt, Expression of Ia antigens on T lymphocytes, *in*: "Regulatory T Lymphocytes," B. Pernix, H. J. Vogel, eds., Academic Press, New York (1980).

43. A. M. Pullen, P. Marrack, and J. Kappler, The T cell repertoire is heavily influenced by tolerance to polymorphic self-antigens, *Nature* 335:796 (1988).

44. D. E. Speiser, R. Schneider, H. Hengartner, H. R. MacDonald, and R. Zinkernagel, Clonal deletion of self-reactive T cells in irradiation bone marrow chimeras and neonatally tolerant mice. Evidence for intercellular transfer of Mls[a], *J. Exp. Med.* 170:595 (1989).

45. S. R. Webb, J. Hu Li, I. MacNeil, P. Marrack, J. Sprent, and D. B. Wilson, T cell receptors for responses to Mls determinants and allo-H-2 determinants appear to be encoded on different chromosomes, *J. Exp. Med.* 161:269 (1985).

46. S. R. Webb, A. Okamoto and J. Sprent, Analysis of T hybridomas prepared from a T cell clone with three specificities. Recognition of self + X and allo-H-2 determinants segregates from recognition of Mls[a] determinants, *J. Immunol* 141:1828 (1988).

47. M. E. Katz and C. A. Janeway, Jr., The immunobiology of T cell responses to Mls-locus disparate stimulator cells. II. Effects of Mls-locus-disparate stimulator cells on cloned, protein antigen-specific, Ia-restricted T cell lines, *J. Immunol.* 134:2064 (1985).

48 A. M. Pullen, W. Potts, E. K. Wakeland, J. Kappler, and P. Marrack, Surprisingly uneven distribution of the T cell receptor Vβ repertoire in wild mice, *J. Exp Med.* 171:49 (1990).

MULTIPLE MECHANISMS OF T CELL TOLERANCE TO Mls-1[a]

Marcia A. Blackman,[*][+] John W. Kappler,[*][+][@] and Philippa Marrack[*][+][@][$]

*Howard Hughes Medical Institute at Denver, [+]Division of Basic Immunology, Department of Medicine, [#]Division of Basic Sciences, Department of Pediatrics, National Jewish Center for Immunology and Respiratory Medicine, Denver, Colorado
[$]Department of Biochemistry, Biophysics and Genetics, [@]Department of Microbiology, Immunology and Medicine, University of Colorado Health Sciences Center, Denver, Colorado

A fundamental role of the immune system is to identify and destroy foreign antigens (Ag). A critical requirement of this function is the ability to distinguish between what is foreign, and should be eliminated, and what is "self" and should be tolerated. This lack of response to self is termed self-tolerance. Three mechanisms have been proposed for maintaining self tolerance-clonal deletion, or elimination of self-reactive clones, clonal anergy, or functional inactivation of self-reactive clones, and suppression, or negative regulation of self-reactive clones.

Recent advances in our understanding of the structure of the T cell receptor for antigen (TcR), antigen processing and presentation on molecules encoded by the major histocompatibility complex (MHC), and the structure of MHC class I and class II (by inference) molecules has resulted in our current concept of TcR recognition of Ag/MHC. The $\alpha\beta$ TcR is a heterodimeric membrane-bound molecule with an antigen combining site that is formed by interaction of the variable components of the alpha and beta chains.[1] The TcR binds an antigenic peptide that is bound in a groove of the MHC molecule formed by two alpha helices and a beta-pleated sheet.[2,3,4] The TcR has only a short cytoplasmic tail, but signaling to the interior of the cell is mediated, at least in part, by CD3, a complex of five molecules (γ, δ, ϵ, ζ, η) with which the TcR associates in the membrane.[5] The TcR also associates with the accessory or co-receptor molecules, CD4 and CD8. These molecules bind monomorphic determinants on MHC (class II and class I, respectively), and serve to increase the avidity of TcR binding to Ag/MHC, and are also implicated in signaling.[6] The participation of other cell surface molecules in signaling, such as CD45, has also been described.[7] Thus, the binding of Ag/MHC by the TcR on the surface of the T cell triggers a cascade of signaling events that results in changes in gene expression culminating in the effector function appropriate to the type of T cell triggered.

Despite the fact that recognition of conventional Ag/MHC, i.e., a peptide bound in the MHC groove, is mediated by the combining site of the TcR that is shaped by variable region contributions from both receptor chains, superantigens have been described in which recognition is dominated by the $V\beta$ element alone, such that all T cells expressing a particular $V\beta$ share specificity for a given superantigen. For example, CD4[+]$V\beta$8.1-bearing T cells almost always recognize Mls-1[a] + class II, regardless of the other components of the receptor.[8] A panel of self and foreign superantigens has

Mechanisms of Lymphocyte Activation and Immune Regulation III
Edited by S. Gupta *et al.*, Plenum Press, New York, 1991

159

been described and their corresponding pattern of $V\beta$ reactivity in both mouse and man has been defined.[9] Monoclonal antibodies against the panel of $V\beta$s have been generated, and provide valuable tools for monitoring a population of T cells with shared recognition properties. This experimental approach has been used to examine mechanisms of T cell tolerance. For example, what happens to the $V\beta8.1$-bearing T cells in a mouse that expresses Mls-1[a]? This question is the topic of the studies presented here.

Clonal deletion has recently been described as a major mechanism of T cell tolerance in both normal mice, using the $V\beta$/superantigen approach described above,[8,10-13] as well as in TcR transgenic mice.[14-16] Specifically, with regard to tolerance to Mls-1[a], $V\beta8.1$,[8] $\beta6$[12] and $\beta9$-bearing[17] T cells, which have been shown to recognize Mls-1[a], are virtually eliminated from the periphery of Mls-1[a]-expressing mice. Also, in a TcR β chain transgenic mouse, most $M\beta8.1^+$ T cells were eliminated from the periphery when the transgenic mouse expressed Mls-1[a].[18] Analysis of the thymus of these mice showed that clonal elimination occurred during T cell development in the thymus. Thus, there were significant numbers of $V\beta8.1^+$ immature thymocytes (CD4CD8 "double positive"), but almost no $V\beta8.1^+$ mature thymocytes (CD4 or CD8 "single positive"), indicating that clonal deletion occurred at a specific stage of thymocyte maturation, sometime during the transition from immature to mature thymocytes. Clonal deletion acting on double-positive thymocytes is consistent with the absence of all $V\beta8.1$-expressing peripheral T cells, rather than just the $CD4^+$ cells which are capable of Mls-1[a] recognition. That clonal deletion acts on double positive thymocytes has also been more directly shown by introduction of monoclonal antibodies against CD4 or class II MHC to block the direct molecular interactions required for clonal deletion during thymocyte maturation in the presence of self-antigen, both *in vitro* and *in vivo*.[19-21]

In summary, clonal deletion has been demonstrated in naturally occurring tolerance (i.e., where expression of the self antigen is not manipulated) in both normal and transgenic mice, as well as during *in vitro* exposure of thymocytes to antigen in organ culture[22] and *in vivo* introduction of self antigen by injection from birth.[23] Clonal deletion of double positive, immature thymocytes appears to be the predominant method of tolerance to self antigens that are expressed in the thymus.

Table 1. Analysis of Reactivity of T Cell Hybridomas from Mls-1[b] $V\beta8.1$ Transgenic Mice

		Responsive to stimulation with:			
$V\beta8.1$ TcR$^+$	CD4$^+$	$\alpha V\beta8.1$	αCD3	SEB	Mls-1[a]
29(100%)	29(100%)	27(93%)	27(93%)	28(97%)	12(41%)

$V\beta8.1$ and CD4 expression was determined by staining cells with biotinylated KJ16[32] ($V\beta8.1$- and 8.2-specific) and F23.2 [33] ($V\beta8.2$-specific) antibodies, followed by PE-avidin, and FITC-conjugated GK1.5(CD4-specific), respectively. Responsiveness of T cell hybridomas was determined by measuring IL-2 production after 24 h. of stimulation, as follows. Anti- $V\beta8$ and anti-CD3[34] antibodies were immobilized by overnight incubation in the cold in 96-well microtiter plates (100μg/ml), followed by extensive washing. 1 x 10[5] hybridoma cells were added to the wells, incubated for 18-24 hours, and the supernatants were tested for IL-2 content, as previously described.[35] SEB (Staphylococcal enterotoxin B) reactivity was measured by stimulation of 10[5] hybridoma cells with 10μg/ml SEB in the presence of 10[5] CH12 antigen presenting cells. Reactivity to Mls-1[a] was determined by measuring IL-2 production in response to 10[5] CBA/J (Mls-1[a]) spleen cells.

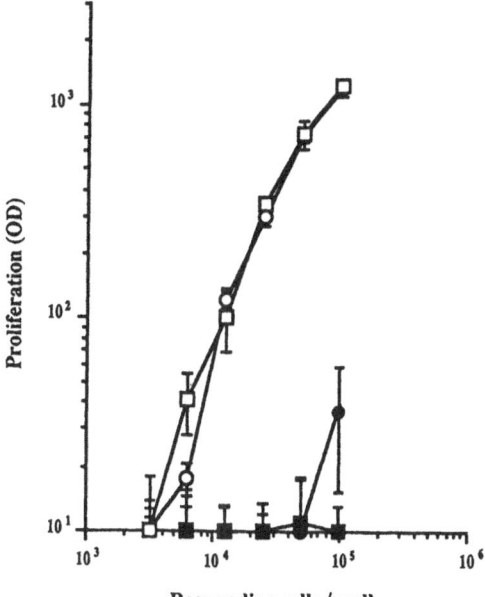

Figure 1. CD4$^+$ peripheral T cells from
Mls-1$_b$ β transgenic mice respond
to Mls-1^a, whereas CD4$^+$ peripheral
T cells from Mls-1^a β transgenic
mice are tolerant to Mls-1^a. *In
vitro* proliferation of CD4$^+$ peripheral
T cells from Mls-1^a (●,■) and
Mls-1^b (○,□) β transgenic mice
was measured in response to Mls-1^a,
expressed by 5 x 10^5 mitomycin
C-treated CBA/J spleen cells. CD4$^+$
cells were purified by panning on anti-
CD8 (53.6)-coated plates. Each point
represents the mean of triplicate
determinations, +/- SEM.

During our studies on tolerance, we generated a TcR β chain transgenic mouse
expressing a Vβ8.1-containing β chain that was cloned from a strongly-Mls-1^a-reactive
T cell hybridoma, 3DT-52.5.[24] The β chain construct was a 26 kb genomic construct
containing approximately 9 kb of 5' flanking sequence and 7 kb of 3' flanking sequence,
including the TcR β chain enhancer. When tested by transfection in a β-negative
variant of 3DT-52.5 that had lost Mls-1^a reactivity, the transfected β chain restored
Mls-1^a reactivity.[25] The transgenic founder mouse was H-2^k and Mls-1^b. The
transgene (approximately 16 copies) was expressed in > 95% of peripheral T cells, exhibited
allelic exclusion, and was expressed in a tissue-specific fashion.[26] Peripheral T cells
from these mice were highly reactive to Mls-1^a (Figure 1a). However, surprisingly, only
approximately 40% of CD4$^+$ Vβ8.1$^+$ T cell hybridomas generated from these transgenic
mice were reactive to Mls-1^a *in vitro* (Table 1), despite the fact that they were
responsive to immobilized anti Vβ8 and anti-CD3 antibodies, as well as the bacterial
enterotoxin, SEB (Staphylococcal enterotoxin B), a foreign superantigen that stimulates
Vβ8.1-bearing T cells. Thus, it appears likely that this particular β chain is
unusual in that it is very dependent upon its alpha chain association for Mls-1^a
reactivity (see discussion).

Table 2. T Cell Anergy in Vβ8.1 Transgenic Mice

Stimulus	Proliferation (Relative Proliferative Units)[a]	
	Mls-1 expression of transgenic host	
	b	a/b
Mls-1[a]	81.5 +/- 14.6	2 +/-0.5
SEB	59.4 +/- 3.2	5.3 +/-0.1
αVβ8	128.5+/-8.1	26.2 +/-2.8
αCβ	100 +/- 7.4	10.8+/-0.6
αCD3	63.8 +/-4.4	7.2+/-1.4
ConA	115.4+/-13.4	17.6+/-1.8
PMA + ionomycin	145.4+/-18.3	117.1+/-19

CD4+ peripheral T cells were purified from Mls-1[a]-expressing and non-expressing transgenic mice and tested for their ability to respond when stimulated with anti-Vβ8, anti-Cβ and anti-CD3$_\varepsilon$ antibodies immobilized to plastic plates (see legend to Table 1), to Mls-1[a] presented by CBA/J mitomycin C-treated spleen cells, to 10μg/ml SEB, to 4 μg/ml ConA or 0.5μM ionomycin and 10^{-8}M PMA. All stimulation, with the exception of PMA and ionomycin, was done in the presence of 5 x 10^5 inactivated CBA/Ca filler cells.

[a] Relative proliferative units were determined as follows. Two-fold dilutions of responding cells (1 x 10^5 to 4 x 10^3) were stimulated as indicated for 72-96 hours. Proliferation was measured with an MTT/ELISA assay[36]. The data were expressed on a log/log plot, standardized to the curve obtained with anti-Cβ stimulation of the CD4+ T cells from Mls-1[b] β transgenic mice(100 units), and expressed as relative proliferative units.

The founder mouse was then crossed with CBA/Ca (H-2[k], Mls-1[b]) and CBA/J (H-2[k], Mls-1[a]) to study tolerance to Mls-1[a] in this mouse. The expression of Mls-1[a] in the transgenic mice had surprisingly little effect on peripheral expression of Vβ8.1-bearing T cells. Density of TcR and accessory molecule cell surface expression was unaffected, and numbers of thymocytes and peripheral T cells in the two lines of mice were comparable. However, two differences were noted. There was a reproducibly lower percent of transgene expression in the CD4+ peripheral T cells (80-92% compared with >95% in the Mls-1[b] β transgenic mice) whereas the transgene expression in the CD8+ T cells was >95% in both lines of mice. Also, the CD4:CD8 ratio of peripheral T cells was significantly skewed in the Mls-1[a] transgenic mice (0.4 ± 0.02 compared with 0.9 ± 0.03 and 0.85 ± 0.02 for Mls-1[b] transgenic and non-transgenic mice, respectively. (The ratio was not determined for the Mls-1[a] normal mice, because all the Vβ8.1-bearing T cells were deleted in these mice.)

Also in contrast to the situation described for normal mice, analysis of transgene expression in the thymus of adult mice revealed little evidence for clonal deletion. The number of mature Vβ8.1+ thymocytes was only slightly reduced (<20%). Despite the lack of evidence for massive clonal deletion, these mice were tolerant to Mls-1[a], as measured in *in vitro* in a mixed lymphocyte assay against Mls-1[a]-expressing, mitomycin-treated spleen cells (Figure 1b).

Further analysis of the responsiveness of peripheral T cells in the Mls-1[a] transgenic mice revealed a significant reduction of responsiveness to a panel of stimuli (Table 2). We interpret this reduction in responsiveness to indicate that some cells were responding to stimulation normally, whereas others were not responding, rather than a reduction in responsiveness of all cells. This was confirmed by the analysis of the ability of individual cells to flux calcium in response to TcR litigation. Results showed that approximately 50% of the cells failed to increase intracellular calcium levels in the Mls-1^a-expressing β transgenic mice.[27] The cells were capable of proliferation in response to stimulation with PMA and ionomycin, however, confirming that they were viable. We believe that the cells capable of proliferation represent the non-Mls-1^a-reactive Vβ8.1$^+$ T cells, whereas the non-responsive Vβ8.1$^+$ peripheral T cells have receptors capable of Mls-1^a recognition and hence have been inactivated. Lack of responsiveness of self-reactive T cells could be mediated by clonal anergy or suppression. In order to distinguish between these two possibilities, we mixed the T cells from Mls-1^a and Mls-1^b β transgenic mice in a 50:50 ratio and tested the proliferative response. The results failed to reveal any evidence for the presence of suppressor cells,[27] arguing against a mechanism of suppression and implicating clonal anergy to account for the tolerance.

Recently, clonal anergy has been shown to be the mechanism of tolerance in several experimental systems. Evidence exists that clonal deletion does not occur extrathymically, so clonal anergy may be a common mechanism of peripheral tolerance in cases where self-reactive thymocytes sneak through the thymus or when self antigen is introduced experimentally, or are expressed in the periphery late in life, after a significant number of T cells have matured. First, adult injection of Mls-1^a-bearing T cells resulted not in the elimination of Vβ6 cells, but in their functional inactivation.[28] Also, IE-reactive Vβ11 and Vβ 17 T cells that encounter IE expressed abnormally in the periphery of transgenic mice are not deleted, but are rendered anergic.[29]

In our experiments, anergy appears to be a peripheral mechanism. Only CD4$^+$ cells are inactivated, and analysis of thymocyte reactivity fails to reveal evidence for anergy.[26] However, a recent study reports that anergy may be mediated by a radio-resistant component of the thymus, the thymic epithelium.[30] These conclusions remain to be confirmed with thymic chimeras. We cannot eliminate the possibility that in our system cells are being anergized at a late stage of thymocyte development and being immediately exported to the periphery rather than accumulating as a stable thymic population. However, the fact that anergy is a peripheral mechanism in our system is supported by the finding that we can reproduce the anergy in Mls-1^bβ transgenic mice by a single intravenous injection of Mls-1^a-bearing spleen cells.[27]

The surprising finding is that anergy is a major mechanism of tolerance of Vβ8.1 cells to Mls-1^a in our transgenic mice, whereas, in normal mice,[8-12] as well as in another Vβ8.1 transgenic mouse,[18] tolerance of Vβ8.1 T cells to Mls-1^a is mediated (almost) exclusively by clonal deletion. We don't think that this reflects an abnormality of Mls-1^a expression in these mice, because the non-transgenic littermates are capable of early and efficient deletion of both Vβ6$^+$ and Vβ8.1$^+$ (Mls-1^a-reactive) thymocytes.[27] It is likely that the difference in mechanisms of tolerance is due to receptor affinity. As stated previously, the T cell hybridoma from which the transgenic β chain was cloned was strongly Mls-1^a-reactive. However, it is likely that this β chain is very dependent upon the alpha chain with which it pairs for Mls-1^a reactivity, because less than 50% of T cell hybridomas generated from the Mls-1^b-expressing transgenic mice, when paired with a normal complement of α chains, were Mls-1^a-reactive. The Mls-1^a reactivity of this β chain is not dependent upon Vα usage, however. Analysis of Vα usage in a panel of T cell hybridomas failed to reveal a simple correlation between Vα usage and Mls-1^a reactivity.[31] Thus Mls-1^a reactivity is likely to be controlled by the pairing of α chains expressing a particular pattern of Jα or a specific pattern of junctional diversity.

A final question concerns the mechanism of anergy in these cells. Our data show that the anergy is not due to a reduction of TcR or accessory molecule density, because levels of expression of these molecules in peripheral T cells are comparable in anergized (Mls-1^a-expressing) and non-anergized (Mls-1^b-expressing) transgenic mice. Second, the anergy is not due the uncoupling of the TcR and CD3, because direct stimulation with antibodies to anti-CD3$_\epsilon$ does not bypass the non-responsiveness. Third, we cannot overcome the anergy with increasing concentration of antigen.[27] However, we can overcome the anergy by stimulation with PMA and ionomycin, agents which activate protein kinase C and increase intracellular calcium directly, thus bypassing the TcR and CD3. Therefore, the block in signaling is localized to early receptor/CD3 signaling events.

REFERENCES

1. Reviewed in P. Marrack and J. Kappler, The T cell receptor, *Science* 238:1073 (1987).
2. P. J. Bjorkman, M. A. Saper, B. Samraoui, W. S. Bennett, J. L. Strominger, and D. C. Wiley, Structure of the human class 1 histocompatibility antigen, HLA-A2, *Nature* 329:506 (1987).
3. P. J. Bjorkman, M. A. Saper, B. Samraoui, W. S. Bennett, J. L. Strominger, and D. C. Wiley, The foreign antigen binding site and T cell recognition regions of class I histocompatibility molecules, *Nature* 329:512 (1988).
4. J. H. Brown, T. Jardetzky, M. A. Saper, B. Samraoui, P. J. Bjorkman, and D. C. Wiley, A hypothetical model of the foreign antigen binding site of class II histocompatibility molecules, *Nature* 332:845 (1988).
5. Reviewed in A. Weiss, J. Imboden, K. Hardy, B. Manger, C. Terhorst, and J. Stobo, The role of the T3/antigen receptor complex in T-cell activation, *Ann. Rev. Immunol.* 4:593 (1986).
6. Reviewed in C. A. Janeway, Jr., T cell development: accessories or coreceptors, *Nature* 335:208 (1988).
7. E. A. Clark and J. A. Ledbetter, Leukocyte cell surface enzymology:CD45(LCA, T200) is a protein tyrosine phosphatase, *Immunol. Today* 10:255 (1989).
8. J. W. Kappler, U. Staerz, J. White, P. C. Marrack, Self-tolerance eliminates T cells specific for Mls-modified products of the major histocompatibility complex, *Nature* 332:35 (1988).
9. Reviewed in P. Marrack and J. W. Kappler, The staphylococcal enterotoxins and their relatives, *Science*, in press (1990).
10. J. Kappler, N. Roehm, P. Marrack, T cell tolerance by clonal elimination in the thymus, *Cell* 49:273 (1987).
11. A. M. Pullen, P. Marrack, J. W. Kappler, The T-cell repertoire is heavily influenced by tolerance to polymorphic self antigens, *Nature* 335:796 (1988).
12. H. R. MacDonald, R. Schneider, R. K. Lees, R. C. Howe, H. Acha Orbea, H. Festenstein, R. M. Zinkernagel, and H. Hengartner, T-cell receptor $V\beta$ use predicts reactivity and tolerance to Mls[a]-encoded antigens, *Nature* 332:40 (1988).
13. J. Bill, O. Kanagawa, D. Woodland, and E. Palmer, The MHC molecule I-E is necessary but not sufficient for clonal deletion of $V\beta 11$-bearing T cells, *J. Exp. Med.* 169:1405 (1989).
14. P. Kisielow, H. Bluthmann, U. D. Staerz, M. Steinmetz, and H. von Boehmer, Tolerance in T-cell-receptor transgenic mice involves deletion of nonmature $CD4^+8^+$ thymocytes, *Nature* 333:742 (1988).
15. W. Sha, F. Nelson, R. Newberry, D. Kranz, J. Russell, and D. Loh, Positive and negative selection of an antigen receptor on T cells in transgenic mice, *Nature* 336:73 (1988).
16. L. Berg, B. Fazekas de St. Groth, A. Pullen, and M. Davis, Phenotypic differences between $\alpha\beta$ versus β T-cell receptor transgenic mice undergoing negative selection, *Nature* 340:559 (1989).
17. M. P. Happ, D. L. Woodland, and E. Palmer, A third T-cell receptor β-chain variable region gene encodes reactivity to Mls-1[a] gene products, *Proc. Natl. Acad. Sci. USA* 86:6293 (1989).
18. H. Pircher, T. W. Mak, R. Lang, W. Ballhausen, E. Ruedi, H. Hengartner, R. M. Zinkernagel, and K. Burki, T cell tolerance to Mls[a]-encoded antigens in T cell receptor $V\beta 8.1$ chain transgenic mice, *EMBO J.* 8:719 (1989).
19. B. J. Fowlkes, R. H. Schwartz, and D. M. Pardoll, Deletion of self-reactive thymocytes occurs at a $CD4^+CD8^+$ precursor stage, *Nature* 334:620 (1988).
20. M. McDuffie, N. Roehm, J. Kappler, and P. C. Marrack, Involvement of major histocompatibility products in tolerance induction in the thymus, *J. Immunol.* 141:1840 (1988).

21. H. R. MacDonald, H. Hengartner, and T. Pedrazzini, Intrathymic deletion of self-reactive cells prevented by neonatal anti-CD4 antibody treatment, *Nature* 335:174 (1988).

22. T. Finkel, J. Cambier, R. Kubo, W. Born, P. Marrack, and J. Kappler, The thymus has two functionally distinct populations of $\alpha\beta^+$ T cells: one population is deleted by ligation of $\alpha\beta$ TCR, *Cell* 58:1047 (1989).

23. H. R. MacDonald, T. Pedrazzini, R. Schneider, J. A. Louis, R. M. Zinkernagel, and H. Hengartner, Intrathymic elimination of Mls^a-reactive (Vβ6) cells during neonatal tolerance induction of Mls^a-encoded antigens, *J. Exp. Med.* 167:2005 (1988).

24. M. A. Blackman, J. Yague, R. Kubo, D. Gay, C. Coleclough, E. Palmer, J. Kappler, and P. Marrack, The T cell repertoire may be biased in favor of MHC recognition, *Cell* 47:349 (1986).

25. M. Blackman and H. G.-Burgert, unpublished observations.

26. M. A. Blackman, H. G.-Burgert, D. L. Woodland, E. Palmer, J. W. Kappler, and P. Marrack, A role for clonal inactivation in T cell tolerance to Mls-1^a *Nature* (1990, in press).

27. M. A. Blackman, unpublished observations.

28. H.-G. Rammensee, R. Kroschewski, B. Frangoulis, Clonal anergy induced in mature Vβ^+ T lymphocytes on immunizing Mls-1^b mice with Mls-1^a-expressing cells, *Nature* 339:541 (1989).

29. L. C. Burkly, D. Lo, O. Kanagawa, R. L. Brinster, and R. A. Flavell, T cell tolerance by clonal anergy in transgenic mice with non-lymphoid expression of MHC class II IE, *Nature* 342:564 (1989).

30. F. Ramsdell, T. Lantz, B. J. Fowlkes, A nondeletional mechanism of thymic self tolerance, *Science* 246:1038 (1989).

31. M. Blackman and D. L. Woodland, unpublished observations.

32. K. Haskins, C. Hannum, J. White, N. Roehm, R. Kubo, J. Kappler, and P. Marrack, The major histocompatibility complex-restricted antigen receptor on T cells. VI. An antibody to a receptor allotype, *J. Exp. Med.* 160:452 (1984).

33. U. D. Staerz, H.-G. Rammensee, J. D. Benedetto, M. J. Bevan, Characterization of a murine monoclonal antibody specific for an allotypic determinant on T cell antigen receptor, *J. Immunol.* 134:3394 (1985).

34. O. Leo, M. Foo, D. H. Sachs, L. E. Samelson, and J. A. Bluestone, Identification of a monoclonal antibody specific for a murine T3 polypeptide, *Proc. Nat. Acad. Sci. USA* 83:767 (1987).

35. J. Kappler, B. Skidmore, J. White, and P. Marrack, Antigen-inducible, H-2-restricted, interleukin 2-producing T cell hybridomas. Lack of independent antigen and H-2 recognition, *J. Exp. Med.* 153:119 (1981).

36. T. Mosmann, Rapid colorimetric assay for cellular growth and survival: application to proliferation and cytotoxicity assays, *J. Immunol. Methods* 65:55 (1983).

INDUCTION AND MAINTENANCE OF ANERGY IN MATURE T CELLS

Marc K. Jenkins,* Daniel Mueller,[+] Ronald H. Schwartz,[+]
Simon Carding,[†] Kim Bottomley,[†] Miguel J. Stadecker[++],
Kevin B. Urdahl,* and Steven D. Norton*

*University of Minnesota
 Department of Microbiology
 Minneapolis Minnesota
[+]National Institutes of Health
 National Institute of Allergy and Infectious Disease
 Laboratory of Cellular and Molecular Immunology
 Bethesda, Maryland
[†]Yale University School of Medicine
 Howard Hughes Medical Institute
 Section of Immunobiology
 New Haven, Connecticut
[++]Tufts University School of Medicine
 Department of Pathology
 Boston, Massachusetts

INTRODUCTION

Three models have been proposed to explain the inability of T lymphocytes to respond to self-antigens (reviewed in [1]): (a) self-reactive T cells are present but are prevented from functioning by suppressor T cells (suppression); (b) self-reactive T cells are present but have been functionally inactivated following interaction with host antigens (clonal anergy); and (c) self-reactive T cells are physically deleted (clonal deletion). A growing body of conclusive evidence indicates that clonal deletion is a major mechanism of tolerance induction for those antigens expressed in the thymus, the site of T cell development.[2-5] It is difficult, however, to understand how this mechanism could account for tolerance to tissue-specific antigens, expressed in low amounts outside of the thymus. A potential resolution to this paradox may be found in recent studies that provide *in vivo* evidence for a nondeletional mechanism of clonal anergy[6-8] that may operate outside of the thymus. The characteristics of the anergy observed *in vivo* are strikingly similar to those described for the induction of unresponsiveness *in vitro* for type 1 CD4[+] T cell clones. Here we review our results on the induction of anergy in T cell clones and present new data on the mechanism by which it is maintained.

INDUCTION OF CLONAL ANERGY *IN VITRO*

Optimal production of IL-2 and subsequent proliferation by murine Th1 clones appears to depend on two events: (a) occupancy of the T cell receptor (TCR) by antigenic peptides bound to class II major histocompatibility complex-encoded molecules (Ia) expressed on the surface of antigen-presenting cells (APC); and (b) reception of a non-specific costimulatory signal from the APC (reviewed in [9]). Additionally, evidence from several labs now

Mechanisms of Lymphocyte Activation and Immune Regulation III
Edited by S. Gupta *et al.*, Plenum Press, New York, 1991

167

Table 1. Activation Events Associated with Normal or Anergic Th1 Clones

Response	Normal T Cells			Anergic T Cells
	Normal APC + Antigen	Fixed APC + Antigen	Anti-CD3	Normal APC + Antigen
Proliferation:	++	--	+ or −	+ or −
Plus Allogeneic Splenocytes	++	++	++	?
Lymphokines:				
IL-2	++	--	+	−
IL-3	++	+	+	+
IFN-γ	++	+	++	+
Biochemical Events:				
IP production	++	+	++	++
PKC activation	++	+	++	?
Calcium increase	++	+	++	++
ζ-phosphorylation	++	+	++	?
Surface Molecules:				
IL-2R				++
TCR				++
CD4				++
LFA-1				++
ICAM-1				++

Results from previous studies (9, 10, 12, 17-19, 21) are summarized here. The magnitude of the indicated responses is shown as maximal (++), less than 20% of maximal (+), or undetectable (--).

suggests that TCR occupancy in the absence of functional APC renders Th1 clones anergic to restimulation with antigen and normal APC for long periods of time (greater than 7 days). Anergy has been achieved by stimulating Th1 cells with minimal antigenic peptides and Ia$^+$ cells that do not support T cell proliferation: chemically modified[10] or heavily irradiated resting B cells,[11] purified Ia molecules incorporated into planar membranes,[12] pancreatic β cells expressing a transgenic Ia molecule,[13] or haptenated Ia$^+$ keratinocytes.[14] Anergy has also been induced by stimulating class II MHC$^+$ human T cell clones with antigenic peptides alone.[15] In each case, the induction of anergy depends on the appropriate antigenic peptide and allelic form of the Ia molecule demonstrating a requirement for presentation of antigen to the TCR. Furthermore, agents that directly stimulate the TCR complex (Con A, anti-TCR, or anti-CD3 antibodies) all induce anergy in the absence of accessory cells.[16-18]

The activation events that occur when Th1 clones are incubated with an anergy-inducing stimulus (fixed resting B cells and antigenic peptide or immobilized anti-CD3 antibody alone) are summarized in Table 1. Consistent with their failure to induce T cell proliferation, antigenic peptide and fixed resting cells do not cause detectable IL-2 secretion.[19] Suboptimal amounts of IL-3 and IFN-γ are produced and IL-2 receptor surface expression is weakly increased providing further evidence that the TCR is occupied. At the biochemical level, antigenic peptide and fixed resting B cells activate the suboptimal hydrolysis of phosphatidyl inositol bisphosphate yielding inositol trisphosphate, an activator of intracellular calcium increase, and diacylglycerol, an activator of protein kinase C.[17] Based on the suboptimal nature of these responses it is not surprising that significant, albeit suboptimal increases in intracellular calcium and protein kinase C-mediated phosphory-

CD3-ζ is also observed. The suboptimal activation events are consistent with the idea that the TCR is occupied less well by antigenic peptide bound to fixed resting B cells, probably as a result of fixation-related damage to Ia molecules and/or accessory molecules. Although permissive, suboptimal TCR occupancy is not a requirement since anergy is very efficiently induced by Con A or anti-CD3 antibody (Table 1) which both cause nearly maximal TCR-associated biochemical events.[17,18] Anergy induction and the aforementioned suboptimal lymphokine production can be mimicked by calcium ionophore alone, suggesting that the critical biochemical event is an increase in intracellular calcium.[19] Since anergy cannot be induced in the presence of cycloheximide,[12] it is possible that the calcium increase activates the synthesis of new proteins responsible for the maintenance of the anergic state.

Allogeneic accessory cells (that cannot be recognized by the TCR on the responding T cells) allow proliferation in response to antigenic peptide and fixed resting B cells that express the appropriate Ia molecule.[20,21] Subsequent anergy is also prevented.[21] This costimulatory function is provided most efficiently by T cell-depleted low density cells (activated B cells, macrophages, and dendritic cells), less well by resting B cells and not at all by resting T cells. It is likely that the delivery of costimulation involves physical contact between the responding T cell and the accessory cell based on the observation that allogeneic accessory cells fail to provide costimulation if they are separated from the responding T cells by a semipermeable membrane that allows the diffusion of macromolecules but prevents cell-cell contact. Provision of costimulation does not augment inositol phosphate generation, calcium increase, or PKC activation indicating that a biochemical pathway distinct from inositol phospholipid hydrolysis is responsible.[17,20]

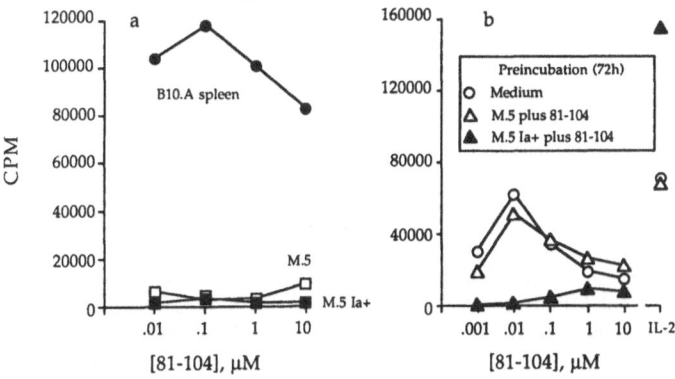

Figure 1. (a) A pigeon cytochrome c/Eβ^k:Eα^k molecule-specific Th1 clone (A.E7) was incubated for 3 days with the indicated concentrations of the relevant antigenic peptide, pigeon cytochrome c fragment 81-104, and irradiated splenic B10.A APC (closed circles), uninduced M.5 cells (open squares) or 4 day IFN-γ-treated (60 U/ml) Eβ^k:Eα^{k+} M.5 cells (closed squares). Tritiated thymidine was included during the last 16 hours of the culture period. The incorporation of tritiated thymidine into DNA was quantitated by liquid scintillation counting. Results are expressed as counts per minute (CPM).
(b) A.E7 cells were incubated for 3 days with medium alone (closed circles) or with 10 μM pigeon cytochrome c fragment 81-104 and uninduced M.5 cells (open triangles) or 4 day IFN-γ-treated Eβ^k:Eα^{k+} M.5 cells (closed triangles). The T cells were reisolated and stimulated with irradiated splenic B10.A APC and the indicated antigen concentrations or with exogenous IL-2 alone. T cell proliferation was measured as described in the legend to Figure 1a.

All of the regimens described above for inducing anergy require that accessory cells be modified in some way or that they be absent. This raises the question of whether unmodified costimulation-deficient accessory cells exist naturally *in vivo*. Epithelial cells are potential candidates. To address this, we examined the APC function of the M.5 thyroid-derived epithelial cell line. Although these cells express high levels of Ia molecules on their surface following treatment with IFN-γ, they fail to stimulate a primary allogeneic mixed lymphocyte reaction.[22] As shown in Figure 1a, they also fail to stimulate proliferation by a pigeon cytochrome c-specific T cell clone in the presence of the relevant antigenic peptide although they do induce IL-3 production (not shown). Most importantly, the T cells become anergic, failing to respond when stimulated with splenic APC and antigen (Figure 1b). These results demonstrate that anergy can be induced by an unmodified APC. Recent results suggest that Ia$^+$ thymic epithelial cells also fail to stimulate antigen-specific proliferation by Th1 clones.[23]

Based on these results, IL-2 production and T cell proliferation occur as a result of biochemical second messengers generated by occupancy of two surface receptors: the TCR and the putative surface receptor for a costimulatory ligand on the APC. APC populations vary in their capacity to provide the costimulatory activity, perhaps explaining the dramatic differences in APC potency among various cells that express similar numbers of Ia molecules. Anergy, in contrast, occurs as a result of TCR occupancy and the associated increase in intracellular calcium unopposed by the biochemical signals generated by the costimulating receptor/ligand interaction.

MAINTENANCE OF ANERGY

The molecular mechanism that prevents anergic Th1 clones from proliferating in response to antigenic peptides and normal APC is unknown. Several reasonable possibilities, however, have been ruled out. Zanders et al.[24] showed that shortly after exposure to antigenic peptide in the absence of accessory cells human T cell clones expressed reduced TCR levels potentially explaining their anergic condition. We and others,[12,18,25] however, have shown that although the TCR may be transiently modulated, TCR levels return to normal at later times when the cells are fully unresponsive. As summarized in Table 1, anergic T cells also express normal levels of other surface molecules known to be important for T cell activation including CD4, LFA-1, and ICAM-1.

Anergy is also not explained by an inability to respond to IL-2 (Table 1, Figure 1b), suggesting that the IL-2 receptor signals normally.[10-12,14-19,21] This also demonstrates that Th1 clones are not killed, at least in the short-term, by exposure to anergy-inducing conditions. It should be pointed out, however, that potent anergy-inducing stimuli such as

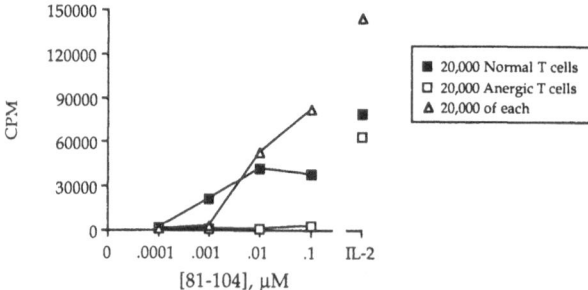

Figure 2. A.E7 cells were incubated either with medium or 1.5 γM calcium ionophore (ionomycin) alone for 40 hours. The cells were washed, rested for 24 hours and restimulated either separately or as an equal mixture with splenic B10.A APC and the indicated antigen concentrations or with exogenous IL-2 alone. T cell proliferation was measured as described in the legend to Figure 1.

Table 2. Anergy Cannot be Explained by a Reduction in Inositol Phosphate Production

T cells	Proliferation (CPM)		Inositol Phosphates (CPM)	
	Medium	10 μM Antigen	Medium	10 μM Antigen
Normal	197	14,345	79	3,803
Anergic	358	3,851	62	3,633

Anergy was induced by preincubating A.E7 cells with $E\beta^k$:$E\alpha^k$ molecule containing planar lipid membranes and 10 μM pigeon cytochrome c fragment 81-104 for 24 hours as described previously (12). Normal controls were A.E7 cells preincubated in medium alone. The T cells from both groups were reisolated and rested for 6 days in the absence of antigen. Some of the normal and anergic A.E7 cells were labeled with ^{3}H-myoinositol for the last 24 hours of the 6 day rest period. The incorporation of label was similar in both groups. Proliferation (by unlabeled A.E7 cells) or inositol phosphate production (by labeled A.E7 cells) was measured at 72 hours and 1 hour, respectively, as previously described (19). The amount of labeled inositol phosphates produced was quantitated by liquid scintillation counting.

immobilized anti-CD3 antibody can lead to cell death after several days. This may be related to the TCR-mediated growth inhibition and cell death observed for T cell hybridomas[26] and immature thymocytes.[27] Indeed, several other parallels exist between these phenomena and anergy, including susceptibility to blockage by cyclosporine A.[21,27,28]

As shown in Figure 2, anergic T cells are not very suppressive, i.e., an equal mixture of anergic and rested Th1 cells proliferated in response to antigen and normal APC. The greater than normal response achieved by the mixture at high antigen dose is probably a result of the anergic T cells responding to IL-2 produced by the normal T cells. It is conceivable, how ever, that at limiting antigen concentrations anergic T cells could bind to antigen/Ia molecule complexes, preventing their recognition by normal T cells. This form of competition may explain the reduced proliferation exhibited by the mixture at 0.001 μM antigen (Figure 2).

Although a reduction in the number of TCR is unlikely to account for anergy, it was possible that TCR on anergic T cells do not transduce signals properly. To test this, a Th1 clone in its normal or anergic state was tested for antigen-specific proliferation and inositol phosphate generation seven days after the induction of anergy. The inositol phosphate assay appears to be a direct measure of TCR occupancy independent of the costimulatory signals required for IL-2 production.[17] As shown in Table 2, stimulation of normal T cells with syngenic APC and 10 μM antigen resulted in large increases in proliferation at 72 hours and inositol phosphate production at one hour. As expected, stimulation of anergic T cells resulted in a greatly reduced proliferative response. These cells, however, generated control levels of inositol phosphate (Table 2) and increases in intracellular calcium (data not shown). It should be noted, however, that at early times after exposure to antigenic peptide and either fixed resting B cells (inactivating conditions) or normal splenic APC (activating conditions) the dose response of inositol phosphate production to antigen restimulation was shifted to higher antigen concentrations (data not shown). This transient desensitization is probably related to the temporary refractoriness of T cell clones following normal antigenic stimulation.[29] Therefore, it is unlikely that proliferative anergy is related to a failure of the TCR to generate biochemical second messengers, at least those related to inositol phospholipid hydrolysis.

Table 3. Lymphokine Production by Normal and Anergic Th1 Clones.

| | IL-2 | | IFN-γ | | IL-3 |
| | Protein (Units) | mRNA (% cells[+]) | Protein (Units) | mRNA (% cells[+]) | Protein (Units) |
T cells[a]					
Normal	<1	4	12	6	<1
Normal + Antigen	41	56	625	72	160
Anergic	<1	8	<5	6	<1
Anergic + Antigen	<1	10	175	57	20

Lymphokine Production[b]:

[a]Anergy was induced by preincubating Th1 clones for 24–48 hours with antigenic peptide and fixed syngeneic resting B cells (10) or Ia molecule-containing planar membranes (12) or with calcium ionophore alone (19).

[b]As previously described (19), lymphokine proteins were measured at 24 hours from supernatants of cultures containing normal or anergic Th1 clones stimulated with irradiated splenic APC and 10 µM pigeon cytochrome c fragment 81-104. Units are defined as the reciprocal of the supernatant dilution that yielded 50% of the maximal response of the relevant indicator cell line. Mean values for 3-6 experiments are shown. Lymphokine mRNA was quantitated by in situ hybridization (37) in normal or anergic Th1 cells stimulated for 6 hours with irradiated splenic APC and 10 µM pigeon cytochrome c fragment 81-104. Positive cells are defined as those possessing greater than 5 silver grains. Mean values from 2 experiments are shown.

T cell anergy can be attributed to a lack of lymphokine production, particularly the production of IL-2 (Table 3). The inability of anergic T cells to secrete detectable IL-2 protein into the supernatant correlates with a deficit in antigen-inducible IL-2 mRNA production as measured by *in situ* hybridization. The production of IFN-γ and IL-3 by anergic T cells is reduced, but easily detectable. The partial reduction in IFN-γ secretion is paralleled by a partial reduction in the number of cells producing IFN-γ mRNA (Table 3) and in the amount of mRNA produced by responding cells (data not shown). These results demonstrate that anergy is at the level of lymphokine production and rule out aberrant lymphokine consumption, degradation, or responsiveness as a mechanism.

CONCLUSION

The following is a potential molecular scenario for the induction and maintenance of T cell anergy. TCR occupancy would result in the generation of inositol phosphates which in turn would cause an increase in intracellular calcium from internal and external sources. This calcium increase would trigger the synthesis of new negative regulatory proteins that could act by blocking the synthesis or function of critical nuclear DNA binding proteins known to positively regulate the IL-2 enhancer.[30] Thus, anergic T cells could express normal levels of TCR and other activation molecules and engage these receptors normally upon interaction with antigen-bearing APC leading to the normal generation of TCR-associated biochemical second messengers. Due to the action of the negative regulatory proteins, however, these cells would be unable to produce the functional positive transcriptional regulators normally induced by these second messengers. Provision of the costimulatory signal could prevent the synthesis of the negative regulators or lead to their dilution as the T cells divide in response to the IL-2 they produce. The latter possibility is favored by the observation that stimulating already anergic T cells with IL-2 partially reverses the anergy.[31]

The type of anergy described here is a potential explanation for earlier observations of experimentally-induced T cell unresponsiveness following injection of adult animals with antigen-coupled lymphocytes or their membranes (reviewed in [9]). Most of these studies employed the intravenous route of administration, a route that might bypass normal deposition in accessory cell rich lymph nodes, and antigen coupling methods (heavy haptenation or carbodiimide crosslinkers) that would be expected to compromise accessory cell function. Recently it has been shown that adult mice rendered specifically-unresponsive following intravenous injection of Mls-disparate splenocytes contain normal numbers of T cells expressing normal levels of Mls-specific TCR.[8] These cells fail, however, to produce IL-2 when stimulated *in vitro*, suggesting that they are anergic. The limited expression of Mls antigens to B cells, cells known to be poor providers of costimulatory signals,[21] may be related to the ease of inducing this type of anergy without modifying the injected splenocytes.

Results derived from two independent sets of transgenic mice suggest that anergy can be a mechanism of self-tolerance during T cell development *in vivo*. Transgenic mice expressing a foreign Ia molecule only on β cells of the pancreas fail to develop autoimmune diabetes.[32] Surprisingly, these mice were tolerant to the transgenic Ia molecule as assessed by a mixed lymphocyte reaction even though this molecule was not expressed in the thymus. Normal numbers of T cells expressing TCR specific for the transgenic Ia molecule could be detected in these mice.[7] In contrast to normal non-transgenic T cells from the same strain, T cells from the β cell transgenic mice did not proliferate when stimulated with antibodies against the TCR known to be specific for the transgenic Ia molecule, demonstrating that these T cells were anergic. This conclusion is further supported by the finding that the pancreatic β cells expressing the transgenic Ia molecule failed to stimulate proliferation by an appropriate peptide-specific T cell line and instead rendered the line unresponsive to restimulation with peptide and splenic APC.[13] Mice expressing transgenic foreign class I MHC molecules are also tolerant to the transgene product although T cells specific for the transgenic MHC molecule are present.[33] It should be noted that, although several other studies on transgenic MHC molecule expression in the pancreas showed no evidence for autoimmune rejection of the pancreas, no transgene product-specific tolerance was observed.[34,35] These differences may be explained by tolerance to pancreatic cell peptide/ Ia molecule complexes but not spleen cell peptide/Ia molecules complexes present on the APC used in the mixed lymphocyte reaction.

Lethally-irradiated F_1 mice (AxB) reconstituted with parental (A) bone marrow are tolerant to the MHC antigens expressed on either parent (A or B).[6] These chimeric mice express strain B MHC antigens only on radioresistant host elements (including the thymic epithelium) and not on radiosensitive bone marrow-derived cells which are rapidly replaced by strain A cells. Since bone marrow-derived cells are thought to be required for clonal deletion,[36] how do these mice develop tolerance to B MHC antigens? Recent results demonstrate that these mice contain normal numbers of T cells expressing TCR specific for one of the strain B MHC molecules ruling out clonal deletion as a mechanism.[6] These T cells, however, fail to produce IL-2 when stimulated with APC from strain B or with antibodies specific for their TCR. Like anergic T cell clones, T cells from the chimeras will proliferate, albeit suboptimally, in response to strain B MHC antigens if exogenous IL-2 is provided. The observation that thymic epithelial cells fail to stimulate proliferation by antigen-specific Th1 clones[23] raises the possibility that anergy in the chimeras is induced by the thymic epithelium, the only tissue expected to constitutively express strain B Ia molecules.

Although these studies on manipulated animals provide conclusive evidence that anergy can be a mechanism of T cell unresponsiveness *in vivo*, they do not address the question of whether anergy contributes to self-tolerance induction in the normal individual. This is difficult to test in normal animals in that there is no current method to functionally or physically detect anergic T cells without foreknowledge of their antigenic specificity. Therefore, we can only speculate on the physiological role of T cell anergy. Anergy may be a mechanism for inducing peripheral tolerance to antigens not expressed in the thymus or may complement clonal deletion as a mechanism of intra-thymic tolerance. It is difficult to understand how anergy could be induced in CD4[+] T cells in the peripheral lymphoid tissues because Ia molecules are expressed primarily on stimulatory "professional" APC and not on parenchymal cells of the type that have been shown to lack costimulatory function. It is possible, however, that this function could be served by epithelial cells that express Ia molecules in response to IFN-γ at a site of inflammation. Alternatively, since thymic epithelial cells constitutively express Ia molecules and appear to lack costimulatory function, it is possible that anergy is induced intra-thymically in self-reactive T cells that: (a) escape clonal deletion because they do not encounter bone-marrow derived APC, or (b) recognize epithelial cell, but not hematopoietic cell, peptides bound to MHC.

ACKNOWLEDGMENT

This work was supported in part by NIH grant AI-27998.

REFERENCES

1. G. J. V. Nossal, Cellular mechanisms of immunologic tolerance, *Annu. Rev. Immunol.* 1:33 (1983).
2. J. W. Kappler, N. Roehm, and P. Marrack, T cell tolerance by clonal elimination in the thymus, *Cell* 49:273 (1987).
3. J. W. Kappler, U. Staerz, J. White, and P. Marrack, Self-tolerance eliminates T cells specific for Mls-modified products of the MHC, *Nature* 332:35 (1988).
4. H. R. MacDonald, R. Schneider, R. K. Lees et al., T-cell receptor Vβ use predicts reactivity and tolerance to Mls[a]-encoded antigens, *Nature* 332:40 (1988).
5. P. Kisielow, H. Bluthman, U. Staerz, M. Steinmetz, and H. von Boehmer, Tolerance in T-cell-receptor transgenic mice involves deletion of nonmature CD4[+]8[+] thymocytes, *Nature* 333:742 (1988).
6. F. Ramsdell, T. Lantz, and B. J. Fowlkes, A nondeletional mechanism of thymic self-tolerance, *Science* 246:1083 (1989).
7. L. C. Burkly, D. Lo, O. Kanagawa, R. L. Brinster, and R. A. Flavell, T-cell tolerance by clonal anergy in transgenic mice with nonlymphoid expression of MHC class II I-E, *Nature* 342:564 (1989).
8. H. G. Rammensee, R. Kroschewski, and B. Frangoulis, Clonal anergy induced in mature Vβ6[+] T lymphocytes on immunizing Mls-1[b] mice with Mls-1[a] expressing cells, *Nature* 339:541 (1989).
9. D. L. Mueller, M. K. Jenkins, and R. H. Schwartz, Clonal expansion versus functional clonal inactivation: a costimulatory signalling pathway determines the outcome of T cell antigen receptor occupancy, *Annu. Rev. Immunol.* 7:445 (1989).
10. M. K. Jenkins and R. H. Schwartz, Antigen presentation by chemically modified splenocytes induces antigen-specific T cell unresponsiveness *in vitro* and *in vivo*, *J. Exp. Med.* 165:302 (1987).
11. J. D. Ashwell, M. K. Jenkins, and R. H. Schwartz, Effect of gamma radiation on resting B lymphocytes. II. Functional characterization of the antigen-presentation defect, *J. Immunol.* 141:2536 (1988).
12. H. Quill and R. H. Schwartz, Stimulation of normal inducer T cell clones with antigen presented by purified Ia molecules in planar lipid membranes: specific induction of a long-lived state of proliferative nonresponsiveness, *J. Immunol.* 138:3704 (1987).

13. J. Markmann, D. Lo, A. Naji, R. D. Palmiter, R. L. Brinster, and E. Heber-Katz, Antigen-presenting function of class II MHC expressing pancreatic beta cells, *Nature* 336:476 (1988).

14. A. A. Gaspari, M. K. Jenkins, and S. I. Katz, Class II MHC-bearing keratinocytes induce antigen-specific unresponsiveness in hapten-specific Th1 clones, *J. Immunol.* 141:2216 (1988).

15. J. R. Lamb, B. J. Skidmore, N. Green, J. M. Chiller, and M. Feldmann, Induction of tolerance in influenza virus-immune T lymphocyte clones with synthetic peptides of influenza hemagglutinin, *J. Exp. Med.* 157:1434 (1983).

16. K. Tomonari, T cell receptor expressed on an auto-reactive T-cell clone, clone 4. I. Induction of various T-receptor functions by anti-T idiotypic antibodies, *Cell. Immunol.* 96:147 (1985).

17. D. L. Mueller, M. K. Jenkins, and R. H. Schwartz, An accessory cell-derived costimulatory signal acts independently of protein kinase C activation to allow T cell proliferation and prevent the induction of unresponsiveness, *J. Immunol.* 142:2617 (1989).

18. M. K. Jenkins, C. Chen, G. Jung, D. L. Mueller, and R. H. Schwartz, Inhibition of antigen-specific proliferation of Type I T cell clones following stimulation with immobilized anti-CD3 monoclonal antibody, *J. Immunol.* 144:16 (1990).

19. M. K. Jenkins, D. M. Pardoll, J. Mizuguchi, T. M. Chused, and R. H. Schwartz, Molecular events in the induction of a nonresponsive state in IL-2-producing helper T-lymphocyte clones, *Proc. Natl. Acad. Sci. USA* 84:5409 (1987).

20. E. R. Nisbet-Brown, J. W. W. Lee, R. K. Cheung, and E. W. Gelfand, Antigen-specific and nonspecific mitogenic signals in the activation of human T cell clones, *J. Immunol.* 138:3713 (1987).

21. M. K. Jenkins, J. D. Ashwell, and R. H. Schwartz, Allogeneic non-T spleen cells restore the responsiveness of normal T cell clones stimulated with antigen and chemically modified antigen-presenting cells, *J. Immunol.* 140:3324 (1988).

22. M. E. Stein and M. J. Stadecker, Characterization and antigen-presentation function of a murine thyroid-derived epithelial cell line, *J. Immunol.* 139:1786 (1987).

23. R. G. Lorenz and P. M. Allen, Thymic cortical epithelial cells lack full capacity for antigen presentation, *Nature* 340:557 (1989).

24. E. D. Zanders, J. R. Lamb, M. Feldmann, N. Green, and P. C. C. Beverly, Tolerance of T-cell clones is associated with membrane antigen changes, *Nature* 303:625 (1983).

25. K. Tomonari, Down-regulation of the T cell receptor by a mitogenic anti-Thy-1 antibody, *Eur. J. Immunol.* 18:179 (1988).

26. J. D. Ashwell, R. E. Cunningham, P. D. Noguchi, and D. Hernandez, Cell growth cycle block of T cell hybridomas upon activation with antigen, *J. Exp. Med.* 165:173 (1987).

27. Y. Shi, B. M. Sahai, and D. R. Green, Cyclosporin A inhibits activation-induced cell death in T-cell hybridomas and thymocytes, *Nature* 339:625 (1989).

28. M. Mercept, A. M. Weissman, S. J. Frank, R. D. Klausner, and J. D. Ashwell, Activation-driven programmed cell death and T cell receptor $\zeta\eta$ expression, *Science* 246:1162.

29. D. Wilde and F. Fitch, Antigen-reactive cloned helper T cells. I. Unresponsiveness to antigenic restimulation develops after stimulation of cloned helper T cells, *J. Immunol.* 132:1632 (1984).

30. G. R. Crabtree, Contingent genetic regulatory events in T lymphocyte activation, *Science* 243:355 (1989).

31. G. Essery, M. Feldmann, and J. R. Lamb, IL-2·can prevent and reverse antigen-induced unresponsiveness in cloned human T lymphocytes, *Immunology* 64:413 (1988).

32. D. Lo, L. C. Burkly, G. Widera, C. Dowing, R. A. Flavell, R. D. Palmiter, and R. L. Brinster, Diabetes and tolerance in transgenic mice expressing class II molecules in pancreatic beta cells, *Cell* 53:159 (1988).

33. G. Morahan, J. Allison, and J. F. A. P. Miller, Tolerance of class I histocompatibility antigens expressed extrathymically, *Nature* 339:622 (1989).

34. J. Bohme, K. Haskins, P. Stecha, W. van Ewjik, M. LeMeur, P. Gerlinger, C. Benoist, and D. Mathis, Transgenic mice with I-A on islet cells are normoglycemic but immunologically intolerant, *Science* 244:1179 (1989).

175

35. J. Miller, L. Daitch, S. Rath, and E. Selsing, Tissue-specific expression of allogeneic class II MHC molecules induces neither tissue rejection nor clonal inactivation of alloreactive T cells, *J. Immunol.* 144:334 (1990).

36. J. Sprent, D. Lo, E. K. Gao, and Y. Ron, T cell selection in the thymus, *Immunol. Rev.* 101:173 (1988).

37. S. R. Carding, J. West, A. Woods, and K. Bottomly, Differential activation of cytokine genes in normal CD4-bearing T cells is stimulus dependent, *Eur. J. Immunol.* 19:231 (1989).

CLONING AND CHARACTERIZATION OF A PROTEIN BINDING TO THE Jκ

RECOMBINATION SIGNAL SEQUENCE OF IMMUNOGLOBULIN GENES

Yasushi Hamaguchi, Norisada Mastunami, Yoshiki Yamamoto, Kogo Kuze,
Kenji Kangawa,* Hisayuki Matsuo,* Masashi Kawaichi, and Tasuku Honjo

Department of Medical Chemistry
Kyoto University Faculty of Medicine
Kyoto, Japan
*Department of Biochemistry
 Miyazaki Medical College
 Miyazaki, Japan

SUMMARY

A protein with molecular weight of 60,000 that binds to the recombination signal sequence (RS) of the immunoglobulin Jκ segment was purified from the nuclear extract of a murine pre B cell line 38B9. This binding protein was found in lymphoid cell lines but not in non-lymphoid cell lines. The Kd value of the Jκ RS binding protein to the Jκ RS was 1 nM. The cDNA clone (RBP-2) was isolated based on partial amino-acid sequence of this protein. This cDNA encodes 526 amino-acid residues, and its sequence does not show extensive overall homology with any known proteins, but displays an interesting homology to a 40-residue region that is conserved among a subset of site specific recombinase (integrase family).

INTRODUCTION

Recombination of the immunoglobulin gene takes place at the early phase of B lymphocyte differentiation, that is, in close association with the commitment of the stem cell to the B lymphocyte lineage. The variable region genes of immunoglobulin heavy chains as well as light chain are encoded by three (or two) separate germline DNA elements. During B lymphocyte differentiation, these three segments are assembled into complete V gene exon by site specific recombination. This V-(D)-J recombination that joins variable (V), diversity (D), and joining (J) segments contributes to amplification of the diversity of antibody repertoire. Comparison of nucleotide sequences of the flanking regions of the V, D and J segments has shown that two common blocks of nucleotide sequence are conserved. They are a heptamer CACTGTG and a T-rich nonamer GGTTTTTGT which are separated by spacer sequences of either 12 or 23 bases.[1,2]

V-(D)-J recombinase activity can be assayed in cells through the use of introduced recombination substrates which contain unrearranged V, D or J segments flanking selectable marker genes.[3,4] Such experiments showed that the recombinase is targeted to the respective gene coding segments by similar recombination recognition sequences, heptamer and nonamer sequences, that flank V, D and J coding segments of immunoglobulin or T cell receptor. Furthermore, those coding segments were shown to be not necessarily required for V-(D)-J recombination.[5] It is, therefore, reasonable to assume that the DNA-binding protein that would recognize the conserved recombination signal sequence might be involved in the recombinational machinery.

Mechanisms of Lymphocyte Activation and Immune Regulation III
Edited by S. Gupta *et al.*, Plenum Press, New York, 1991

177

```
5'     GATCCGGTTTTTGTACAGCCAGACAGTGGAGTACTACCACTGTGG
3'        GCCAAAAACATGTCGGTCTGTCACCTCATGATGGTGACACCCTAG        J_k

5'     GATCCGGTTTTTGTACAGCCAGACAGTGGAGTACTACCACTcTGG
3'        GCCAAAAACATGTCGGTCTGTCACCTCATGATGGTGAgACCCTAG      J_k1P3

5'     GATCCtagcatgtaACAGCCAGACAGTGGAGTACTACtgtccgaG
3'        GatcgtacatTGTCGGTCTGTCACCTCATGATGacaggctCCTAG    J_k79M

5'     GATCCtagcatgtaACAGCCAttgcatcGAGTACTACtgtccgaG
3'        GatcgtacatTGTCGGTaacgtagCTCATGATGacaggctCCTAG    J_k79SM

5'     ctagaCACAGTGatacaaatcataACAtAAACCt
3'        tGTGTCACtatgtttagtatTGTaTTTGGagatc              V_k

5'     ctagatcgtcgaatacaaatcataggacctcgat
3'        tagcagcttatgtttagtatcctggagctagatc              V_k79M
```

Figure 1. Nucleotide sequence of RS DNA probes. Jκ sequence was synthesized according to the published mouse sequence Jκ₁[22]. Jκ 79M are completely replaced by other sequences. Jκ 79SM is the same as Jκ79M except that the spacer sequence has 7-base replacement. Vκ sequence was sequenced according to the published sequence of mouse Vκ41[23]. The heptamer and nonamer sequences are completely substituted in Vκ 79M. Lower cases represent substituted bases relative to the Jκ RS. Heptamer and nonamer sequences are underlined. These nucleotides were cloned into the plasmid vector pUC19 and digested with HindIII and EcoRI. The fragments were labeled with α-^{32}P-dATP.

To dissect the enzymatic machinery responsible for V-(D)-J recombination, we set out to purify a RS binding protein from an Abelson leukemic virus transformed pre B cell line (38B9) that continues to rearrange the immunoglobulin gene *in vitro*.[6] We report here that purification of the Jκ RS binding protein RBP-2 protein of 60 kD that specifically interacts with the heptamer signal sequence and that isolation of a complementary DNA clone (RBP-2) based on the partial amino acid sequence of this protein.

RESULTS

Purification and Characterization of Jκ RS Binding Protein

The Jκ RS and its mutant RS probes together with the Vκ probe (Figure 1) were used to find a protein that specifically interacted with the Jκ RS. Several bands with altered mobilities were shown in the gel retardation assay of Jκ RS probe but not with other probes including Jκ 79 M, Jκ 79SM, and Vκ RS probes. So we set out to purify the protein that could produce the specific gel retardation band of the Jκ RS probe. Purification was accomplished by two different affinity columns; heparin agarose and Jκ RS probe coupled agarose. Each column chromatography was repeated twice. The detailed purification procedure was described previously.[7] The overall purification of this protein was estimated to be about 17,000 fold. The purity of the final preparation was assessed by SDS-polyacrylamide gel electrophoresis and silver staining. The final fraction contained an essentially homogenous 60 kD protein as shown in Figure 2. The relative activities of the Jκ RS binding protein in the second DNA affinity column chromatography

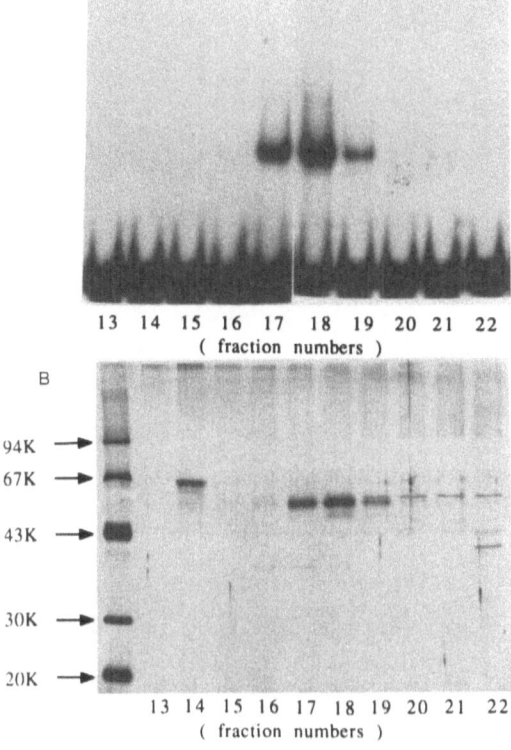

Figure 2. DNA-affinity column chromatography
of Jκ RS binding protein. A. Gel retardation
assay of the second DNA-affinity column fractions (1 μ each) with ³²P-labeled Jκ RS probe.
Gel retardation assay was done as described
previously. Only 10 out of 50 fractions were
assayed. Fractions 17-19 contained the protein
binding to the Jκ RS probe. B. 20 μl each
of fractions 13-22 were electrophoresed in 10%
acrylamide gels and stained with silver.

coincided well with the relative amounts of the 60 kD protein. The specificity of DNA
sequences to which the purified protein bound was examined by the gel retardation assay using
various mutant probes (Figure 3). The protein-DNA complex was detected only with the Jκ
RS probe but not with other DNA fragments. However, the Jκ RS binding protein could also
bind to other mutant DNA probes when their concentrations were increased.

To study further the affinity of the Jκ RS binding protein to various DNA probes, we
carried out kinetic analyses. The saturation curve and the Scatchard plot indicate that the
Kd value of the J RS binding protein to the Jκ RS probe were 1.0 nM in the absence of
poly(dI)poly(dC) (Figure 4). The dissociation constants of the Jκ RS binding protein to
the other DNA probes were quantitated in the presence of poly(dI)poly(dC), which reduces
nonspecific binding at high concentrations of DNA probe. The Kd values for the Jκ 79M and
Vκ probes in the absence of poly(dI)poly(dC) are preliminary because of the appearance of
several other bands. The Kd values of the other mutant Jκ and Vκ RS probes were
roughly at least one order higher than that of the Jκ RS probe as summarized in Table 1.
In order to study the precise sequence of the Jκ RS probe where the Jκ RS binding
protein interacts, we carried out a DNase I footprinting analysis of the DNA-protein complex.
This experiment indicated that the "GTG" sequence in the heptamer sequence interacted with the

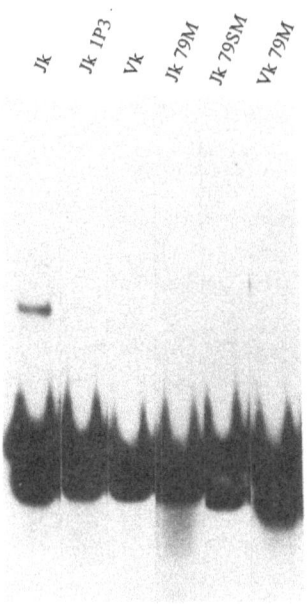

Figure 3. Gel retardation assay
with various RS probes.
[32]P labeled RS probes
described in Figure 1 were
bound to 1 μ of DNA
affinity column fraction in
the presence of 1 μg of
poly(dI)poly(dC). DNA
probes are indicated at
the top.

Jκ RS binding protein. The results are in general agreement with a previous mutagenesis experiment that revealed several important bases required for in vivo rearrangement in a pre-B cell line.[5] A marked reduction of the affinity by a single base substitution of the third G in the heptamer (Jκ IP3 probe) is also consistent with the conclusion that the GTG sequence of the heptamer interacts with the Jκ RS binding protein.

Crude nuclear extracts from lymphoid and non-lymphoid cell lines were prepared, and tested for their ability to produce the specific complex with the Jκ RS probe. Extracts from a murine pro-B (LyD9)[8] and a murine pro-T (FTD11)[9] cell lines contained the Jκ RS binding activity as shown in Figure 5. Other lymphoid cell lines that contain similar binding activity include a human leukemic T cell line (Jurkat), an Epstein-Barr virus-transformed B cell line (CESS), a human HTLV-I-transformed T cell line (MT-1) and a murine leukemic B cell line (BCL1). In contrast, extract from non-lymphoid cell line L tk[−] as well as NIH-3T3, SC-1, WEHI-3 and COS-7 cells did not contain the Jκ RS-specific binding activity.

Isolation and Characterization of RBP-2 Gene

The purified preparation was then subjected to trypsin digestion and the resulting peptides were fractionated by reverse-phase HPLC. Oligodeoxyribonucleotide probes were synthesized on the basis of partial amino acid sequences of the tryptic digested peptide and used to screen a cDNA library derived from 38B9 poly(A)[+] RNA. We have isolated two cDNA clones (RBP-1 and RBP-2) out of 1 x 10[6] colonies of the cDNA library.[10] They were identical except that RBP-2 has an extra 550-bp sequence at the 5' end and utilizes the 140 bp downstream poly(A) addition site. Nucleotide sequence determination of the

Table 1. Kd Values of Various Probes to Jκ RS Binding Protein

| | Kd X 10^9 | |
Probes	+poly(dI)poly(dC)	-poly(dI)poly(dC)
Jκ	10	1
Jκ 1P3	66	-
Jκ 79M	300	(95)
Vκ wild	330	(110)

Numbers in parentheses are preliminary (See text).

cDNA revealed an open reading frame that encodes 526 amino acid residues (Figure 6). All the 23 partial amino acid sequences determined were identified in this amino acid sequence deduced from the cDNA sequence. The calculated molecular weight of the RBP-2 protein is 57,000 (including the initiating methionine), in reasonable agreement with the value (60 k Dalton) estimated by SDS-polyacrylamide electrophoresis.

To confirm that the cDNA clones isolated encode the Jκ RS binding protein, COS-7 cells were transfected with RBP-2 cDNA in Okayama-Berg expression vector (Figure 7A). After 48 hr culture, RBP-2 and mock transfected COS-7 cells were harvested. Their nuclear extracts were purified in the Jκ RS DNA-coupled affinity chromatography and assayed for the binding activity to the Jκ RS probe in a gel retardation assay. As shown in Figure 7B, the binding activity was detected in the extract of RBP-2-transfected COS-7 cells but not of mock transfected COS-7 cells. Furthermore, a marked reduction of the affinity was observed by using a mutant probe (Jκ 1P3 probe). These results strongly indicate that the RBP-2 cDNA sequence encodes the Jκ RS-binding protein.

Computer survey did not pick up any other proteins that have extensive homology with the overall sequence of the RBP-2 protein. The protein does not seem to have any known DNA binding domains such as zinc finger, helix-turn-helix, or leucine zipper structure. However, the RBP-2 protein contains the sequence similar to the 40-residue motif of the integrase family that includes λ phage integrase, fimbriae switch recombinase of *E. coli* and yeast FLP recombinase (Figure 8, discussed later).[11]

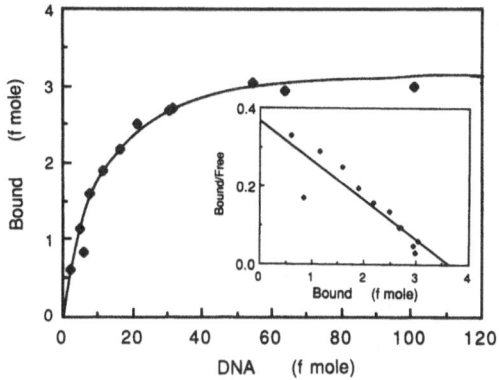

Figure 4. Saturation curve and Scatchard plot of association of the Jκ RS binding protein with Jκ RS in the absence of poly(dI)poly(dC). 50 fmoles of DNA-affinity column II fraction were used. The amounts of free and bound DNA were estimated by a scintillation counting and plotted. Insert, Scatchard plot.

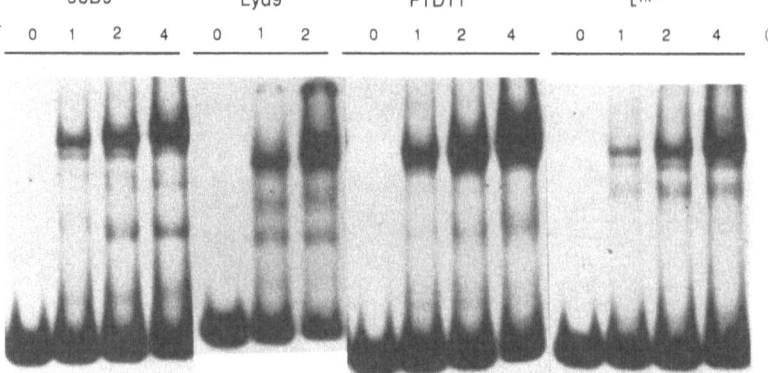

Figure 5. Tissue specificity of the Jκ RS binding protein. Jκ RS binding activity was examined in nuclear extracts from various cell lines with ^{32}P-labeled Jκ RS probe. Various amounts of nuclear extracts were incubated with Jκ probe in the presence of 1 μg of poly(dI)poly(dC) and electrophoresed. Pre-B cell line 38B9, pro-B cell line LyD9, and pro-T cell line FTD 11 are lymphoid cells, and L tk- is a non-lymphoid cell.

DISCUSSION

First we have purified the RBP-2 protein almost to homogeneity. The molecular weight of this protein was 60k Dalton. The Kd value of this protein to the Jκ probe was 1 nM. The RBP-2 protein has a very strict sequence specificity as a single base substitution in the heptamer sequence greatly reduced its binding to this protein. The heptamer and nonamer sequences of the Jκ and Vκ probes are identical except for one base substitution (T → A) in the nonamer sequence. This base substitution is rather frequent in the nonamer sequence and is known to be neutral to the V-J recombination frequency in vivo.[5] Nonetheless, the RBP-2 protein has a very weak affinity for the Vκ sequence, indicating that the length of the spacer is important for binding recognition.

The fact that T cell receptor genes introduced into the pre-B cell line 38B9 can undergo D-J rearrangement[12] implies that the same recombination machinery is involved in the antigen receptor gene rearrangement in T and B cells. Since th RS sequence of the Jκ segment is almost identical to those of the T cell receptor V and D, and immunoglobulin V_H, Vλ and J_H segments,[13,14] it is likely that the RBP-2 protein, which was detected also in T lineage cells, binds to these RS sequences as well.

Site-specific recombinases can be categorized into at least two families, based on their protein structure and chemistry of strand breakage.[15] One of these is the resolvase/invertase family and the other family is called the integrase family. Structural comparison of the integrases showed that these proteins share a highly conserved 40-residue motif.[11] The amino acid sequence deduced from the cDNA sequence has shown that the RBP-2 protein has the sequence similar to the 40-residue motif of the integrase. In the 40-residue motif, three residues (His[1], Arg[4] and Tyr[38]) are perfectly conserved among all the 18 integrase proteins so far sequenced. The conserved tyrosine residue of FLP[16,17] and λ phage integrase[18] is involved in the catalytic reaction of the integrase. The RBP-2 protein does conserve the His[1] and Tyr[38] residues but not the Arg[4] residue. In addition, Leu[11] is more than 60% conserved among the other recombinases, but is not conserved in the RBP-2 protein. Nonetheless, the RBP-2 protein contains two highly (more than 68%) conserved residues (Gly[15] and His[40]) and seven moderately (more than 21%) conserved residues within the 40-residue motif. If we regard residues in the same exchange group as conserved, 21 out of 40 residues of the motif region of the RBP-2 protein are more than 32% conserved among all the 19 proteins. This homology suggests that RBP-2 protein might be involved in V-(D)-J recombination as recombinase. Among the other 18 integrases, the RBP-2 protein has the highest homology with the P4 integrase (Figure 8B). The 40-residue motifs of these two proteins share 9 identical residues and 11 residues of the same family. Furthermore, the

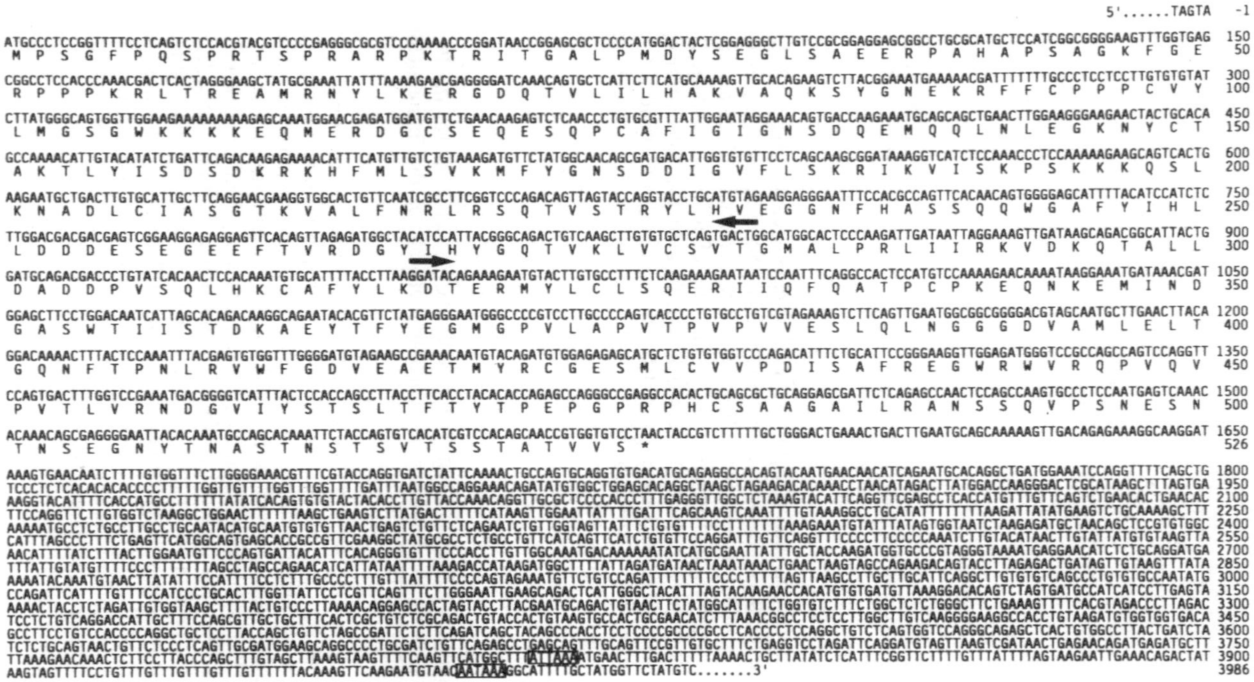

Figure 6. Nucleotide and predicted amino acid sequence of RBP-2 protein. Poly(A) addition signal sequences were boxed. Asterisk indicates termination codon. The boundaries of the integrase motif region (Figure 8) are indicated by two horizontal arrows.

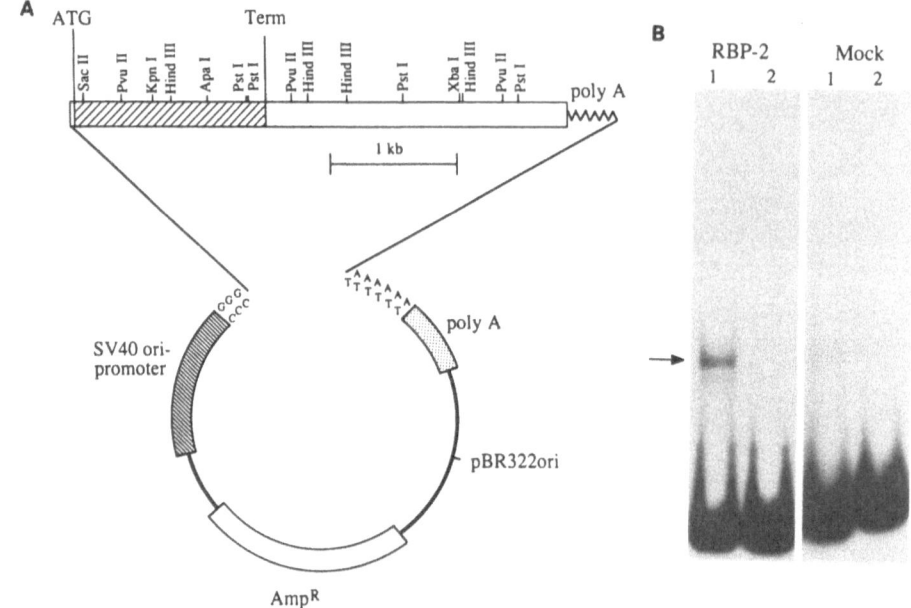

Figure 7. A. Restriction map and structure of RBP-2 cDNA clone. We used the original RBP-2 clone in Okayama-Berg expression vector. Solid and zig-zag lines indicate vector and poly(A) tails, respectively. Black and open rectangles indicate coding and untranslated regions, respectively. B. Expression of RBP-2 cDNA in COS-7 cells. Nuclear extracts from RBP-2 and mock transfected COS-7 cells were purified in a DNA affinity column and were assayed for the binding activity to the Jκ RS probe (lane 1) or its single-base substitution mutant probe, Jκ 1P3 (lane 2) in a gel retardation assay. Term, termination site.

first 16 of the recognition sequence "GAGTCCGGCCTTCGGCACCA" for the P4 integrase 19 has some homology with the sequence of heptamer-nonamer of the immunoglobulin gene, that is, "GAGTC(A)CGGTTTTTGT".

In vitro studies of site-specific recombination indicate that the reaction is essentially catalyzed by a single protein that seems to be required in stoichiometric rather than catalytic amounts although additional accessory proteins are required in some cases.[15] It is, therefore, reasonable to assume that the RBP-2 protein may be a recombinase of V-(D)-J recombination although other accessory binding proteins like integration host factor[15] might be required for efficient recombination. Actually, presence of the RBP-2 protein in the cells (CESS, BCL1, MT-1 and Jurkat) which have finished DNA rearrangement of either immunoglobulin or T cell receptor gene implicates that another regulatory molecule may be required for rearrangement of antigen receptor genes. The previously reported RS binding proteins like Vκ RS binding protein[20] and nonamer binding protein[21] might be candidates for accessory protein(s).

ACKNOWLEDGMENTS

This investigation was supported by Tokubetsu Suishin grant from the Ministry of Education, Science and Culture of Japan. We are grateful to Ms. J. Kuno for her excellent technical assistance and to Ms. K. Hirano for her help in preparation of the manuscript. We thank Ono Pharmaceutical Co. Ltd. and The Green Cross Corporation for their help in large scale preparations of 38B9 cells.

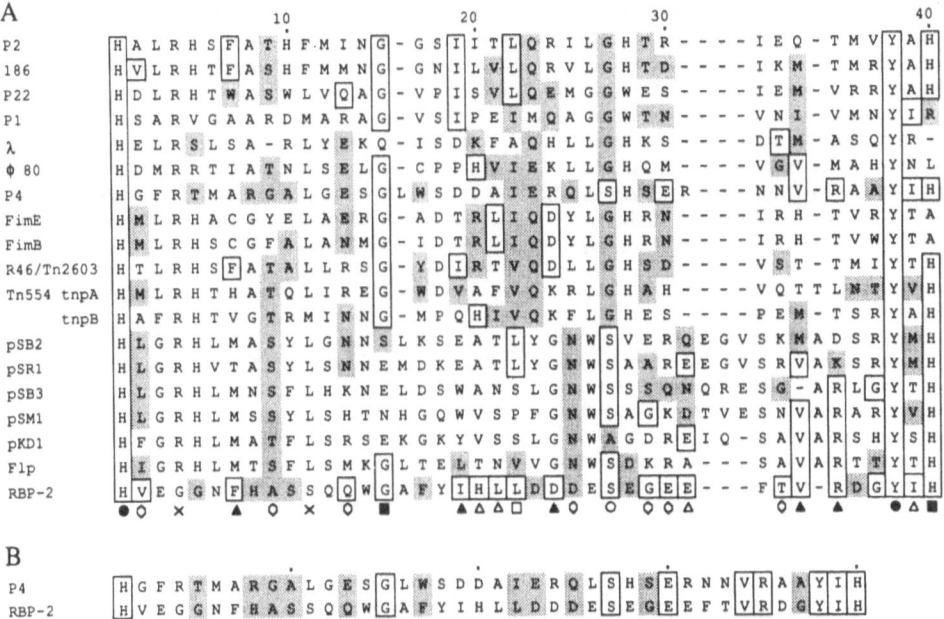

Figure 8. Comparison of the 40-residue motif sequences of the RBP-2 protein and recombinases of the integrase family. A. Alignment with all the known sequences. B. Alignment with P4 integrase. Only residues that are conserved between the RBP-2 protein and any of the other proteins are boxed. Residues that belong to the same exchange groups with those of RBP-2 are shaded. Extents of conservations are indicated by closed and open symbols at the bottom, excluding and including, respectively, residues of the same exchange group. Circles, 100%; rectangles, >68%; triangles, >21%. Residues with less than two identical members but more than 7 members of identical exchange groups are indicated by open hexagons. Residues that are conserved >60% among the other recombinases, but not conserved in RBP-2, are shown by crosses. Origins of integrase sequences are listed in reference 10.

REFERENCES

1. P. Early, H. Huang, M. Davis, K. Calame, and L. Hood, An immunoglobulin heavy chain variable region gene is generated from three segments of DNA: V_H, D and J_H, *Cell* 19:981 (1980).

2. H. Sakano, R. Maki, Y. Kurosawa, W. Roeder, M. Weigert, and S. Tonegawa, Two types of somatic recombinations are necessary for the generation of complete immunoglobulin heavy chain genes, *Nature* 286:676 (1980).

3. S. Lewis, A. Gifford, and D. Baltimore, Joining of Vκ to Jκ gene segments in a retroviral vector introduced into lymphoid cells, *Nature* 286:676 (1980).

4. T. K. Blackwell and F. W. Alt, Site-specific recombination between immunoglobulin D and J_H segments that were introduced into the genome of a murine pre-B cell line, *Cell* 37:105 (1984).

5. S. Akira, K. Okazaki, and H. Sakano, Two pairs of recombination signals are sufficient to cause immunoglobulin V-(D)-J joining, *Science* 238:1134 (1987).

6. F. Alt, N. Rosenberg, S. Lewis, E. Thomas, and D. Baltimore, Organization and reorganization of immunoglobulin genes in A-MuLV-transformed cells: rearrangement of heavy but not light chain genes, *Cell* 27:381 (1981).

7. Y. Hamaguchi, N. Matsunami, Y. Yamamoto, and T. Honjo, Purification and characterization of a protein that binds to the recombination signal sequence of the immunoglobulin Jκ segment, *Nucleic Acids Res.* 17:9015 (1989).

8. R. Palacios and M. Steinmetz, IL3-dependent mouse clones that express B-220 surface antigen, contain Ig genes in germ-line configuration, and generate B lymphocytes *in vivo*, *Cell* 41:727 (1985).

9. J. Pelkonen, P. Sideras, H.-G. Rammensee, K. Karjalainen, and R. Palacios, Thymo-cyte clones from 14-day mouse embryos. I. State of T cell receptor genes, surface markers, and growth requirements, *J. Exp. Med.* 166:1245 (1987).

10. N. Matsunami, Y. Hamaguchi, Y. Yamanoto, K. Kuze, K. Kangawa, H. Matsuo, M. Kawaichi, and T. Honjo, A protein binding to the Jκ recombination sequence of immunoglobulin genes contains a sequence rerated to the integrase motif, *Nature* 342:934 (1989).

11. P. Argos, A. Landy, K. Abremski, J. B. Egan, E. Haggard-Ljungquist, R. H. Hoess, M. L. Kahn, B. Kalionis, S. V. L. Narayana, L. S. Pierson III, N. Sternberg, and J. M. Leong, The integrase family of site-specific recombinases: regional similarities and global diversity, *EMBO J.* 5:433 (1986).

12. G. Yancopoulos, T. K. Blackwell, H. Suh, L. Hood, and F. W. Alt, Introduced T cell receptor variable region gene segments recombine in pre-B cells: evidence that B and T cells use a common recombinase, *Cell* 44:251 (1986).

13. J. Kavaler, M. M. Davis, and Y. Chien, Localization of a T-cell receptor diversity-region element, *Nature* 310:421 (1984).

14. E. A. Kabat, T. T. Wu, M. Reid-Miller, H. M. Perry, and K. S. Gottesman, "Sequences of Proteins of Immunological Interest, Ed. 4." NIH, Bethesda (1987).

15. N. L. Craig, The mechanism of conservative site-specific recombination, *Annu. Rev. Genet.* 22:77 (1988).

16. R. L. Parsons, P. V. Prasad, R. M. Harshey, and M. Jayaram, Step-arrest mutants of FLP recombinase: implications for the catalytic mechanism of DNA recombi-nation. *Mol. Cell. Biol.* 8:3303 (1988).

17. P. V. Prasad, L.-J. Young, and M. Jayaram, Mutations in the 2-μm circle site-specific recombinase that abolish recombination without affecting substrate recognition, *Proc. Natl. Acad. Sci. USA* 84:2189 (1987).

18. C. A. Pargellis, S. E. Nunes-Düby, L. Moitoso de Vargas, and A. Landy, Suicide recombination substrates yield covalent λ integrase-DNA complexes and lead to identification of the active site tyrosine, *J. Biol. Chem.* 263:7678 (1988).

19. L. S. Pierson III, and M. L. Kahn, Integration of satellite bacteriophage P4 in *Escherichia coli*: DNA sequences of the phage and host regions involved in site-specific recombination. *J. Mol. Biol.* 196:487 (1987).

20. R. J. Aguilera, S. Akira, K. Okazaki, and H. Sakano, A pre-B cell nuclear protein which specifically interacts with the immunoglobulin V-J recombination sequences. *Cell* 51:909 (1987).

21. B. D. Halligan and S. V. Desiderio, Identification of a DNA binding protein that recognizes the nonamer recombinational signal sequence of immunoglobulin genes, *Proc. Natl. Acad. Sci. USA* 84:7019 (1987).

IN SITU DETECTION OF STAGE-SPECIFIC GENES AND ENHANCERS IN B CELL DIFFERENTIATION VIA GENE-SEARCH RETROVIRUSES

William G. Kerr, Garry P. Nolan, Jeffrey B. Johnsen, and Leonard A. Herzenberg

Department of Genetics
Stanford University
Stanford, California

We demonstrate that infection of an LPS-responsive pre-B cell line with transcriptionally-defective retroviruses containing a reporter gene (*lacZ*) can result in viral integrations where expression of *lacZ* is differentiation stage-dependent. Because expression of *lacZ* is dependent upon flanking cellular sequences these retroviral integrations represent *in situ* gene fusions with cellular enhancers (Enhsr1) and genes (Gensr1) which are either induced or repressed during LPS-stimulated differentiation. One of the well-documented effects of LPS upon pre-B cells is the induction of κ light chain transcription via NF-κB. The identification of LPS-stimulated gene repression during B cell differentiation indicates that LPS has multiple effects upon gene expression during the pre-B to B cell transition. The identification of cellular enhancers and genes which are downregulated during the transition from the pre-B to the B cell stage indicates that other transcription factors, in addition to NF-κB, are required for this step in differentiation. Finally, we present some initial experiments which indicate the gene-search retroviruses can introduce expression of *lacZ* into normal hematopoietic cells *in vitro* and *in vivo*.

INTRODUCTION

The differentiation of hematopoietic stem cells to committed B cell progenitors and then to immunoglobulin-secreting plasma cells occurs in a step-wise fashion.[1] These stages of differentiation represent discrete points where intrinsic (e.g., Ig rearrangement), or extrinsic (e.g., stromal cells, T cells) mechanisms can influence the maturation of B-lineage cells. Because B-lineage cells must satisfy different requirements at different stages in their maturation, the expression of stage-specific gene products presumably allows B-lineage cells to respond to the intrinsic or extrinsic signals controlling their differentiation.

In Figure 1 we show a schematic depicting the step-wise differentiation of B-lineage cells and indicate some of the known genes or gene products which are expressed in a differentiation stage-specific fashion. It is difficult to know if any of these genes are involved in deciding whether a cell should commit to the next step in differentiation. Some gene products may influence the decision by providing information relevant to the status of the cell (e.g., pseudo-light chain proteins)[2] or others may be involved in implementing the differentiation decision (e.g., J-chain).[3] The gene products responsible for decision-making during differentiation of B cells have yet to be identified. Although B cell differentiation is one of the best understood mammalian differentiation pathways, we probably know only a fraction of the proteins which are necessary for their ordered maturation.

Mechanisms of Lymphocyte Activation and Immune Regulation III
Edited by S. Gupta *et al.*, Plenum Press, New York, 1991

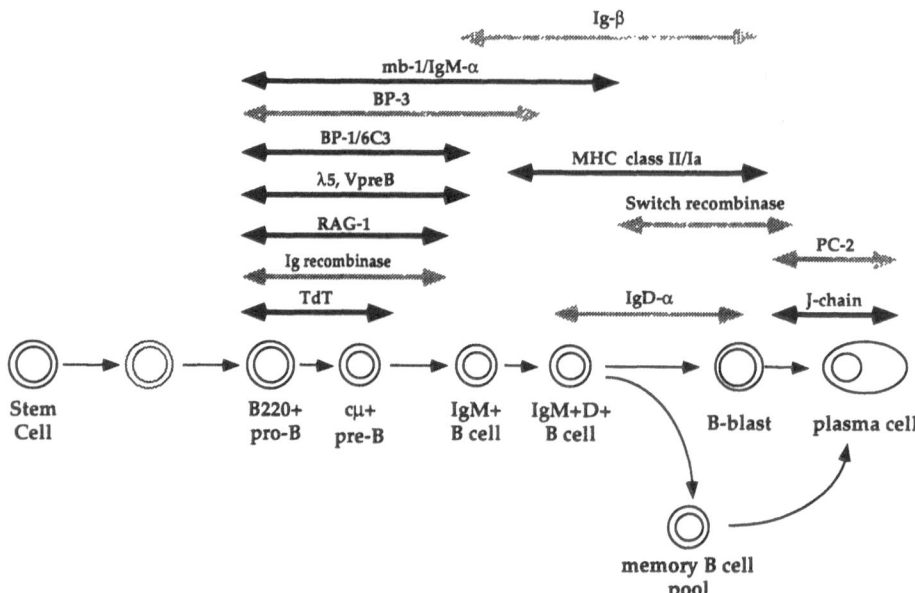

Figure 1. Stage-specific gene expression during murine B cell differentiation. This schematic depicts the step-wise maturation of B-lineage cells with the order of differentiation indicated by thin black arrows. Above specific stages in B-lineage differentiation we indicate genes (thick black arrows) or gene products (thick gray arrows) which are expressed in these specific stages of B cell differentiation. Gene products are distinguished from genes because coding sequences for these proteins have not yet been identified and consequently the evidence for their stage-specific expression is based on biochemical or immunochemical data. This figure represents a compilation of data from references listed below in addition to interpretation by the author. For references concerning expression of: BP-1/6C3, BP-3, Ia, PC-2 see [1]; switch recombination see [25]; Ig recombination see 6; TdT see 7; RAG-1 see [28]; J-chain see [3]; λ5, VpreB, and mb-1 see [17]; IgM-α, IgD-α, and Ig-β see [29-31].

Here we describe our continuing effort to detect and identify stage-specific genes in the differentiation of B-lineage cells. Our approach to this problem relies on forming *in situ* gene fusions with cellular genes via self-inactivating retroviruses containing *lacZ* reporter gene constructs.[4,5] Activation of the *lacZ* reporter gene contained within the transcriptionally-active provirus requires integration in either transcriptionally-active chromatin (Enhsr1) or in an intron of a transcriptionally-active gene (Gensr1). Others have utilized reporter gene constructs containing *lacZ* to identify developmentally-regulated chromatin and genes in mammalian embryonic development.[6,7] Transposable elements containing *lacZ* have been used to detect developmentally regulated chromatin domains in *Drosophilia*.[8,9,10]

RESULTS

Models for Expression of LacZ *via the Gene-Search Retroviruses*

In Figures 2 and 3 we illustrate how the gene-search retroviruses self-inactivate their ability to drive transcription from their 5' LTR. This approach was originally described by Yu et al..[11] When copying viral RNA into DNA, reverse transcriptase uses the 3' U3 as the

Enhsr1

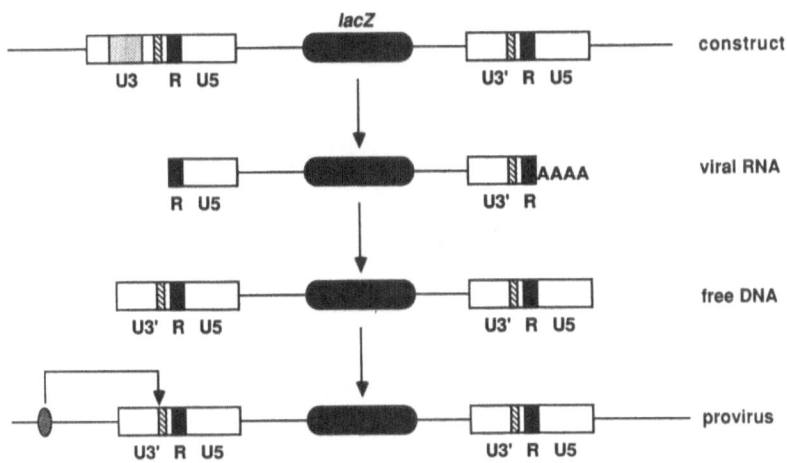

Figure 2. Schematic depiction of Enhsr1 and model for *lacZ* expression.
This figure depicts how a *lacZ*-encoding provirus which lacks the
Moloney leukemia virus enhancer region (stippled box) would be generated from the Enhsr1 construct. The provirus generated from Enhsr1 will
lack the viral enhancer region (stippled box) but still retain CAAT and
TATA box motifs (hatched box). This defective proviral transcriptional
unit can be activated by an enhancer element (stippled oval) in the flanking, endogenous chromatin.

template for the U3 region in both the 5' and 3' viral LTR.[12] When a mutation or deletion
is made in the 3' U3 of a retroviral construct, infection of target cells will generate
proviruses which have a different transcriptional phenotype than the parent construct. The
gene-search retroviruses exploit this feature of retroviral biology since they generate
proviruses which are transcriptionally-incompetent relative to the parental construct.

In the case of Enhsr1, the retrovirus construct contains a deletion of the Moloney enhancer region in the 3' LTR (Figure 2). Following entry of Enhsr1 viral particles into target
cells, a provirus will be generated which will lack the Moloney enhancer region in either
LTR. The only transcription control elements remaining in the LTR will be the CAAT
and TATA motifs of the Moloney promoter. Because the promoter elements are no longer
associated with the viral enhancer, the *lacZ* gene must rely on transcription control ele
ments in the flanking cellular chromatin for efficient expression. Thus *in situ* gene fusions
with Enhsr1 represent a fusion between transcriptional regulatory elements, presumably
enhancers, which are capable of activating transcription at the Moloney promoter. Therefore,
in situ gene fusions with Enhsr1 represent a fusion of cellular transcriptional control
elements and the Moloney promoter region.[4]

In *lacZ*+ cells with integrations of the Gensr1 provirus a transcriptional and translational fusion between a cellular gene and *lacZ* is generated (Figure 3). To accomplish this we
developed the *lacZ* reporter construct, AcLac.[5] In AcLac the translation initiation codon
(ATG) of *lacZ* has been replaced by the splice-acceptor region of the Moloney *env* gene.
Because the splice-acceptor sequences are in frame with *lacZ*, AcLac will form a
translational fusion with a cellular gene if it is spliced to a coding exon. The
splice-acceptor of *env* is predicted to have three separate splice acceptance points which
each break the translational codon at one of three possible positions in the triplet.[13,14]
Because of this feature, Gensr1 can potentially form a translational fusion with any
cellular gene that contains an intron following a coding exon.[5]

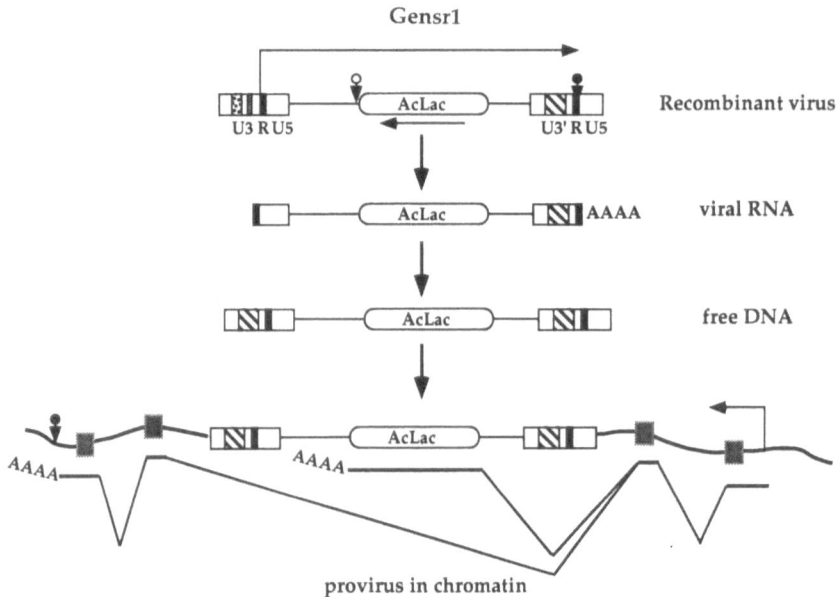

Figure 3. Model for Gensr1. A flow chart which depicts the generation of Gensr1 viral RNA, reverse transcription to a free DNA intermediate and subsequent proviral integration in an intron of a transcriptionally-active gene. A striped rectangle in the viral LTR indicates that the LTR contains a deletion of both the Moloney enhancer and promoter regions which results in a pro-virus which is transcriptionally-inert. Horizontal arrows indicate transcriptional orientation, note that transcriptional orientation of AcLac is opposite that of the retroviral construct and consequently we have inserted a unidirectional polyA site downstream of AcLac (vertical arrow with empty circle) which will function as the polyA cleavage and addition site for the *in situ* fusion with a cellular gene.

Enhsr1 Integrations Identify Cellular Transcription Control Elements That Confer Differentiation Stage-Specific Expression on LacZ

By infecting the pre-B cell line, 70Z/3, in either the uninduced or the induced state, we have derived a series of 70Z/3-Enhsr1 clones where *lacZ* expression is differentiation stage-specific.[4,5] We found that expression of *lacZ* is repressed in three clones (Figure 4)

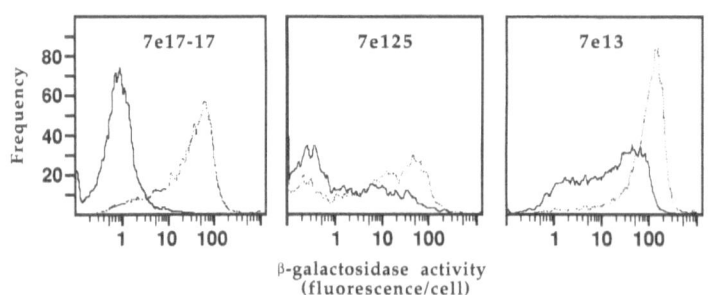

Figure 4. 70Z/3-Enhsr1 *lacZ*[+] integrants which are repressed during LPS-stimulation. Analysis of cells cultured in LPS is represented as a solid line while cells cultured in normal medium are represented as a broken line. See [32] for details on FACS-GAL analysis and [5] for details concerning analysis of 70Z/3-Enhsr 1 clones.

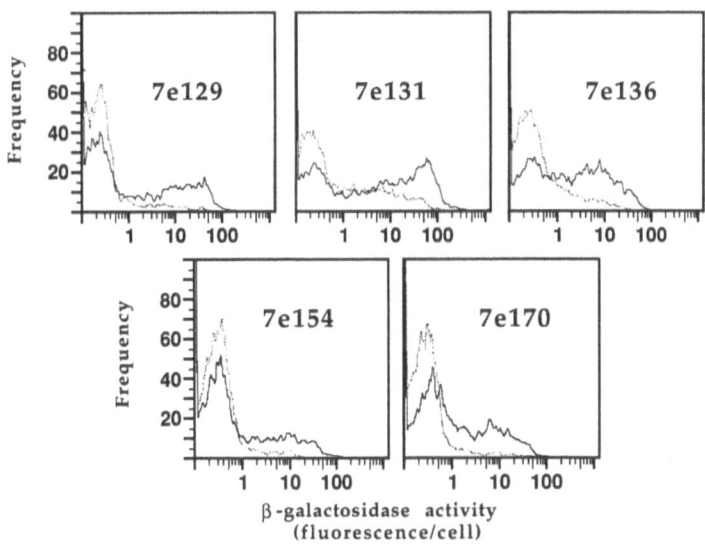

Figure 5. 70Z/3-Enhsr1 integrants which are induced by LPS stimulation. As in Figure 4, solid line represents LPS analysis and broken line normal medium analysis.

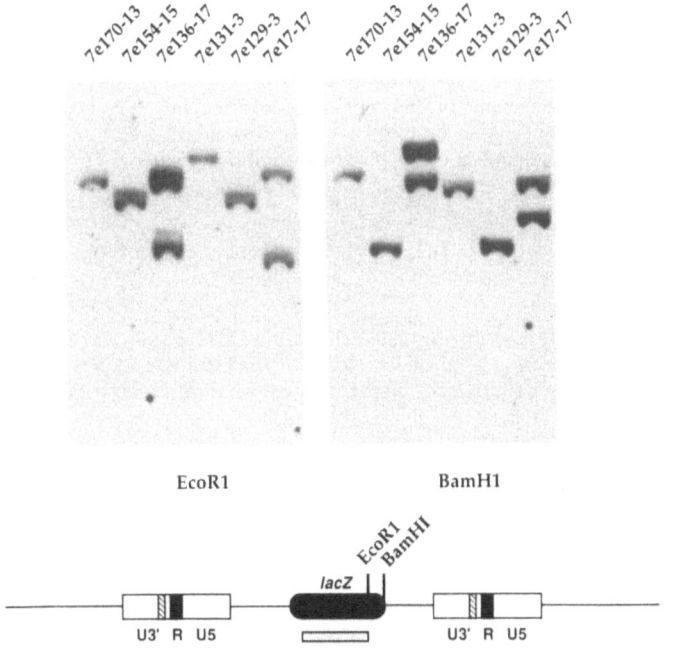

Figure 6. LPS-inducible Enhsr1 integrations represent independent proviral integrations. Southern blot analysis of genomic DNA from 70Z-Enhsr1 clones digested with either EcoRI or BamHI and hybridized with a *lacZ* probe (gray rectangle below provirus).

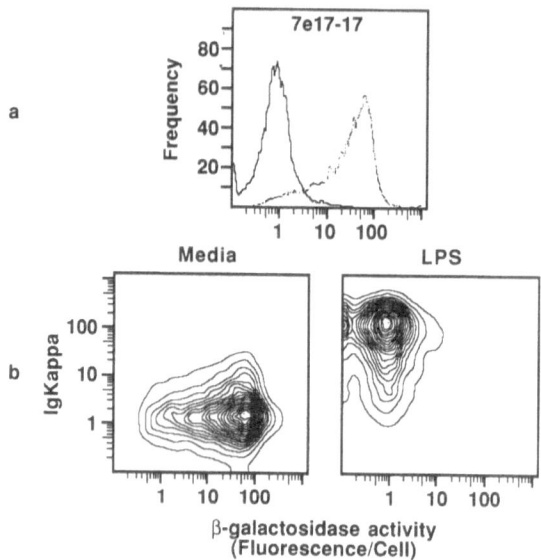

Figure 7. Regulation of *lacZ* and κ light chain
expression following LPS-induction of the
Enhsr1 clone, 7e17-17 cells. Equal numbers of
7e17-17 cells were either cultured in normal
media or in media containing 10 μg/ml of
LPS for 24 hours. The cells were then stained
as described above and analyzed on the FACS.
(a) Histograms representing β-galactosidase
activity of 7e17-17 cells after 24 hours in
LPS-containing media (solid line) or in normal
media (broken line). (b) Dual parameter FACS
analysis of κ light chain expression (Texas
Red) vs. β-galactosidase activity
(Fluorescein) in 7e17-17 cells cultured in
normal media or in LPS-containing media. For a
more detailed description of combining surface
staining with the FACS-GAL technique see [5].

induced in five clones (Figure 5) during LPS-stimulated differentiation of 70Z/3 cells. These
eight clones were derived from a total of 411 *lacZ*[+] 70Z/3-Enhsr1 clones and were the
only clones found to undergo significant changes in expression during LPS-stimulated
differentiation.

Because the five LPS-inducible clones shared a similar pattern of β-galactosidase expression upon LPS induction we felt that they may represent retroviral integrations in an
identical site. Recent work has demonstrated that retroviral integration can be non-random
and in some cases independent proviruses can integrate repetitively at an identical
location.[15] We performed Southern blot analysis to examine the integration diversity among
the five LPS-inducible clones (Figure 6). Because the restriction enzymes EcoRI and BamHI
are only present once within the proviral DNA, these enzymes will generate *lacZ*-hybridizing
restriction fragments unique to each proviral integration. Because the restriction fragments
hybridizing with *lacZ* in Figure 6 are not of identical size in the five LPS- inducible
clones (7e129-3, 7e131-3, 7e136, 7e154-15, 7e170), we conclude that they represent independent integrations of Enhsr1. Although the integrations are not identical by conventional
Southern blotting, we cannot exclude the possibility that they have integrated within the same
chromosomal locus.

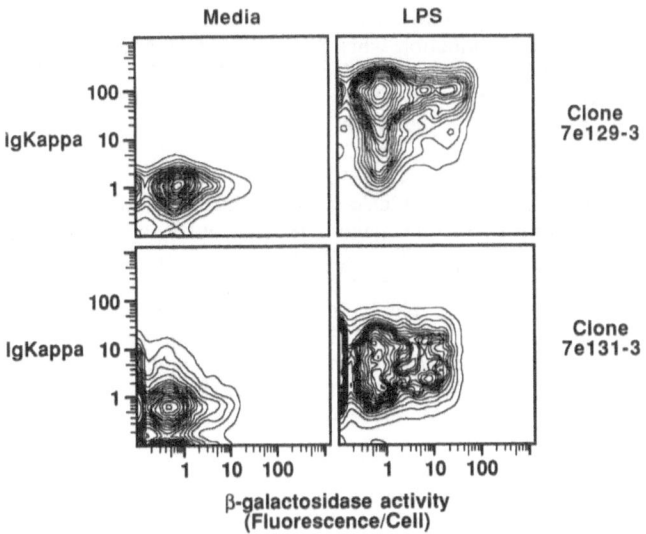

Figure 8. B-cell stage specific expression of *lacZ* in the Enhsr1
clones, 7e129-3 and 7e131-3. Cells were cultured, stained
and analyzed in a similar fashion to 7e17-17 cells in Figure 7.

To confirm that the Enhsr1 cells were differentiating in response to LPS we examined the
expression of the *lacZ* via FACS-GAL in combination with immunofluorescent staining for
surface IgMκ. This dual parameter FACS analysis revealed that the Enhsr1 clones were differentiating in response to LPS by expressing surface IgMκ while either repressing *lacZ*
expression (7e17-17; Figure 7) or inducing *lacZ* expression (7e129-3, 73131-3; Figure 8).
It is clear from this analysis that expression of *lacZ* in the repressible clone, 7e17-17,
is reciprocally regulated with respect to differentiation of 70Z/3 cells. *LacZ* is
expressed by the 7e17-17 cells in the absence of LPS when the cells are IgMκ⁻ (the pre-B
phenotype); however, after culture in LPS for 24 hr the 7e17-17 cells acquire the B cell
phenotype (IgMκ⁺) and became *lacZ*⁻. This analysis confirms the differentiation
stage-specific expression of *lacZ* in this Enhsr1 clone. In this situation we presume that
the Enhsr1 provirus has integrated near a pre-B specific enhancer which controls the expression of a gene(s) that is shut off when the cell becomes a B cell. This explanation assumes
the loss of positive regulation during LPS stimulation; however, active negative regulation,
such as a silencer element,[16] is an equally plausible explanation.

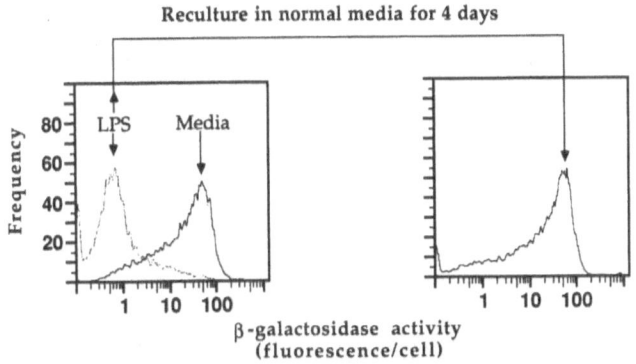

Figure 9. Repression of *lacZ* in 7e17-17 cells is relieved
when LPS is removed. Cells which had been cultured
for 24 hours in LPS were washed two times, recultured
in normal medium and analyzed by FACS-GAL as before.

Dual parameter FACS analysis of the LPS-inducible Enhsr1 clones revealed that there are two distinct phenotypes of *lacZ* induction when *lacZ* expression and differentiation are analyzed simultaneously (Figure 8). In some clones (e.g., 7e129-3) the $lacZ^+$ cells tend to be found among the differentiated cells (IgMκ^+) within the clone while in another clone the $lacZ^+$ cells are equally represented in the IgMκ^+ and IgMκ^- portion of the clone (7e131-3).

Presumably the cellular enhancers which are controlling expression of *lacZ* in Enhsr1 clones also control expression of a cellular gene. Because multi-parameter FACS analysis allows us to determine the phenotype of the $lacZ^+$ cells within these clones we can make some predictions about the role of the associated cellular gene. Since differentiated cells IgMκ^+) are enriched among the $lacZ^+$ cells in 7e129-3 while there is no enrichment of IgMκ^+ cells among the $lacZ^+$ cells in 7e131-3, we would propose that expression of the cellular gene associated with $lacZ^+$ expression in 7e129-3 is related to the differentiation of 70Z/3 cells. The cellular enhancer controlling *lacZ* in 7e131-3 may control a gene whose expression is responsive to the mitogenic effects of LPS.

To test the linkage between LPS-induced repression of *lacZ* and differentiation in 7e17-17 cells we asked whether the repression could be relieved by removal of LPS. In Figure 9 we show *lacZ* expression is recovered by LPS-treated 7e17-17 cells when they are cultured in the absence of LPS. The repression of *lacZ* is completely reversed since β-galactosidase β-gal) activity in the post-LPS 7e17-17 cells is equivalent to that of 7e17-17 cells which were never treated with LPS. It is interesting that the recovery of expression requires a period of 4 days while repression requires only 20-24 hr. The repression of *lacZ* and its reversibility in 7e17-17 indicate that LPS can also mediate repression of genes in pre-B cells. This would suggest that although LPS can mediate activation of genes in lymphoid[17] and myeloid[18] cells it may also have a repressive effect upon gene expression in B cell differentiation. One of the well-documented effects of LPS in pre-B cells is the activation of κ light chain transcription via the transcription factor, NF-κB.[19,20] Since NF-κB is thought to be responsible for the induction of gene expression, it presumably is incapable of mediating gene repression in LPS-treated pre-B cells as is seen with repression of *lacZ* in 7e17-17 cells. If NF-κB is incapable of mediating repression then multiple regulatory pathways must be triggered during LPS-stimulated differentiation of pre-B cells to B cells.

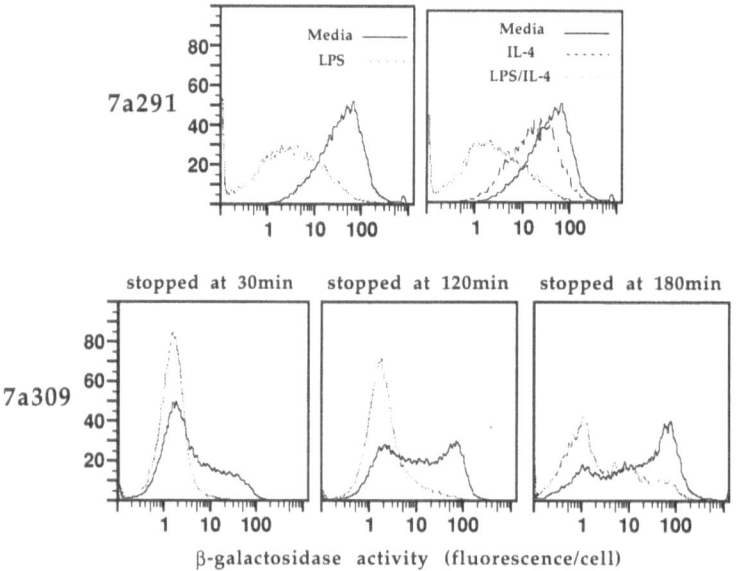

β-galactosidase activity (fluorescence/cell)

Figure 10. Gensr1 integrants are differentially-regulated during LPS stimulation of 70Z/3 cells. For 7a309, solid line is FACS-GAL analysis of LPS-stimulated cells and broken line is FACS-GAL analysis of cells in normal medium.

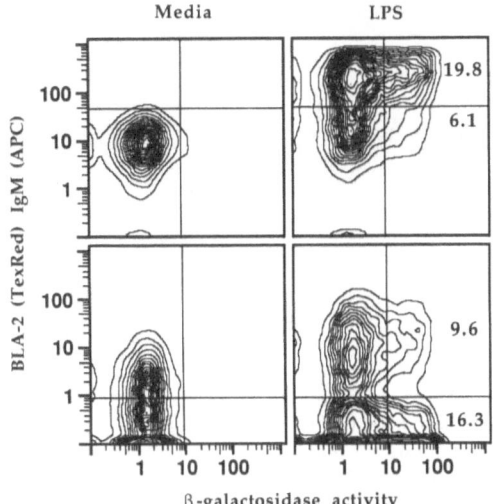

Media   LPS

β-galactosidase activity

Figure 11. LPS stimulation of 7a309 cells results in *lacZ*[+] cells which are enriched for "differentiated" cells (IgMκ[+]) rather than "activated" cells (BLA-2[+]). Simultaneous analysis of BLA-2 (Texas Red), IgM (Allophycocyanin) and β-galactosidase activity (Fluorescein) in 7a309 cells that were cultured for 24 hours with or without 10 μg/ml LPS.

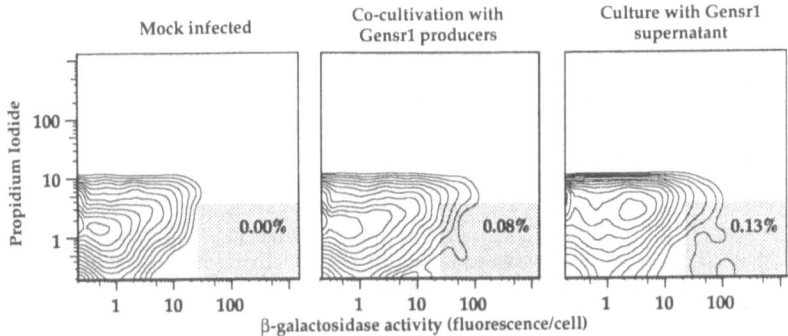

β-galactosidase activity (fluorescence/cell)

Figure 12. Transduction of *lacZ* expression into hematopoietic progenitor cells via Gensr1. Day 14 fetal liver cells were plated at 5 x 106 cells/ml either on an irradiated monolayer of Gensr1 producers (Ψ2/A8) or in Gensr1 supernatant with 4μg/ml polybrene for 24 hours. The mock-infected cells were cultured at an equivalent cell density in polybrene for an equal amount of time. The cells were then analyzed by FACS-GAL. The propidium iodide positive cells were hardware gated and thus excluded from the data collection. Data were collected on over 200,000 cells via Electric Desk run on a DEC 3600 computer. The percentage of β-gal positive cells within the gates (grayed rectangle) was estimated by Electric Desk run on a DEC 6310 computer. The percentage of β-gal[+] cells is indicated within the gray rectangles of each logarithmic 50% plot.

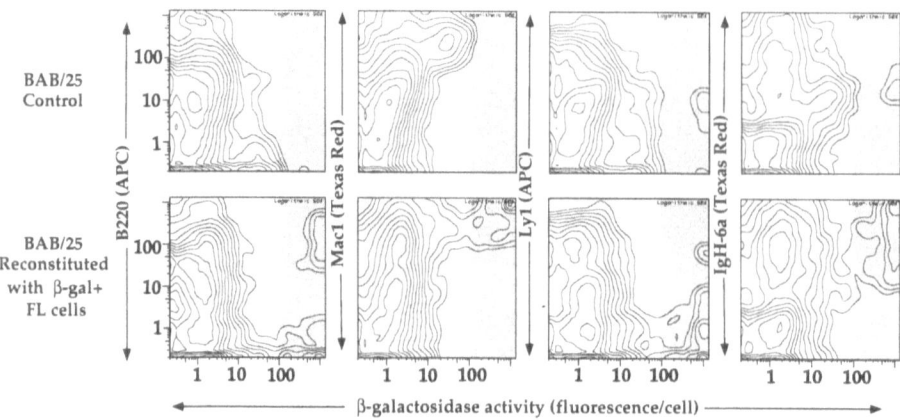

Figure 13. *In vivo* expression of *lacZ* following reconstitution with β-gal⁺ fetal liver
cells obtained by infection with the Enhsr1 retrovirus. Multiparameter FACS
analysis of hematopoietic lineage markers (y-axis) versus β-gal activity (x-axis) of
PerC cells from a BAB/25 host (IgH-6b) reconstituted with 5,000 β-gal⁺ BALB/C
(IgH6a) fetal liver cells following co-cultivation with Enhsr1 producer cells. There
were 0.05% β-gal⁺ cells in the control BAB/25 PerC that was examined on the same
day while there were 0.63% β-gal⁺ cells in the BAB/25 host reconstituted with
lacZ⁺ fetal liver cells. The stains were B220 (APC) vs. Mac1 (TexRed) vs. β-gal
and Ly-1 (APC) vs. IgH6a (TexRed) vs. β-gal.

Gensr1 Integrations Identify Cellular Genes Which are Repressed or Induced During
LPS-Induced Differentiation

The Enhsr1 integrations which were differentially regulated during LPS induction of
70Z/3 cells proved that *in situ* gene fusions could place a reporter gene (*lacZ*) under
the control of endogenous transcription control elements.[4,5] Although isolation and
characterization of the cellular gene associated with these native control elements is
possible, this task could prove laborious when one considers the large distances over which
enhancers can act. Therefore, to more readily identify genes associated with *lacZ*
integrations that are differentially regulated we have developed the splice acceptor *lacZ*
reporter construct, AcLac, as well as a retrovirus that contains AcLac, Gensr1.[5] Identification and characterization of the cellular gene associated with expression of *lacZ* in
Gensr1 integrants is amenable to conventional cloning techniques for two reasons: (1)
lacZ is fused to the transcript of the cellular gene, (2) β-galactosidase is fused to
the product of the cellular gene. This permits the associated gene to be identified by
probing cDNA libraries with Gensr1 sequences or oligonucleotides predicted from amino
terminal sequencing of the β-galactosidase fusion protein.

In Figure 10 we illustrate two Gensr1 integrants which are regulated during LPS-
stimulated differentiation of 70Z/3 cells. In 7a291, expression of *lacZ* is substantially
repressed after exposure of the cells to LPS. In addition we find that IL-4 has a mildly
repressive effect on expression of *lacZ* in 7a291. However, there appears to be no
synergistic effect when 7a291 is treated with both LPS and IL-4. In 7a309 expression of
lacZ is induced in a portion of the cells after exposure to LPS. We show the analysis of
β-galactosidase activity at three different times of incubation. The longest incubation
period (180 min) indicates that a majority of the cells in 7a309 have *lacZ* turned on by
growth in LPS while in cells not treated with LPS, *lacZ* expression is inactive in the
majority of the cells.

We then analyzed the induction of *lacZ* in 7a309 against IgMκ and BLA-2 expression
(Figure 11). As with 7e129-3, 7a309 cells which are induced for *lacZ* expression are two-
fold enriched for IgMκ expression. However, *lacZ*⁺ cells show no enrichment for

196

BLA-2 expression relative to the *lac-Z⁻* cells in 7a309. BLA-2 is a surface antigen found on actively dividing lymphoid and myeloid cells[21] whose surface expression is induced on 70Z/3 cells by LPS treatment (W. Kerr, unpublished). Because differentiated cells (IgMκ$^+$) are enriched among the *lacZ$^+$* cells in 7a309 rather than activated cells (BLA-2$^+$) we would propose that the cellular gene product fused with β-galactosidase in 7a309 is linked to B-cell differentiation rather than activation.

Gene-Search Retroviruses May Be Useful for Progenitor Cell and Gene Expression Studies in Normal Hematolymphoid Differentiation

In addition to studying B cell gene expression with transformed cell lines, we wish to apply the gene-search retroviruses to the study of normal hematopoietic differentiation both *in vivo* and *in vitro*. We have taken some of the first steps down this path. In Figure 12 we demonstrate that early hematopoietic progenitor cells (day 14 fetal liver cells) can express β-galactosidase following infection with Gensr1. Although the efficiency of this procedure is quite low, β-galactosidase$^+$ cells are readily discernible in the Gensr1 infections relative to the mock infected culture.

In another experiment we co-cultured day 14 fetal liver cells from a BALB/C fetus with irradiated Enhsr1 producer cells and sorted β-galactosidase$^+$ cells via FACS-GAL. We obtained approximately 5,000 cells by this procedure and used these cells for reconstitution of a lethally-irradiated BAB/25 allotype congenic host. Two months following irradiation, bone marrow, spleen and peritonal cavity (PerC) cells were analyzed for expression of β-galactosidase versus hematopoietic lineage-specific markers. We did not find a significant percentage of cells expressing β-galactosidase in the spleen and bone marrow of this animal; however, we did find a significant percentage (~0.6%) of cells in the PerC which express β-gal (Figure 13). The β-gal$^+$ cells in the PerC express the donor IgM allotype (IgH-6a) and Mac1. They are Ly1$^-$ and a portion of them express the 6B2 determinant, an epitope of the T200 antigen that is specific to B-lineage cells.[22] From the data in hand we cannot unambiguously assign the β-gal$^+$ cells to a hematopoietic lineage. However, this experiment indicates that the gene-search retroviruses may permit the study of gene expression in normal hematopoietic differentiation. Further experiments are required to determine if this approach is feasible for all arms of the hematopoietic system.

DISCUSSION

In this manuscript we have reviewed the use of gene-search retroviruses for the *in situ* detection of mammalian gene expression. We have utilized these retroviruses to identify genetic loci which respond to LPS-induced differentiation of a pre-B cell line. We have found that LPS mediates both repression and induction of gene expression during pre-B to B cell differentiation. Finally we present some preliminary experiments which indicate that the gene-search retroviruses may permit the study of gene expression during normal hematopoietic differentiation.

We have identified a total of ten gene-search retroviral integrations (both Enhsr1 and Gensr1) where expression of *lacZ* is differentially expressed during LPS stimulation of 70Z/3 cells. Four of these integrations (3-Enhsr1, 1-Gensr1) are repressed during LPS-induced differentiation of 70Z/3 cells while six (5-Enhsr1, 1-Gensr1) are LPS-inducible. Multi-parameter FACS analysis allowed expression of *lacZ* and surface antigens to be monitored simultaneously. This analysis demonstrated that the LPS-repressed Enhsr1 integration (7e17-17) is clearly restricted to the pre-B stage of differentiation. This analysis also indicated that *lacZ* expression in two LPS-inducible integrations (7e129-3, 7a309) is clearly associated with the differentiation of 70Z/3 cells to the B cell stage (surface IgMκ$^+$). From this analysis we predict that the cellular genes linked to these gene-search integrations are associated with a specific stage of B-lineage differentiation.

Our identification of multiple gene-search integrations which are either repressed or induced during the pre-B to B cell transition is not surprising. This transition is arguably the most dramatic step in B-lineage differentiation, because it represents the transition from antigen-independent to antigen-dependent differentiation. Upon becoming a mature B cell, a B-lineage clone is no longer concerned with successfully rearranging the heavy and light chain genes under the restriction of maintaining allelic exclusion but is now ready to respond to antigen, T cells and macrophages in order to participate in an immune response to foreign antigen. Thus, the repression and induction of multiple genes during the transition from the pre-B to the B cell stage is probably a necessary requirement in B cell differentiation. In addition, our identification of loci which are repressed during LPS-induced differentiation of 70Z/3 cells argues for the participation of multiple regulatory circuits in the pre-B to B cell transition, some of which must not involve NF-κB. In support of this hypothesis, others have proposed a regulatory pathway that mediates the induction of κ light chain transcription which is independent of NF-κB.[23,24]

Finally, we are optimistic that the gene-search retroviruses will permit tracking of gene expression during normal lymphoid differentiation. Two different questions concerning lymphoid development can be approached by this technology depending upon the type of locus where the gene-search virus integrates. The first question is: What is the differentiation potential of hematopoietic progenitor cells? If we succeed in integrating a gene-search virus in a constitutively expressed locus (e.g., β-actin) of a hematopoietic stem cell, we can then analyze the differentiation potential of this pluripotent cell via surface staining for hematopoietic lineage markers in concert with FACS-GAL. Second, can we get integration of a gene-search virus in a lineage- or differentiation stage-specific locus? This would provide a molecular tag for this locus permitting its isolation and characterization. *In vitro* culture systems (Dexter or Witte-Whitlock cultures) and *in vivo* models (SCID, W/W^V) exist for the propagation of hematopoietic cells, making this approach technically feasible.

ACKNOWLEDGEMENTS

This work was supported by NIH grants CA42509, HD01287, and CA119512. W. G. K. is a Fellow of the Irvington Institute for Medical Research. W. G. K. also acknowledges the support of the NIH Training Program in Immunology AI-07290.

REFERENCES

1. M. D. Cooper and P. D. Burrows, B-cell differentiation, in: "Immunoglobulin Genes," T. Honjo, F. Alt, and T. Rabbitts, eds., Academic Press Limited, London (1989).
2. W. G. Kerr, M. D. Cooper, L. Feng, P. D. Burrows, and L. M. Hendershot, Mu heavy chains can associate with a pseudo-light chain complex (ΨL) in human pre-B cell lines, *Internat. Immunol.* 1:355 (1989).
3. M. E. Koshland, The immunoglobulin helper: the J chain, in: "Immunoglobulin Genes," T. Honjo, F. Alt, and T. Rabbitts, eds., Academic Press Limited, London (1989).
4. W. G. Kerr, G. P. Nolan, and L. A. Herzenberg, *In situ* detection of transcriptionally-active chromatin and genetic regulatory elements in individual viable mammalian cells, *Immunology* 68 (Suppl. 2):74 (1989).
5. W. G. Kerr, G. P. Nolan, A. T. Serafini, and L. A. Herzenberg, Transcriptionally-defective retroviruses containing *lacZ* for the *in situ* detection of endogenous genes and developmentally-regulated chromatin, *CSHSQB* 54:767 (1989).
6. N. D. Allen, D. G. Can, S. C. Barton, S. Hettle, and M. A. Surami, Transgenes as probes for active chromosomal domains in mouse development, *Nature* 333:852 (1988).
7. A. Gossler, A. L. Joyner, J. Rossant, and W. C. Skarnes, Mouse embryonic stem cells and reporter constructs to detect developmentally regulated genes, *Science* 244:463 (1989).
8. C. J. O'Kane and W. J. Gehring, Detection *in situ* of genomic regulatory elements in *Drosophilia, Proc. Natl. Acad. Sci. USA* 84:9123 (1987).
9. H. J. Bellen, C. J. O'Kane, C. Wilson, U. Grossniklaus, R. K. Pearson, and W. J. Gehring, P-element-mediated enhancer detection: a versatile method to study development in *Drosophilia, Genes and Devel.* 3:1288 (1989).

10. C. Wilson, R. K. Pearson, H. J. Bellen, C. J. O'Kane, U. Grossniklaus, and W. J. Gehring, P-element-mediated enhancer detection: an efficient method for isolating and characterizing developmentally regulated genes in *Drosophilia, Genes and Devel.* 3:1301 (1989).

11. S. F. Yu, T. von Ruden, P. W. Kantoff, C. Garber, M. Sieberg, U. Ruther, W. F. Anderson, E. F. Wagner, and E. Gilboa, Self-inactivating retroviral vectors designed for transfer of whole genes into mammalian cells, *Proc. Natl. Acad. Sci. USA* 83:3194 (1986).

12. H. E. Varmus, Form and function of retroviral proviruses, *Science* 216:812 (1982).

13. R. Mann and D. Baltimore, Varying the position of a retrovirus packaging sequence results in the encapsidation of both unspliced and spliced RNAs, *J. Virol.* 54:401 (1985).

14. P. A. Lazo, V. Prasad, and P. N. Tsichlis, Splice acceptor site for the *env* message of Moloney murine leukemia virus, *J. Virol.* 61:2038 (1987).

15. C. C. Shih, J. P. Stoye, and J. M. Coffin, Highly preferred targets for retrovirus integration, *Cell* 53:531 (1988).

16. A. Winoto and D. Baltimore, $\alpha\beta$ lineage-specific expression of the α T cell receptor gene by nearby silencers, *Cell* 59:649 (1989).

17. F. Melchers, The many roles of immunoglobulin molecules and growth control of the B-lymphocyte lineage, *in* "Immunoglobulin Genes," T. Honjo, F. Alt, and T. Rabbitts, eds., Academic Press Limited, London (1989).

18. A. N. Shakhov, M. A. Collart, P. Vassali, S. A. Nedospasov, and C. V. Jongeneel, κB-type enhancers are involved in lipopolysaccharide-mediated transcriptional activation of the tumor necrosis factor α gene in primary macrophages, *J. Exp. Med.* 171:35 (1990).

19. R. Sen and D. Baltimore, Multiple nuclear factors interact with the immunoglobulin enhancer sequences, *Cell* 46:705 (1986).

20. R. Sen and D. Baltimore, Inducibility of κ immunoglobulin enhancer-binding protein NF-κB by a post-translational mechanism, *Cell* 47:921 (1986).

21. R. R. Hardy, K. Hayakawa, D. R. Parks, L. A. Herzenberg, and L. A. Herzenberg, Murine B cell differentiation lineages, *J. Exp. Med.* 159:1169 (1984).

22. R. L. Coffman and I. L. Weissman, B220, a B cell specific member of the T200 glycoprotein, *Nature* 289:681 (1981).

23. M. Briskin, M. Kuwabara, D. Sigman, and R. Wall, Induction of κ transcription by interferon-γ without activation of NF-κB, *Science* 242:1036 (1988).

24. G. Lee, L. R. Ellingsworth, S. Gillis, R. Wall, and P. W. Kincade, β transforming growth factors are potential regulators of B lymphocytes, *J. Exp. Med.* 166:1290 (1987).

25. T. Honjo, A. Shimuzu, and Y. Yaoita, Constant-region genes of the immunoglobulin heavy chain and the molecular mechanism of heavy chain class switching, *in* "Immunoglobulin Genes," T. Honjo, F. Alt, and T. Rabbitts, eds., Academic Press Limited, London (1989).

26. M. R. Lieber, J. E. Hesse, K. Mizuuchi, and M. Gellert, Developmental stage specificity of the lymphoid V(D)J recombination activity, *Genes and Devel.* 1:451 (1987).

27. S. Desiderio, G. Yancopolous, M. Paskind, E. Thomas, M. Boss, N. Landau, F. Alt, and D. Baltimore, Insertion of N-regions into heavy-chain genes is correlated with expression of terminal deoxynucleotide transferase in B cells, *Nature* 311:752 (1984).

28. D. G. Schatz, M. A. Oettinger, and D. Baltimore, the V(D)J recombination activating gene, RAG-1, *Cell* 59:1035 (1989).

29. J. Hombach, T. Tsubata, L. Leclerq, H. Stappert, and M. Reth, Molecular components of the B-cell antigen receptor complex of the IgM class, *Nature* 343:760 (1990).

30. J. Wienands, J. Hombach, A. Radbruch, C. Riesterer, and M. Reth, Molecular components of the B cell antigen receptor complex of IgD differ partly from those of IgM, *EMBO* 9:449 (1990).

31. J. Chen, A. M. Stall, L. A. Herzenberg, and L. A. Herzenberg, Differences in glycoproteins complexes associated with IgM and IgD on normal murine B cells potentially enable transduction of different signals (1990, submitted).

32. G. P. Nolan, S. Fiering, J. F. Nicolas, and L. A. Herzenberg, Fluorescence-activated cell analysis and sorting of viable mammalian cells based on β-D-galactosidase activity after transduction of *Escherichia coli lacZ*, *Proc. Natl. Acad. Sci. USA* 85:2603 (1988).

B CELL DEVELOPMENT IN FETAL LIVER

Fritz Melchers, Andreas Strasser, Steven R. Bauer, Akira Kudo,
Philipp Thalmann, and Antonius Rolink

Basel Institute for Immunology
Basel, Switzerland

Embryonic development of B lymphocytes in the mouse is remarkably well programmed in time of gestation. A first wave of development in embryonic blood and placenta[1,2] reaches its peak of mitogen-reactive precursors at day 12 of gestation. This is followed by a second wave in fetal liver, which reaches its peak of mitogen-reactive B cells around birth, i.e., at day 19 of gestation.[3,4] This development also occurs when embryonic bodies with blood islands and beating heart cells develop *in vitro* from embryonic stem cells (U. Chen and F. Melchers, manuscript in preparation). Since mitogen-reactive precursors develop from embryonic stem cells with much the same time schedule *in vitro* as they do *in vivo* cell cycle times, numbers of divisions and differentiation steps along the embryonic B lymphocyte development must be tightly controlled. We have investigated B cell development in fetal liver of the mouse on the cellular and molecular level in order to understand how such ordered development in time of gestation is achieved.

A series of B lymphocyte lineage markers and functions are used to study B cell developments. A protein plaque assay[5] can be used to quantitate the number of Ig-secreting cells which develop at the end of the B lymphocyte lineage pathways upon stimulation, usually by antigens or mitogens. Mitogen-reactive, surface Ig (sIg) positive precursors can be evaluated with this plaque test in cultures with the polyclonal activator lipopolysaccharide (LPS) and rat thymus cells as "filler" cells which allow the quantitation of the number of mitogen-reactive B cells in limiting dilution analysis with efficiencies of plating the reactive precursors near 100%.[6] SIg-positive B cells and B lineage cells expressing intra-cytoplasmic μ heavy (H) chains can be quantitated by immunofluorescence. Cells expressing certain genes of interest, as mRNA molecules, can be quantitated by *in situ* hybridizations with radioactive gene probes, as long as the expression is in the range of 50 mRNA molecules per cell or higher.[7] Detection and quantitation of cells expressing Ig genes and pre B cell-specific genes such as V_{preB} and λ [8-10] are particularly interesting at early stages of B cell development. Earlier stages of B lineage cells are furthermore identifiable by the monoclonal antibody (mAb) G-5-2 which recognizes a 76kd glycoprotein[11] recently identified as an alkaline phosphatase.[12] The mAb G-5-2 allows enrichment of precursor B cells by fluorescence-activated cell sorting (FACS). Enrichment of G-5-2 positive cells leads to concomitant enrichment of cells expressing V_{preB} and λ_5 and, not in early but in later stages of pre B cell development, μH chain expressing cells.[11,13,14] Therefore, the mAb G-5-2 appears to recognize cells committed to the B lineage before and after μH chain expression.

The three Ig gene loci are rearranged successively.[15] First, D_H segments are rearranged with J_H segments, then V_H segments with $D_H J_H$ segments followed by V_K to J_K and, finally, $V\lambda$ to $J\lambda$. Rearranged chromosomes can be distinguished from nonrearranged, germ-line

Mechanisms of Lymphocyte Activation and Immune Regulation III
Edited by S. Gupta *et al.*, Plenum Press, New York, 1991

201

chromosomes in single cells by the polymerase chain reaction (PCR) (D. Haasner and F. Melchers, in preparation).

Our study of B cell development in fetal liver from day 12 until day 19 of gestation makes use of these methods. It has been observed early on[3,4] that pre B cell development uses the same time schedule *in vitro* as it does *in vivo*. As a consequence, fetal liver cells taken at different days of gestation into tissue culture will complete their programs of development with the same timed programs as if they had remained *in vivo* in fetal liver.

A more thorough reinvestigation of this development of precursor B cells in fetal liver, and in bone marrow *in vitro* has clarified several issues. It has become evident that pre B to B cell development depends on interactions of these precursors with stromal cells of the microenvironment of B cell-generating organs such as fetal liver and bone marrow.[16-30] Stromal cell lines have been developed which support this B cell development. Interleukin (IL-) -7 has been implicated as an important factor in this development.[31] The *in vitro* conditions for precursor cell differentiation have allowed a change from progenitors to B cells within several weeks. It has remained difficult to quantitate the number of progenitors and precursors at a given stage of B cell differentiation in these cultures.

We enriched G-5-2 positive cells from day 13, 14 and 15 of gestation from fetal liver by FACS and cultured them at medium (2×10^5/ml) to high (2×10^6/ml) densities.[14] They did not develop into sIg$^+$ mitogen-reactive cells within a week. At day 16 of gestation fetal liver cells became sIg$^+$ and can be cultured with LPS and rat thymus cells to develop into clones of IgM PFC at high frequencies. G-5-2 positive cells from day 16 to 19 of gestation were around 5% sIg positive, and one in 30 to 100 cells was a precursor of a mitogen-reactive B cell developing into a clone of IgM PFC. Thus, there was a rapid change between day 15 and 16 of gestation in fetal liver *in vivo* from undetectable levels of sIg$^+$ cells to 5% sIg$^+$ cell in the G-5-2 positive population, with a concomitant change of mitogen-reactive cells from less than 1 in 3000 to 1 in 30 to 100.

Since unseparated fetal liver cells of early (13 to 15) days of gestation (containing all possible cells including stromal cells) could become sIg$^+$ and mitogen-reactive *in vitro* if they were kept at high densities in tissue culture beyond the time equivalent to day 16 of gestation, we attempted to support the *in vitro* development of G-5-2 positive precursors of day 13 and 14 fetal liver with irradiated embryonic stromal cell layers, or with irradiated stromal cell lines capable of induction to B cell differentiation, such as the ST-2 cell line.[25,26,30] Indeed, when the day 13 or 14 G-5-2$^+$ cells were cultured on layers of stromal cells of either type, sIg$^+$ cells appeared *in vitro* at a time equivalent to day 16 of gestation, and the cells became stimulatable to clones of IgM PFC in cultures containing LPS and rat thymus cells. Again, within 24 hours of culture at the time equivalent to day 16 of gestation, 4 to 11% of the cultured G-5-2$^+$ cells became sIg$^+$ and the frequencies of mitogen-reactive precursors changed from less than 1 in 3000 to 1 in 30 to 100. Beyond these 24 hours there was no further increase in the number of sIg$^+$ or of mitogen-reactive precursors in the time equivalent to day 16 to 19 of gestation. Addition of IL-1 to IL-7 in all possible combinations to these cultures did not change the outcome of the experiment. The number of cells in culture increased no more than a factor of 4 between day 13 or 14 and the time experiment to day 17 or 18, when the cells were cultured on stromal cell layers. This indicates that proliferation of precursors is limited under these conditions, and may not exceed 3 to 4 divisions. In summary, these results show that stromal cells can provide the environment for early progenitors to develop into B cells within 3 days without extensive proliferation. Exogenously added IL-7 appears not to influence this development.

The development appears to occur in one synchronous wave in which all cells are at the same stage of development at any given time of gestation, and this wave remains synchronous when continued *in vitro* on stromal cells. Our results also suggest that preB cell development occurs in two phases. A first phase between day 13 and 16 is stromal cell-dependent. At this stage Ig rearrangements occur. Monoclonal antibodies which inhibit adherence of progenitors to stromal cells inhibit B cell development in this phase.[14] This is followed by a second phase between day 16 and 19 which is stromal cell-independent, where cells are already sIg$^+$ and which ends in the cells becoming mitogen-reactive at day 18 to 19.[4,14]

When G-5-2$^+$ cells of day 14 fetal liver were grown in the presence of different interleukins on a series of established stromal cell lines, we found that the preadipose cell line MC3T3-G2/PA6 (called PA6)[27-29] propagated continuous cell proliferation in the presence of IL-7 (Melchers et al., in preparation). With a cell cycle time of around 24 hours G-5-2$^+$ cells grew within 3 weeks to 10^8 times the original number of cells. The proliferating cells were found to be B220$^+$, BP-1$^+$, sIg$^-$ and could not be induced by LPS in rat thymus cells to IgM PFC. When these proliferating cells were transferred to ST-2 stromal cells (which are capable of inducing B cell differentiation), the G-5-2$^+$ cells showed only limited proliferation even in the presence of exogenously added IL-7. Within 3 days, however, 5-10% of the cells cultured on ST-2 became sIg$^+$ and mitogen-reactive. At the same time they lost their capacity to grow on PA-6 in the presence of IL-7. We think that the G-5-2$^+$ cells growing on PA-6 plus IL-7 are committed B progenitors which can be induced to rearrange and express Ig genes when put in contact with ST-2 stromal cells. We estimate that around 40% of the G-5-2$^+$ cells of day 14 are PA-6/IL-7-reactive cells, while in day 15 and 16 fetal liver, none were detectable in these growth assays.

The following picture of B cell development in fetal liver emerges. Development from pluripotent stem cells reaches a stage of B lineage-commitment at around day 12 to 14 of gestation where B220$^+$, BP-1$^+$, G-5-2$^+$, sIg$^-$ progenitors exist which are capable of extensive, if not unlimited, proliferation in a fashion of B lineage-committed stem cells. For symmetric divisions into two identical progenitor daughter cells they need the contact to stromal cells of type I (PA-6) and IL-7 as cytokine. When the same cells come in contact with stromal cells of type II (represented by ST-2), they undergo a limited number of divisions (estimated between 2 and 5) and differentiate into sIg$^+$ cells. During embryonic development in fetal liver, this occurs synchronously for all cells from day 13 onwards. Between day 13 and 16 all Ig loci are rearranged in a successive fashion. Induction of these rearrangements needs the contact with stromal cells of type II. Since not all rearrangements are expected to be productive, due to the use of pseudogenes, and as a consequence of out-of-frame rearrangements, only a part of all cells emerging in this differentiation are sIg$^+$ and mitogen-reactive. Over 90% of all differentiating pre B cells may have chromosomes with non-productively rearranged Ig genes (either H and/or L chains). This number is at least within the range of frequencies of productive versus nonproductive rearrangements expected from a stochastic process of rearrangement in which possible corrections from nonproductive to productive rearrangements are not the rule.

It is tempting to speculate that the four successive Ig gene rearrangements ($D_H \rightarrow J_H$, $V_H \rightarrow D_H J_H$, $V_K \rightarrow J_K$, $V\lambda \rightarrow J\lambda$) are carried by four critical cell divisions that progenitor B cells carry out in contact with stromal cells of type II. We have proposed[13] that the two pre B cell-specific genes V_{preB} and λ_5 produce proteins, which form a L-chain-like complex which can associate with μH chains in pre B cells. These and other possible complexes of V_{preB} and λ_5 with yet unidentified H-chain-like proteins make contact on the surface of pro and pre B cells with distinct stromal cell determinants. One of these contacts always only induces one round of division, the next successive Ig gene rearrangement and, with the product of this rearrangement, a new complex on the surface of the pre B cell ready for a different contact with stromal cells. These hypotheses are now testable with B progenitor cell lines and stromal cell lines.

Continuously growing progenitor B cell lines are ideal targets for transfection of genes with possible functions in B cell development and growth control. They open new and exciting experimental possibilities to understand the cellular and molecular controls in B cell responses.

ACKNOWLEDGMENTS

The Basel Institute for Immunology was founded and is supported by F. Hoffman-La Roche Ltd., CH-4002 Basel.

REFERENCES

1. F. Melchers, Murine embryonic B lymphocyte development in the placenta, *Nature* 277:219 (1979).

2. F. Melchers and J. Abramczuk, Murine embryonic blood between day 10 and 13 of gestation as a source of immature precursor B cells, *Eur. J. Immunol.* 10:763 (1980).

3. F. Melchers, B lymphocyte development in fetal liver. I. Development of reactivities to B cell mitogens *in vivo* and *in vitro, Eur. J. Immunol.* 7:476 (1977).

4. F. Melchers, B lymphocyte development in fetal liver. II. Frequencies of precursor B cells during gestation, *Eur. J. Immunol.* 7:482 (1977).

5. E. Gronowicz, A. Coutinho, and F. Melchers, A plaque assay for all cells secreting Ig of a given type or class, *Eur. J. Immunol.* 6:588 (1976).

6. J. Andersson, A. Coutinho, W. Lernhardt, and F. Melchers, Clonal growth and maturation to immunoglobulin secretion *in vitro* of every growth-inducible B lymphocyte, *Cell* 11:27 (1977).

7. C. N. Berger, *In situ* hybridization of immunoglobulin-specific RNA in single cells of the B lymphocyte lineage with radio-labelled DNA probes, *EMBO J.* 5:85 (1986).

8. N. Sakaguchi and F. Melchers, λ_5, a new light-chain-related locus selectively expressed in pre-B lymphoyctes, *Nature* 324:579 (1986).

9. A. Kudo, N. Sakaguchi, and F. Melchers, Organization of the murine Ig-related λ_5 gene transcribed selectively in pre-B lymphocytes, *EMBO J.* 6:103 (1987).

10. A. Kudo and F. Melchers, A second gene, V_{preB} in the 15 locus of the mouse, which appears to be selectively expressed in pre-B lymphocytes, *EMBO J.* 6:2267 (1987).

11. A. Strasser, PB76: a novel surface glycoprotein preferentially expressed on mouse pre-B cells and plasma cells detected by the monoclonal antibody G-5-2, *Eur. J. Immunol.* 18:1803 (1988).

12. C. Marquez, A. de la Hera, E. Leonardo, L. Pezzi, A. Strasser, and A. C. Martínez, Identity of PB76 differentiation antigen and lymphocyte alkaline phosphatase, *Eur. J. Immunol.* 20:947 (1990).

13. F. Melchers, A. Strasser, S. R. Bauer, A. Kudo, P. Thalmann, and A. Rolink, Cellular stages and molecular steps of murine B cell development, *Cold Spring Harb. Symp. Quant. Biol.* 54:183 (1990).

14. A. Strasser, T. Rolink, and F. Melchers, One synchronous wave of B cell development in mouse fetal liver changes at day 16 of gestation from dependence to independence of a stromal cell environment, *J. Exp. Med.* 170:1973 (1989).

15. S. Tonegawa, Somatic generation of antibody diversity, *Nature* 302:575 (1983).

16. P. W. Kincade, Experimental models for understanding B lymphocyte formation, *Adv. Immunol.* 41:181 (1987).

17. P. W. Kincade, G. Lee, C. W. Paige, and M. P. Scheid, Cellular interactions affecting the maturation of murine B lymphocyte precursors *in vitro, J. Immunol.* 127:255 (1981).

18. J. W. Coleman, J. H. K. Yeung, M. D. Tingle, and B. K. Park, Enzyme-linked immunosorbent assay (ELISA) for detection of antibodies to protein-reactive drugs and metabolites: criteria for identification of antibody activity, *J. Immunol. Meth.* 88:37 (1986).

19. C. Whitlock, K. Denis, D. Robertson, and O. N. Witte, *In vitro* analysis of murine B-cell development, *Ann. Rev. Immunol.* 3:213 (1985).

20. G. H. Gisler, A. Söderberg, and M. Kamber, Functional maturation of murine B lymphocyte precursors. II. Analysis of cells required from the bone marrow microenvironment, *J. Immunol* 138:2433 (1987).

21. P. L. Witte, M. Robertson, A. Henley, M. G. Low, D. L. Stiers, S. Perkins, R. A. Fleischman, and P. W. Kincade, Relationship between B-lineage lymphocytes and stromal cells in long-term bone marrow cultures, *Eur. J. Immunol.* 17:1473 (1987).

22. T. Kinashi, K. Inaba, T. Tsubata, K. Tashiro, R. Palacios, and T. Honjo, Differentiation of an interleukin 3-dependent precursor B cell clone into immunoglobulin-producing cells *in vitro, Proc. Natl. Acad. Sci. U. S. A.* 85:4473 (1988).

23. L. E. Pietrangeli, S.-I. Hayashi, and P. W. Kincade, Stromal cell lines which support lymphocyte growth: characterization, sensitivity to radiation and responsiveness to growth factors, *Eur. J. Immunol.* 18:863 (1988).

24. R. Palacios, S. Stuber, and A. Rolink, Epigenetic influences of bone marrow stroma and liver stroma on the developmental potential of LY1 plus pro-B lymphocytes, *Eur. J. Immunol.* 19:347 (1989).

25. S. I. Nishikawa, M. Ogawa, S. Nishikawa, T. Kunnisada, and H. Kodama, B lymphopoiesis on stromal cell clone: stromal cell clones acting on different stages of B cell differentiation, *Eur. J. Immunol.* 18:1767 (1988).

26. M. Ogawa, S. Nishikawa, Y. Ikuta, F. Yamamura, M. Naito, K. Takahashi, and S. I. Nishikawa, B cell ontogeny in murine embryo studied by a culture system with the monolayer of a stroma cell clone, ST2: B cell progenitor develops first in the embryonal body rather than in the yolk sac, *EMBO J.* 7:1337 (1988).

27. H. A. Kodama, Y. Amagai, H. Koyama, and S. Kasai, A new preadipose cell line derived from newborn mouse calvaria can promote the proliferation of pluripotent hemopoietic stem cells *in vitro, J. Cell. Physiol.* 112:89 (1982).

28. H. A. Kodama, H. Sudo, H. Koyama, S. Kasai, and S. Yamamoto, *In vitro* hemopoiesis within a microenvironment created by MC3T3-G2/PA6 preadipocytes, *J. Cell. Physiol.* 118:233 (1984).

29. H. A. Kodama, H. Hagiwara, H. Sudo, Y. Amagai, T. Yokota, N. Arai, and Y. Kitamura, MC3T3-G2/PA6 preadipocytes support *in vitro* proliferation of hemopoietic stem cells through a mechanism different from that of interleukin 3, *J. Cell. Biol.* 129:20 (1986).

30. T. Sudo, M. Ito, Y. Ogawa, M. Lizuka, A. Kodama, T. Kumisada, S.-I. Hayashi, M. Ogawa, K. Sakai, S. Nishikawa, and S. I. Nishikawa, Interleukin 7 production and function in stromal cell-dependent B cell development, *J. Exp. Med.* 170:338 (1989).

31. A. E. Namen, S. Lupton, K. Kjerrild, J. Urignall, D. Y. Mochizuki, A. Schmierer, B. Mosley, C. J. March, D. Urdai, S. Gillis, D. Cosman, and R. G. Goodwin, Stimulation of B-cell progenitors by cloned murine interleukin-7, *Nature* 333:571 (1988).

IDENTIFICATION OF COMPONENTS OF THE B CELL ANTIGEN RECEPTOR COMPLEX

Michael Reth, Jürgen Wienands, Takeshi Tsubata , and Joachim Hombach

Max-Planck-Institut for Immunology
Freiburg, Germany

INTRODUCTION

Immunoglobulins (Ig) are well-known as antibody molecules which are present in the serum of all vertebrate species. The mouse has five classes of Ig (IgM, IgD, IgG, IgA and IgE) and most of these antibody classes have different effector function during an immune response.[1] Heavy chains of each Ig class exist also in a membrane-bound form[23] which is expressed on the B cell surface as part of the antigen receptor. During B cell development the different antigen receptors are expressed in an ordered fashion.[4] Immature B cells carry only surface Ig (sIg) of the IgM class[5] while mature B cells co-express sIgM and sIgD.[6,7] Upon activation some B cells are generated which after a class switch express sIgG, sIgA or sIgE. Cells with these antigen receptors are dominant in the pool of memory B cells.

In the presence of appropriate T cell help the B cells are activated by the cross-linking of their antigen receptor, which results in B cell proliferation and differentiation. Cross-linking of the B cell antigen receptor is coupled to the generation of the second messengers diacylglycerol and inositol-1, 4, 5-tri-phosphate and to calcium mobilization.[8] How the cross-linked antigen receptor generates the activation signal is presently unknown. The mode of activation was a puzzle so far because membrane-bound Ig only have a short cytoplasmic tail which is 3 amino acids long for IgM and IgD and between 28 to 14 amino acids for the other Ig classes.[9,10]

We have found that Ig heavy chains of class IgM and IgD are associated in the membrane with a heterodimer formed by two transmembrane molecules. These heterodimeric molecules are similar to components of the CD3 complex which are associated with the T cell receptor (TCR) chains. In this report we describe the novel components of the B cell antigen receptor and present a comparative model of the antigen receptors of the B and T cell.

MOLECULAR REQUIREMENTS FOR sIg EXPRESSION CAN BE TESTED IN MYELOMA CELLS

We have previously constructed various μ-vectors and used them to test the regulatory role of μ-chain in pre-B cells.[11] One of these vectors (pSVμm) expresses only the membrane-bound form of μm-chain (μm). Introduction of pSVμm and a light chain vector into pre-B cells and B lymphoma cells results in sIgM expression. However, if the μm-vector is introduced into the λ1 light chain-producing myeloma J558L,[12] membrane-bound IgM molecules are assembled but not transported onto the cell surface.[13,14] Thus the J558L myeloma cells seem to differ from other types of B cells in that they lack one or several factors required for sIg expression. Indeed, membrane-bound IgD molecules are also not expressed on the cell

Mechanisms of Lymphocyte Activation and Immune Regulation III
Edited by S. Gupta *et al.*, Plenum Press, New York, 1991

207

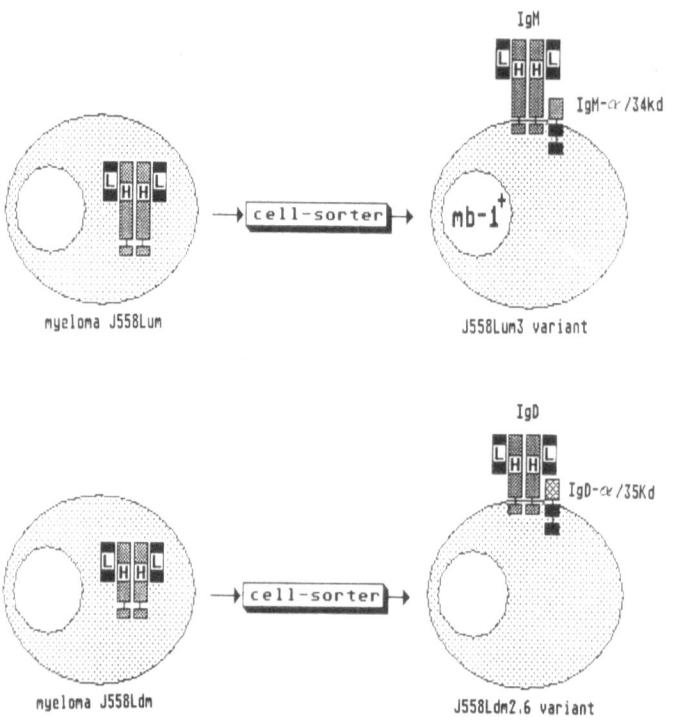

Figure 1. Selection of sIg-positive myeloma variants by cell sorting.

surface of δm-transfected J558L myeloma cells (see below). With a fluorescence-activated cell sorter (FACS) we subsequently isolated sIg$^+$ variants from J558Lμm transfectants (Figure 1). These variant lines, J558Lμm3 and J558Lδm2.6, demonstrate that myeloma cells are in principle able to express sIg and allowed us to analyze the molecular requirement for surface Ig expression.

IN sIg-EXPRESSING B CELLS THE MEMBRANE-BOUND Ig HEAVY CHAINS ARE ASSOCIATED WITH A HETERODIMER

In our comparative analysis of sIg$^-$ and sIg$^+$ myeloma cells we followed the idea that membrane-bound Ig heavy chain could be similar to the TCR chains in that they need associated protein for their proper surface expression. Indeed, on T cells, the antigen receptor chains are non-covalently associated with the components of the CD3 complex.[15,16] In T cells which have loss of one of the CD3 components, the TCR molecules are not transported onto the cell surface.[17]

The complete TCR/CD3 complex can be purified from T cells only if the cells are lysed with a mild detergent like digitonin.[18] To analyze the J558Lμm3 myeloma line for a putative IgM antigen receptor complex we therefore lysed biosynthetically labelled cells with digitonin and isolated their antigen receptor molecules by affinity chromatography. This very efficient purification method could be applied because the heavy chains expressed from our different vectors contained the VH region of the antibody B1-8,[19] and thus together with a λ1 light chain generate a binding site for the hapten 4-hydroxy-3-nitro-phenol (NP). The purified receptor material was analyzed on two-dimensional (nonreducing/reducing) SDS gels. In this analysis we found a 34-kD glycoprotein (IgM-α) which is non-covalently associated with the IgM molecule only in sIgM-expressing myeloma and B lymphoma cells.[20] In the sIgM$^-$ myeloma, J558Lμm, the IgM-α protein was not copurified with the IgM molecule. The IgM-α protein is part of a disulfide-linked dimer because in 2-D gels it lies below the diagonal. IgM-α was first interpreted as a homodimer. However, a close inspection of

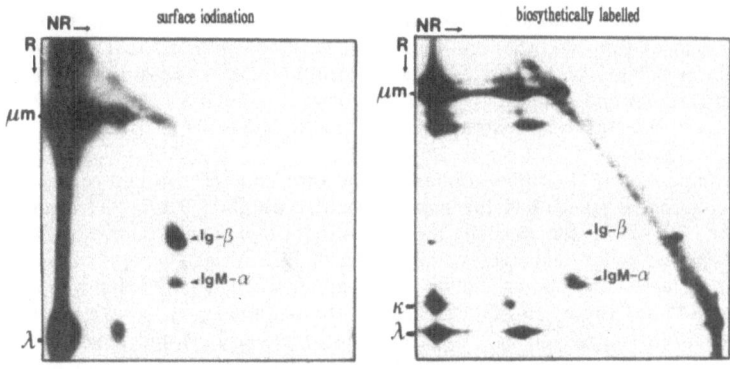

Figure 2. 2D (non-reducing [NR] - reducing [R]) gel analysis of affinity-
purified IgM antigen receptor of the K46λμm transfectant.

the 2-D gels revealed a faint protein band of 39 kD above IgM-α. The 39-kD protein which
we called Ig-β was more prominent in a 2-D gel/analysis of surface-iodinated IgM antigen
receptor.[21] In this analysis Ig-β was beside the Ig chains, μm and λ1, the most dominantly
labelled protein while IgM-α was only weakly labelled (Figure 2).

These data indicate that together with Ig-β, IgM-α forms a heterodimer which is non-
covalently associated with the membrane-bound IgM molecule. The different efficiency
with which these two IgM-associated proteins were labelled by the two labelling procedures
may indicate that inside the cell Ig-β has a slower turnover rate than IgM-α and that Ig-β
is more exposed on the B cell surface than IgM-α.

THE B CELL ANTIGEN RECEPTORS OF CLASS IgM AND IgD DIFFER IN THE α-COMPONENTS OF THEIR RESPECTIVE HETERODIMER

Mature B cell co-express two different classes of antigen receptors, IgM and IgD.[6]
On the surface of these cells IgD is the dominant class and is ten times more abundant than
IgM. The IgD antigen receptor was therefore analyzed next[22] with similar methods to those
used previously for the IgM antigen receptor. A vector specific for the membrane-bound
form of the δ-chain (PSVδm) was introduced either into the myeloma line J558L or into
K46λ, a B lymphoma line which we had previously transfected with a λ1 light chain
vector. After transfection the IgD molecule was expressed on the surface of the K46λδm
but not on the surface of the J558Lδm transfectant. This result suggested that for its
proper expression on the cell surface IgD molecules also need associated proteins, which are
not expressed in the J558L myeloma. Using the FACS to sort the J558Lδm transfectant we

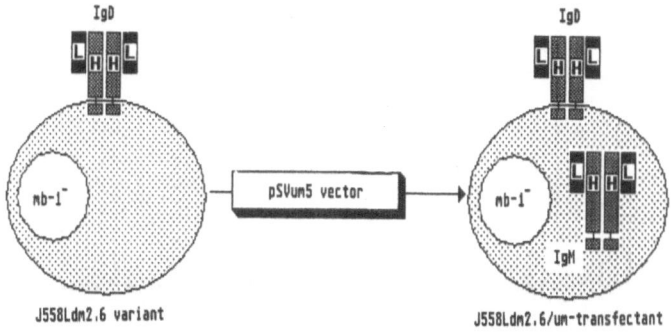

Figure 3. Lack of sIgM expression in the J558Lδm2.6/μm double
transfectant.

obtained sIgD$^+$ variant line J558Lδm2.6 (Figure 3). The IgD antigen receptors of this variant line and that of the two other δm-transfectants (K46λδm and J558Lδm) were analyzed for IgD-associated protein by 2-D gel electrophoresis. A heterodimer consisting of a 35-kD protein (IgD-α) and a 39-kD protein (Ig-β) was co-purified together with the IgD molecules only from the sIgD-expressing cell lines (J558Lδm2.6 and K46λδm).

A direct comparison of the heterodimers of the IgM and IgD antigen receptor showed that the upper component (Ig-β) had the same molecular weight (39 kD). Furthermore, in both receptors the 39-kD protein was the most prominent component of the surface-labelled heterodimer. Together these data suggest that IgM and IgD antigen receptors have the same Ig-β component. The α-components, however, were of different molecular weight for each heterodimer (34 kD versus 35 kD) and seem to be isotype specific. We therefore called the two α-proteins IgM-α and IgD-α. The conclusion that the IgM and the IgD antigen receptors are each associated with their specific heterodimer (IgM-α/Ig-β and IgD-α/Ig-β) is also supported by a transfection experiment in which we introduced the pSVμm vector into the sIgD$^+$ variant J558Lδm2.6 (Figure 3). The obtained double transfectant J558Lδm2.6/μm expresses only the IgD but not the IgM molecules on the cell surface and thus suggests that the two receptors have different requirements for cell surface expression.

IgM-α IS LIKELY TO BE ENCODED BY THE B CELL-SPECIFIC GENE mb-1

The mb-1 gene was isolated by N. Sakaguchi et al.[23] from a subtractive cDNA library which selected for B cell (versus T cell) specific genes. The mb-1 gene encodes a transmembrane protein containing 137-a.a. extracellular portion, a 22-a.a. transmembrane and a 61 a.a. cytoplasmic part. Transcripts of the mb-1 gene are found in pre-B and B cells but not in myeloma cells. This expression pattern of mb-1 was reminiscent of that of sIgM in the different B cell lines we have studied. To elucidate whether there was a correlation between mb-1 and sIgM expression, we tested polyA$^+$ RNA of the different J558L myeloma lines with an mb-1 probe. No transcripts of mb-1 were found in J558L or in J558Lμm. However, the sIgM$^+$ variant line J558Lμm3 contained mb-1 transcripts suggesting that mb-1 encodes a protein required for sIgM expression. This assumption was proven in experiments where we introduced expression vectors of either the mouse or the human mb-1 gene into the J558Lμm line. In the mouse mb-1 transfectants J558Lμm/mb-1m IgM molecules were expressed on the cell surface as strongly as in the J558Lμm3 variant (Figure 4B and D). The sIgM expression in human mb-1 transfectant J558Lμm/mb-1h was 5 to 10 times less than in J588Lμm/mb-1m (Figure 4C). In the mouse mb-1 transfectant J558Lμm/mb-1m the IgM molecules were again associated with the IgM-α heterodimer. We think that it is the IgM-

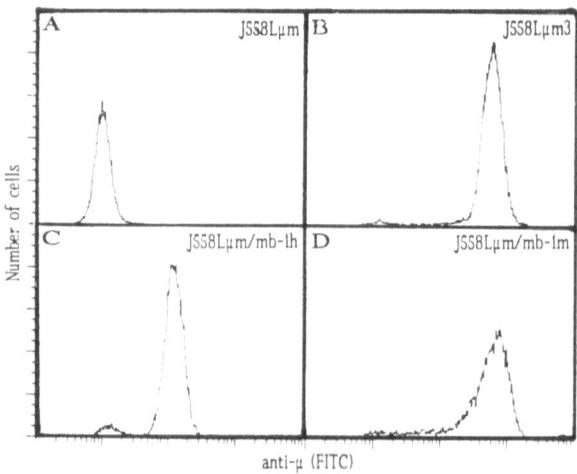

Figure 4. FACScan-analysis of sIgM$^+$ expression in
mb-1 transfectants of J5581μm.

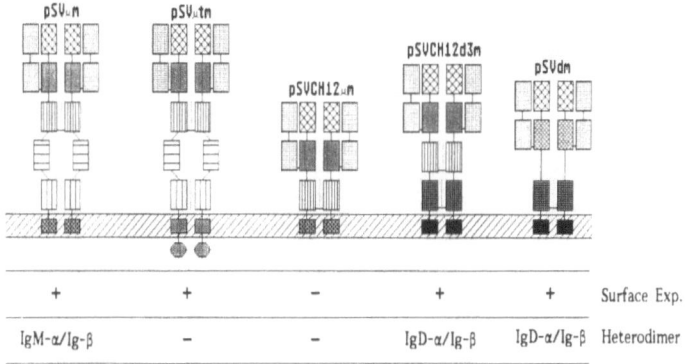

Figure 5. Structural requirement of various μm and δm heavy chains for their expression on the K46λ cell surface and binding to their respective heterodimer.

specific component (IgM-α) which is encoded by the mb-1 gene. First, the presumed molecular weight of the mb-1-encoded protein is closer to that of the α-component than to that of the β-component of the heterodimer. Second, mb-1 transcripts are found only in the sIgM$^+$ J558Lμm3 variant but not in the sIgD$^+$ J558Lδm2.6 variant, indicating that mb-1 encodes an sIgM-specific factor. Indeed, the absence of mb-1 transcripts may explain why IgM molecules are not transported onto the cell surface in J558Lδm2.6/μm double-transfectants (see Figure 3).

STRUCTURAL REQUIREMENTS FOR THE SPECIFIC ASSOCIATION BETWEEN Ig HEAVY CHAIN AND ITS RESPECTIVE HETERODIMER

In K46λμm and K46λδm transfectants, the IgM and IgD antigen receptor is expressed on the cell surface without a previous selection for surface expression. Thus, both types of heterodimers (IgM-α/Ig-β and IgD-α/Ig-β) are constitutively expressed in the K46 B lymphoma. However, each heterodimer is co-purified only together with its respective receptor indicating that they bind the Ig heavy chain of different classes in an isotype-specific fashion. To test what part of the μm or δm chain is involved in this binding we constructed various heavy chain vectors and introduced them into K46λ cells. The pSVμtm vector expresses a chimeric μ-chain in which the transmembrane part of μm was replaced by the transmembrane and cytoplasmic part of the class I H2-K^k molecule. Together with the λl light the μtm chain forms an IgM molecule which is transported on the K46λμtm cell surface without being associated with the IgM specific heterodimer (Figure 5). The pSVCH12μm vector carries a deletion of the CH3 and CH4 exons of the μ gene. This vector expresses a truncated IgM molecule which is not transported onto the cell surface and which does not associate with the heterodimer although it carries the proper μm transmembrane part.

If, however, the CH1 and CH2 domains of the μ chain are connected with the last (C-terminal) C domain (Cδ3) and the transmembrane part of the δm chain, a chimeric μ/δm molecule is expressed on the cell surface of the K46λCH12δ3m transfectant. This chimeric molecule is associated with the IgD specific heterodimer. Together these experiments indicate that both, the last C domain and the transmembrane part, are required for binding of the heavy chain to the heterodimer.

The transmembrane part of the μm and the δm chain have a very similar sequence.[10] The sequences of the last C domains of these two chains (Cμ4 and Cδ3) are evolutionary highly conserved[24] but they differ markedly from each other.[25] Therefore, most likely the membrane-proximal C domains are those which make the isotype specific contact to the respective α-component. Indeed, as judged from the mb-1 sequence, the 137 extracellular amino acids of IgM-α are folded into a C-like domain which may be in direct contact to the Cμ4 domain of the μ-chain (see below).

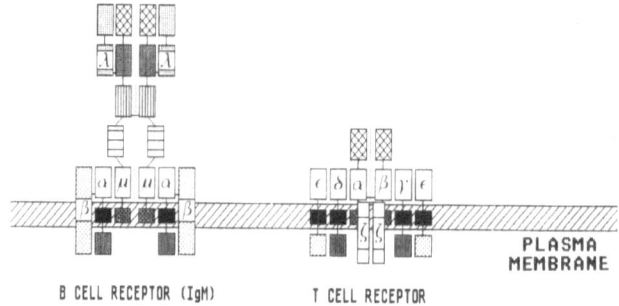

B CELL RECEPTOR (IgM)   T CELL RECEPTOR

Figure 6.   Structural model of B and T cell antigen receptors.

A COMPARATIVE MODEL FOR THE B CELL AND T CELL ANTIGEN RECEPTORS

We have shown that the IgM antigen receptor is a complex of different transmembrane proteins (IgM-α, Ig-β, $\mu2$) and thus similar to the T cell receptor (α, β, γ, δ, ϵ, $\zeta2$). In what follows, we want to compare the two antigen receptors and propose a structural model of these receptors (Figure 6). In the IgM antigen receptor the two μm-chains form a homo-dimer. The μm chains are in contact with each other via the CH_2 and CH_4 domain. This assumption can be made from the μm sequence and the known X-ray structure of the Fc portion of a human IgG2a antibody.[26] The side opposite to the $C\mu4{:}C\mu4$ interface is the most solvent-exposed area of the Fc part of the μ chain. This side also contains several conserved amino acids. We propose that this area of the $C\mu4$ domain is in contact with the IgM-α protein. Because the μm-chain forms a homodimer with two identical outside areas, the IgM-α protein should bind to both sides of the membrane-bound IgM molecule. Thus the IgM-α/Ig-β heterodimer should be present twice in the IgM antigen receptor.

In contrast to the IgM cell antigen receptor heavy chains, the TCR chains are hetero-dimers of either TCR-α/TCR-β or TCR-γ/TCR-δ chains. In the last years several biochemical studies[27,17,28] have found a spatial relationship between the TCR chains and the components of the CD3 complex. These data suggest a physical contact between: TCR-α chain and CD3-δ; TCR-β chain and CD3-γ; CD3-δ and CD3-ϵ as well as CD3-γ and CD3-ϵ. These data can be explained by a model (Figure 6) in which the α-β heterodimer is bound on each side by a non-covalent dimer of δ-ϵ and γ-ϵ. In this model the δ and γ components of the CD3 complex are mediating the specific contact to the α and β TCR-chains and thus would have a similar role to that of the IgM-α component of the B cell antigen receptor. Indeed, the cytoplasmic tails of CD3-γ and CD3-δ share an identical sequence motif with that of the IgM-α component.[29] As seen in our structural model, the antigen receptors of B and T cell would thus have similar structural features, which fits well with their similar functional behavior upon cross-linking. The TCR/CD3 complex, however, contains also a ζ-chain homodimer[30] whose analog we have not yet found in the B cell receptor.

REFERENCES

1.   J. L. Winkelhake, Immunoglobulin structure and effector functions, *Immunochemistry* 15:659 (1978).
2.   F. Alt, A. Bothwell, M. Knapp, E. Siden, E. Mather, M. Koshland, and D. Baltimore, Synthesis of secreted and membrane-bound immunoglobulin mu heavy chains is directed by mRNAs that differ at their 3' ends, *Cell* 20:293 (1980).
3.   P. Early, J. Rogers, M. Davis, K. Calame, M. Bond, R. Wall, and L. Hood, Two mRNAs can be produced from a single immunoglobulin μ gene by alternative RNA processing pathways, *Cell* 20: 313 (1980).
4.   E. Vitteta, E. Puré, P. Isakson, L. Buck, and J. Uhr, The activation of murine B cells: The role of surface immunoglobulins, *Immunol. Rev.* 52:211 (1980).

5. J. F. Kearney, M. D. Cooper, J. Klein, E. R. Abney, R. M. E. Parkhouse, and A. R. Lawton, Ontogeny of Ia and IgD on IgM-bearing B lymphocytes in mice, *J. Exp. Med.* 146:297 (1977).

6. J. W. Goding and J. E. Layton, Antigen-induced co-capping of IgM and IgD-like receptors on murine B cells, *J. Exp. Med.* 144:852 (1976).

7. E. S. Vitetta and J. W. Uhr, Cell surface immunoglobulin. XV. The presence of IgM and an IgD-like molecule on the same cell in murine lymphoid tissue, *Eur. J. Immunol.* 6:140 (1976).

8. J. C. Cambier and J. G. Monroe, B cell activation. V. Differentiation signaling of B cell membrane depolarization, increased I-A expression, G_0 to G_1 transition, and thymidine uptake by anti-IgM and anti-IgD antibodies, *J. Immunol.* 133:576 (1984).

9. Y. Yamawaki-Kataoka, S. Nakai, T. Miyata, and T. Honjo, Nucleotide sequences of gene-segments encoding membrane domains of immunoglobulin γ chains, *Proc. Natl. Acad. Sci. USA* 79:2623 (1982).

10. J. Rogers and R. Wall, Immunoglobulin RNA rearrangements in B lymphocyte differentiation, *Adv. Immunol.* 35:39 (1984).

11. M. Reth, E. Petrac, P. Wiese, L. Lobel, and F. W. Alt, Activation of Vκ gene rearrangement in pre-B cells follows the expression of membrane-bound immunoglobulin heavy chains, *EMBO J.* 6:3299 (1987).

12. V. T. Oi, S. L. Morrison, L. A. Herzenberg, and P. Berg, Immunoglobulin gene expression in transformed lymphoid cells, *Proc. Natl. Acad. Sci. USA* 80:825, 1983.

13. R. Sitia, M. S. Neuberger, and C. Milstein, Regulation of membrane IgM expression in secretory B cells: translational and post-translational events, *EMBO J.* 6:3969 (1987).

14. J. Hombach, F. Sablitzky, K. Rajewsky, and M. Reth, Transfected plasmacytoma cells do not transport the membrane form of IgM to the cell surface, *J. Exp. Med.* 167:652 (1988).

15. L. E. Samelson, J. B. Harford, and R. D. Klausner, Identification of the components of the murine T cell antigen receptor complex, *Cell* 43:223 (1985).

16. H. Clevers, B. Alarcon, T. Wileman, and C. Terhorst, The T cell receptor/CD3 complex: A dynamic protein ensenble, *Ann. Rev. Immunol.* 6:629 (1988).

17. J. S. Bonifacio, C. Chen, J. Lippincott-Schwartz, J. D. Ashwell, and R. D. Klausner, Subunit interactions within the T-cell antigen receptor: Clues from the study of partial complexes, *Proc. Natl. Acad. Sci. USA* 85:6929 (1988).

18. H. C. Oettgen, C. L. Pettey, W. L. Maloy, and C. Terhorst, A T3-like protein complex associated with the antigen receptor on murine T cells, *Nature* 320:272 (1986).

19. M. Reth, G. J. Hämmerling, and K. Rajewsky, Analysis of the repertoire of anti-NP antibodies in C57BL/6 mice by cell fusion. I. Characterization of antibody families in the primary and hyperimmune response, *Eur. J. Immunol.* 8:393 (1978).

20. J. Hombach, L. Leclercq, A. Radbruch, K. Rajewsky, and M. Reth, A novel 34-kd protein co-isolated with the IgM molecule in surface IgM-expressing cells, *EMBO J.* 7:3451 (1988).

21. J. Hombach, T. Tsubata, L. Leclercq, H. Stappert, and M. Reth, Molecular components of the B-cell antigen receptor complex of the IgM class, *Nature* 343:760 (1990).

22. J. Wienands, J. Hombach, A. Radbruch, C. Riesterer, and M. Reth, Molecular components of the B cell antigen receptor complex of class IgD differ partly from those of IgM, *EMBO J.* 9:449 (1990).

23. N. Sakaguchi, S. Kashiwamura, M. Kimoto, P. Thalmann, and F. Melchers, B lymphocyte lineage-restricted expression of mb-1, a gene with CD3-like structural properties, *EMBO J.* 7:3457 (1988).

24. K. E. Bernstein, C. B. Alexander, E. P. Reddy, and R. G. Mage, Complete sequence of a cloned cDNA encoding rabbit secreted μ-chain of V_Ha2 allotype: comparisons with V_Ha1 and membrane μ sequences, *J. Immunol.* 132:490 (1984).

25. E. A. Kabat, T. T. Wu, M. Reid-Miller, H. M. Perry, and K. S. Gottesman, 1987, "Sequences of Proteins of Immunological Interest," U. S. Department of Health and Human Services.

26. J. Deisenhofer, P. M. Colman, O. Epp, and R. Huber, Crystallographic structural studies of a human Fc fragment. II. A complete model based on a Fourier map at 3.5 Å resolution, *Hoppe-Seyler's Z. Physiol. Chem.* 357:1421 (1976).

27. M. B. Brenner, I. S. Trowbridge, and J. L. Strominger, Cross-linking of human T cell receptor proteins: Association between the T cell idiotype β subunit and the T3 glycoprotein heavy subunit, *Cell* 40:183 (1985).

28. F. Koning, W. L. Maloy, and J. E. Coligan, The implications of subunit interactions for the structure of the T cell receptor-CD3 complex, *Euro. J. Immunol.*, 20:299 (1990).

29. M. Reth, Antigen receptor tail clue, *Nature* 338:383 (1989).

30. A. M. Weissman, S. J. Frank, D. G. Orloff, M. Mercep, J. D. Ashwell, and R. D. Klausner, Role of the zeta chain in the expression of the T cell antigen receptor: Genetic reconstitution studies, *EMBO J.* 8:3651 (1989).

DIFFERENCES IN HUMAN B CELL DIFFERENTIATION

Peter D. Burrows,* Hiromi Kubagawa,* Norihiro Nishimoto,* William G. Kerr,#
Gary V. Borzillo,^ Linda M. Hendershot^ , and Max D. Cooper*%

*Division of Developmental and Clinical Immunology
 Departments of Pediatrics, Medicine and Microbiology
 University of Alabama at Birmingham
 Birmingham, Alabama
%The Howard Hughes Medical Institute
 Birmingham, Alabama
#Department of Genetics
 Stanford University School of Medicine
 Stanford, California
^Department of Tumor Cell Biology
 St. Jude Children's Research Hospital
 Memphis, Tennessee

INTRODUCTION

A comparative approach to the study of immune system development has been extremely useful in our understanding of this complex process. For example, it was first experimentally demonstrated in chickens that the immune system had two antigen specific components, T and B cells. This dichotomy was subsequently found to exist in organisms as diverse as humans and frogs.[1,2] Extrapolation of many findings in mammalian B cell differentiation can often be readily made between humans and mice. However, we will discuss several instances where models of B cell development that have been derived from murine systems do not appear to be directly applicable to humans. In some cases the apparent differences may simply be due to the experimental systems employed, but in others there may be fundamental differences between the two species. Since our purpose is not to minimize the value of comparative analysis, recent data concerning early B cell development in humans that most likely is true in mice will also be discussed.

LIGHT CHAIN GENE REARRANGEMENT CAN PRECEDE HEAVY CHAIN GENE REARRANGEMENT IN PRE-B CELLS

The theory that immunoglobulin (Ig) heavy chain rearrangement precedes light chain gene rearrangement derives support from studies in both mice and humans.[3,4] This concept has been developed, mainly by Alt and colleagues,[4] to include a pivotal role for μ chain protein in the regulation of both heavy and light chain gene rearrangement. Thus μ chain synthesized following a productive VDJ rearrangement would prevent further rearrangement of the heavy chain locus, guaranteeing heavy chain allelic exclusion, and would also induce light chain rearrangement. Perhaps the best evidence that μ chain protein can inhibit heavy chain gene rearrangement has come from transgenic mouse studies, where introduction of a rearranged VDJ-Cμ construct into the germline of a mouse severely inhibits rearrangement of the endogenous heavy chain loci.[5] However, the concept that heavy chain gene

Mechanisms of Lymphocyte Activation and Immune Regulation III
Edited by S. Gupta *et al.*, Plenum Press, New York, 1991

rearrangement always precedes and is, in fact, obligatory for light chain gene rearrangement has recently come into question.

We have used Epstein-Barr virus (EBV) to transform B cell progenitors from human bone marrow so that clonal populations would be available for molecular analysis. The ability of EBV to transform these precursors[6,7] has been somewhat puzzling, since the cells are negative by immunofluorescence for the complement receptor 2 (CR2),[8] the receptor for EBV[9] that is readily detectable on B lymphocytes.[8] We have recently demonstrated that the OKB7 anti-CR2 monoclonal antibody, which can block EBV infection of B cells,[9] inhibits EBV-induced proliferation and transformation of human B cell precursors (Burrows and Kuba-gawa, unpublished). Thus these cells must express CR2 at low levels, sufficient for viral entry but insufficient for detection by immunofluorescence.

Cultures of EBV transformed fetal bone marrow cells were initially examined by immunofluorescence for expression of cytoplasmic immunoglobulin. This analysis revealed that cells with pro-B (Ig^-), pre-B (μ^+,LC^-) and B (μ^+,LC^+) cell phenotypes could be immortalized by the virus. Quite unexpectedly, however, cells that produced κ or λ light chains but no Ig heavy chains were also observed.[10] Biochemical and genetic analysis of cloned κ^+, HC^- cell lines confirmed the results of the immunofluorescence assay, and revealed that the κ locus could be rearranged and expressed in cells with germline heavy chain loci and in cells with only DJ_H rearrangements. We have recently sequenced the productive and non-productive κ rearrangements of a clone that had no rearrangements of its heavy chain loci (H. Schroeder et al., unpublished). Interestingly, N sequences were found at the V_κ-J join on the non-productive allele. N sequence addition is thought to be catalyzed by terminal deoxynucleotidyl transferase (TdT), whose expression is restricted to the earliest stages of pre-B cell development.[11,12] Our sequence data would suggest that light chain gene rearrangement can occur very early in differentiation, at the same stage when VDJ_H rearrangement takes place.

The existence of a relatively high frequency of pre-B cells synthesizing a light chain but no heavy chain has thus far been demonstrated only using the EBV transformation system. However, independent evidence that κ gene rearrangement can occur in cells that have not undergone heavy chain gene rearrangement has come from the analysis of Ig gene structure in human B lineage malignancies.[13-15] Recent studies of Abelson murine leukemia virus transformed cell lines indicate that μ chain protein is also not a prerequisite to light chain gene rearrangement in mice.[16] Abelson virus transformants that synthesize a light chain but no heavy chain have been reported although only rarely. These include transformants from both normal bone marrow,[17,18] and from marrow of "leaky" severe combined immunodeficient mice (M. Bosma, personal communication). We would propose that the observed difference in Ig expression by AMuLV versus EBV transformants is an effect of the viruses, and that pre-B cells rearranging light chain before heavy chain are a relatively common feature of both species.

Cloned lines of EBV transformed cells are extremely heterogeneous in terms of morphology and expression of stage-specific B cell differentiation antigens. The cultures include cells with lymphocyte morphology, larger proliferating lymphoblasts, and plasmablasts with abundant cytoplasm and a well-developed endoplasmic reticulum (ER). Like normal plasma cells, the plasmablasts in the transformed clones preferentially express Ig joining (J) chain.[19] EBV can thus induce ongoing morphologic differentiation that reflects in many ways the normal antigen-driven differentiation of a B cell into a plasma cell. Significantly, however, EBV-driven differentiation occurs independently of the Ig gene status of the transformed cell. For example, a subpopulation of cells in a pro-B cell clone with no Ig gene rearrangements can differentiate into "sterile" plasma cells, incapable of Ig production, but able to express other proteins, such as J chain, typical of normal plasma cells. It should be emphasized that the Ig gene status of EBV-transformed cells remains fixed during this remarkable differentiation process.

Abelson virus transformants differ from EBV transformants in several ways. Clones of AMuLV transformed cells are morphologically fairly homogeneous, especially lacking the plasmablasts seen in EBV cell lines. One of the most useful properties of AMuLV transformants has been the capacity of many of them to undergo Ig gene rearrangement *in vitro*.[4,12]

However, in spite of the plasticity of their Ig heavy chain genes, most of these cells remain frozen at the pre-B cell stage of differentiation.[20] We have proposed[10] that it is this block in differentiation seen in AMuLV transformants, in contrast to EBV transformants, that has led to the conclusion that no LC$^+$/HC$^-$ pre-B cells are present in mice. According to this hypothesis, pre-B cells would contain an inhibitor of light chain gene transcription whose expression would cease as the cells progressed to the B cell stage of differentiation. Thus no LC mRNA or protein would be produced by pre-B cells, regardless of the rearrangement status of their Ig light chain genes. EBV, but not AMuLV, can drive cells to the B cell stage and beyond, revealing the previously transcriptionally silent light chain gene arrangements. Studies in mice have revealed an inhibitor of κ transcription, termed IκB, with precisely these characteristics.[21,22] This labile inhibitor apparantly acts indirectly by sequestering the κ transcription activator NF-κB[23] in the cytoplasm. IκB expression would cease just prior the B cell stage of development when complete IgM monomers are expressed on the cell surface. NF-κB would then be free to enter the nucleus where it would induce light chain gene transcription.

It is not clear why pre-B cells would express an inhibitor of light chain gene transcription, particularly since transcription of the germline κ locus is thought to be required for its eventual rearrangement. However, NF-κB is involved in the transcriptional activation of genes other than κ.[23,24] These include IL-2 and its receptor, MHC class I and β_2-microglobulin, a number of viral genes, and the β chain of the T cell receptor. It may be that transcription and expression of these other genes is detrimental to normal pre-B cell development.

EXPRESSION OF THE μ-PSEUDO LIGHT CHAIN COMPLEX BY NORMAL HUMAN PRE-B CELLS

Pre-B cells that express μ chain but no light chain cannot form IgM molecules and thus should lack surface Ig. Certainly by immunofluorescence microscopy pre-B cells are devoid of surface Ig. In addition, chronic treatment of mice with antibodies to μ chain severely depletes the animals of B cells, but has no apparent effect on the pre-B cell compartment.[25,26] Surface expression of μ chains without light chain is prevented by a stringent mechanism that retains incompletely assembled Ig molecules, e.g. μ dimers, in the endoplasmic reticulum (ER). This process is mediated by BiP/GRP78, an ER protein that binds H chains, preventing their transport to the Golgi unless displaced by light chains.[27-29] The finding that certain human μ^+, LC$^-$ pre-B cell leukemias were dimly positive for surface μ chain expression was therefore surprising.[30,31] Recent studies from several laboratories have helped resolve this paradox.[32-36] To briefly summarize, a small fraction of the μ chains in pre-B cells associate with a complex of low molecular weight proteins (16-22 kD) and can be expressed on the cell surface.[34-37] This surrogate light chain complex, termed ΨL in humans and $\omega\iota$ in mice, is encoded by at least two genes whose expression is restricted to the pre-B cell stage of development.[32,33] One of the genes, λ5, has homology at the 3' end to conventional λ LC genes. Another pair of genes, Vpre-B1 and Vpre-B2, are homologous to Ig V region gene segments. Unlike conventional Ig genes, however, the expression of λ5 and Vpre-B does not require rearrangement. Several features of the μ-ΨLC complex with its natural ligand *in vivo* could result in signal transduction.

While these observations suggest an important role for the μ-ΨL complex in the normal development of early B lineage cells, the expression of this protein complex has only been observed on transformed or leukemic pre-B cells. To examine its expression on normal human pre-B cells required the isolation of these cells in high purity, since no monoclonal antibodies (MAb) that react exclusively with μ-ΨL are available. This was accomplished by two color fluorescence activated cell sorting (FACS) using MAbs specific for pan-B lineage antigens (CD19 or CD24) together with MAbs specific for κ and λ light chains. The latter MAbs react with conventional light chains, but not with ΨL. With this combination of antibodies, bone marrow B lymphocytes are positive for both antigens, whereas pre-B cells are CD19$^+$/LC$^-$. Using this strategy we were able to show by both immunofluorescence and immunoprecipitation analyses that normal pre-B cells express the μ-ΨL complex.[43]

The biological function of μ-ΨL has been the subject of some speculation. Successful VDJ$_H$ rearrangement and synthesis of μ chain would lead to expression of this complex on the cell surface. Therefore, it could serve to identify such cells among the majority of B cell precursors that had not yet attempted or completed this process, or had failed. (The latter are thought to constitute a significant percentage of developing pre-B cells.)[12] Since the development of early B lineage cells is dependent on an inductive microenvironment provided in part by stromal cells,[44] the expression of μ-ΨL could serve as a receptor for a complementary ligand on stromal cells. This could allow delivery of a differentiation signal to either inhibit any further heavy chain gene rearrangement, a necessity for allelic exclusion, and/or perhaps induce light chain gene rearrangement. Alternatively, the surface expression of ΨL, which is monomorphic, together with μ, which would contribute tremendous V$_H$ region diversity, could serve as a primitive antigen receptor. Conceivably this could be autoreactive, promoting the expansion or deletion of cells at an early stage in development. The ΨL could function even earlier in B cell differentiation: transcription of germline V$_H$ genes can occur prior to their rearrangement.[4,12] Should these V$_H$ transcripts be expressed as proteins, they could potentially combine with ΨL and be secreted, functioning as part of an immunoregulatory network. Other functions for ΨL have been proposed (e.g. F Melchers, this volume). Clearly, the engineering of a secreted form of μ-ΨL using different μ heavy chains will be necessary to identify the natural ligand(s) of this complex before this issue can be reasonably debated.

REGULATION OF IgD EXPRESSION IN MICE AND HUMANS

Newly formed immature B cells, produced in large numbers in fetal liver and adult bone marrow, express a single Ig isotype, IgM, on their surface. However, with further maturation, the majority of these cells co-express IgD.[45] Following antigen or mitogen-induced terminal differentiation of these mature B cells into IgM secreting plasma cells, the expression of IgD ceases. This type of "isotype switching" differs from the switch to IgG, or IgA, or IgE production in that no DNA rearrangements (see below) are involved, and it is thus reversible. Messenger RNA for μ and δ chain is generated by alternate splicing of pre-mRNA derived from a complex transcriptional unit, VDJ-Cμ-Cδ.[46] Potentially four different types of heavy chain mRNA can be derived from this locus. These would encode the membrane and secretory forms of both μ and δ chains, all of which would contain an identical variable region.

The molecular mechanism for the differentiation-stage-specific regulation of IgD expression has been extremely well studied in mice using both transformed and normal B lineage cells.[47,48] The consensus from these studies is that IgD expression is regulated primarily at the post-transcriptional level. Thus in μ^+/δ^- pre-B cells or in IgM$^+$/IgD$^-$ immature B cells, the δ constant (C) region gene would be transcribed at a level nearly equivalent to that seen in a mature IgM$^+$/IgD$^+$ B cell. However, only in mature B cells would δ mRNA and δ protein be expressed. There is also evidence that in terminally differentiated, IgM-secreting plasma cells the negative regulation of Cδ expression shifts to the transcriptional level, with significant transcription termination upstream of the Cδ-encoding exons.

IgD was the last of the known isotypes to be identified, and its discovery came from the finding of a human myeloma which secreted an immunoglobulin that was not IgM, IgG, IgA, or IgE.[49] IgD was subsequently shown to be expressed on the membrane of the majority of B cells in both humans and mice. A clue that the regulation of IgD expression might differ between mice and men can be found by comparing the serum levels of this isotype in the two species. Although human serum contains low but readily detectable levels of IgD (~39 μg/ml), it was long believed that mouse serum contained no Ig of this isotype. Later studies revealed that mice had very low concentrations of serum IgD, about tenfold less than in man.[50]

We have examined the regulation of IgD expression during human B cell development and indeed find significant differences between mice and humans.[51] A panel of cell lines and malignancies representative of various stages of B cell development was examined for Cδ

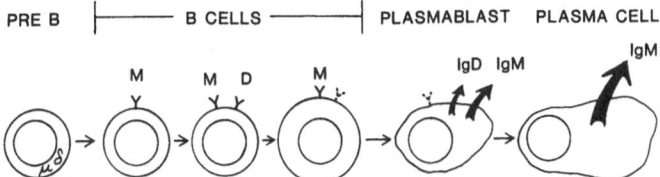

Figure 1. Model for IgD expression during human B cell development. The μ and δ heavy chains are expressed at low levels in pre-B cells and co-expressed on the surface of mature B cells. Both IgD and IgM are to be secreted following antigen- induced proliferation and differentiation, but IgD secretion would cease prior to terminal differentiation into an IgM secreting plasma cell.

transcription using nuclear run-on assays, for δ mRNA by Northern blots, and for δ protein by both immunofluorescence and immunoprecipitation of biosynthetically labeled proteins. The most surprising finding from these studies was that, except in sIgM$^+$/sIgD$^-$ immature B cells, δ mRNA could be readily detected throughout B cell differentiation, from pre-B cells to IgM secreting plasmablasts. This is in striking contrast to mice, as was the finding of both membrane and secretory forms of δ mRNA. The latter has been found at extremely low levels, if at all, in studies of murine δ mRNA.[52] Delta protein was also found whenever δ mRNA was detected. It existed as a short-lived intracellular form in pre-B cells, and as both a membrane and secreted form in IgM secreting plasmablasts. No examples of a truly terminally differentiated IgM secreting plasma cell, e.g. an IgM$^+$ myeloma, were available for analysis. However it seems likely that secretion of IgD must cease prior to this stage of differentiation or else the serum concentration of IgD, even taking its short half life into account, should be much higher. The recent studies of Kuziel et al.[53] are also consistent with this interpretation. Normal human splenic B cells were found to contain both δm and δs mRNA. Stimulation of the cells with the mitogen *Staphylococcus aureus* Cowan I (SA) had little effect on δ mRNA levels; however, stimulation with pokeweed mitogen (PWM) resulted in the disappearance of δ message. SA induces proliferation but very little terminal differentiation of human B cells, whereas PWM can induce terminal differentiation to Ig secreting plasma cells. Thus SA would mimic the early proliferative phases of antigen-induced B cell differentiation, when IgD would be secreted, and PWM would induce differentiation to a stage when secretion of IgD would cease.

Based on these analyses, we have proposed a model for the regulation of human IgD expression depicted in Figure 1. Pre-B cells would express small amounts of both μ and a relatively short lived δ heavy chain. Due in part to its lability, the δ chain in the pre-B cell lines we examined was only detectable by immunoprecipitation after a brief biosynthetic pulse, and not by immunofluorescence or by immunoprecipitation after overnight labeling. The function of δ protein in pre-B cells is unclear, and it is not yet known whether δ can associate with ΨL. However it is known that endogenous VDJ rearrangement is inhibited in transgenic mice harboring a VDJ-Cδ construct[54] as well as it is in similar mice containing a VDJ-Cμ transgene.[5]

Newly-formed human B cells, similar to those in mice, express IgM but not IgD. We have identified two mechanisms for the inhibition of δ expression in these cells. Transcriptional regulation, manifested by transcription termination upstream of Cδ, occurred in one case, while post-transcriptional regulation, in which Cδ is transcribed but no δ mRNA could be detected, occured in another cell line. Given the small sample size it is difficult to predict which is the predominant form of regulation in human IgM$^+$/IgD$^-$ B cells.

IgM secreting plasmablasts, stimulated to differentiate by antigen and T cells, have small amounts of both IgM and IgD on their surface, and can secrete both isotypes.[51] The secretion of IgD, which apparently does not occur to a significant extent in mice, would

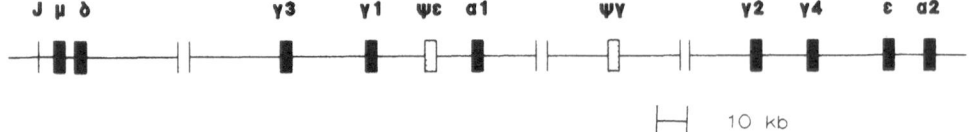

Figure 2. Map of the human immunoglobulin heavy chain constant region locus.

account for the higher serum IgD levels in man. The finding of immunoregulatory T cell bearing receptors for the Fc portion of δ[55] suggests that secreted IgD could have an important function in modulating the immune response.

ISOTYPE SWITCHING IN HUMANS

During their differentiation, some members of a clone of B cells can switch from the production of IgM ± IgD to the production of IgG, IgA, or IgE.[56-59] This allows the same VDJ rearrangement to be expressed with different constant regions, each having different biological effector functions. Isotype switching in both humans and mice is mediated by a DNA deletion event involving all constant region genes upstream of the newly expressed CH gene. (See Figure 2 for a map of the human Ig heavy chain locus.) There continues to be some controversy, however, as to whether DNA deletion is an absolute requirement for the expression of IgG, IgA and IgE.[60-63] A small percentage of B cells express IgG or IgA on their surface. Some of these cells co-express low levels of IgM but, in humans at least, no IgD.[64] The simplest explanation for this phenotype is that the cells have recently switched by DNA deletion, but that residual μ mRNA and protein allows the transient co-expression of IgM and the switched isotype. Other theories hold that DNA deletion has not yet occurred at the B cell stage in differentiation. Several molecular mechanisms have been proposed for this non-deletional type of switch: analogous to the co-expression of IgM and IgD, there could be alternate splicing of long transcripts initiating upstream of the rearranged VDJ_H and reading through the entire IgH locus. "Trans-splicing" of the VDJ exon-encoding RNA onto sterile transcripts (see below) originating from downstream CH genes would also account for this phenotype.[65] Recently Honjo et al. (this volume) have identified an interesting intermediate derived from a cell line that switches *in vitro*. The VDJ_H region was duplicated and inserted upstream of the CH region to which the line eventually switches. We have examined the structure of the IgH locus in a number of sIg$^+$ human B cell malignancies to specifically search for cells that had switched without DNA deletion and have found none.[66,67] In addition, IgA$^+$ B cells isolated from normal human blood show evidence of DNA deletion (A. Radbruch, personal communication). However, it is difficult to exclude formally that there is a transient phase in B cell development when isotype switching could occur without DNA deletion.

Given that the mechanism for isotype switching ultimately, at least, involves DNA deletion, the question arises as to what regulates the production of a particular isotype during an immune response. Is it regulated at the level of the switch itself, or at the cellular level, with selection of switched B cells by T cells and antigen? There is accumulating evidence derived primarily from studies in mice that the switch can be directed.[58,59] Helper T cells, by secreting a given cytokine or combination of them, would instruct a B cell to switch to a particular isotype. This is accomplished by induction of "sterile" transcription of the downstream CH gene. The transcriptionally active and thus "accessible" CH gene would then be available as a substrate for the switch recombination machinery. Other evidence in favor of directed isotype switching has come from the analysis of the non-productive allele in hybridomas that have undergone isotype switching.[68] Switched cells often undergo DNA deletion on both chromosomes, and one of the predictions of the directed switching theory is that the switch should be to the same CH region on both alleles. This was found to be the case in ~60% of the cases, and while it could be argued that 60% is not an overwhelming frequency, the data are consistent with directed class switching.

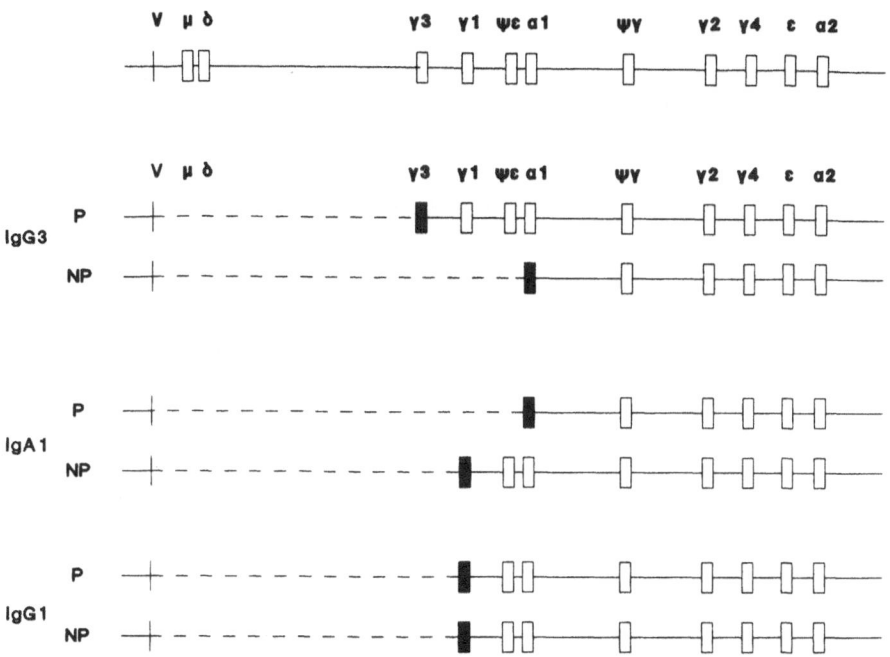

Figure 3. Extent of CH deletions on the productive (P) and non-productive (NP) alleles in human B cell malignancies: evidence for non-directed switch recombination. Switching on the non-productive allele involved both upstream and downstream CH regions, relative to the productive allele. In only twenty percent of the cases was the switch to the same CH gene on both chromosomes.

During an analysis of switch rearrangements in human B lineage malignancies,[66] we had noted that the switch on the non-productive allele showed no evidence of isotype specificity (Figure 3) In only 20% of the cases did the cells switch on the same isotype on both chromosomes. However, there was a tendency for the cells to switch to the nearest neighbor on the non-productive allele. This led us to propose that there was a lack of isotype specificity in the switch rearrangement process in humans. Analysis of a larger number of samples has been consistent with this interpretation[67] (unpublished observation). Again, only 20% of the B cell malignancies had switched to the same isotype on both chromosomes. In 50% of the cases, however, the switch on the non-productive allele was to the nearest 5' CH gene (Figure 4). To determine whether this "geographic specificity" is real or a statistical anomaly will require the analysis of a larger number of samples. It would be of particular interest to examine IgG3 and IgA2 producing cells. The IgG3+ cells should not switch on the non-productive allele, while the IgA2+ cells, which would have the largest choice of upstream CH regions, should preferentially switch to Cε (although see below).

The mechanism by which switching to one isotype on one allele could influence switching to a neighboring isotype on the other allele is not clear. One could envision, for example, that sterile transcripts originating from Cα1 on the productive allele might act in *trans*, forming a heteroduplex with Cα1 on the other chromosome. This might act to influence transcription and switching to the nearest 5' neighbor. However, this model would require that switch recombination occur first on the productive allele. Twenty percent of IgM producing B cell malignancies show evidence for switch recombination on the non-productive allele,[66,67] suggesting that this is not the case.

Further evidence that isotype switching in humans is not directed has come from studies of sterile transcripts in B lineage cell lines and malignancies.[59] Constitutive expression of Cα and multiple Cγ genes was detected in every sIgM+ and sIgM+/sIgD+ sample examined. No sterile transcripts were detected in pre-B cells, and only low levels of

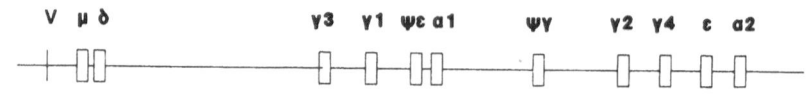

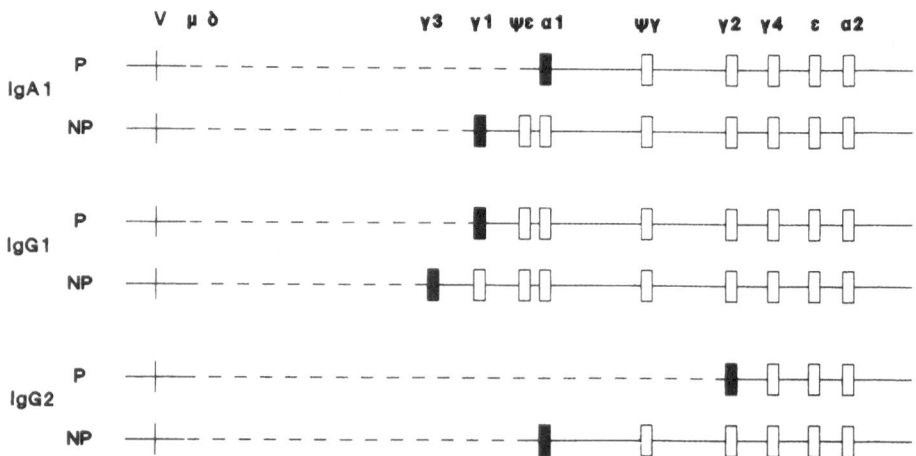

Figure 4. Preferential switching to the nearest 5' neighbor on the non-productive allele in human B cell malignancies. In 50% of cases, the switch on the non-productive (NP) allele occurred to the next CH gene upstream of the one used on the productive (P) allele. The exceptions are Ψε and Ψδ, which contain no switch region and probably cannot participate in isotype switching.

Cα transcription were observed in IgM secreting plasmablasts. Interestingly, Cε was transcriptionally silent throughout B cell development, and thus switching to this isotype may be regulated differently. If sterile transcription is predictive of switch rearrangement potential, then the apparently simultaneous transcription of multiple CH genes in a clonal population of B cells would suggest that switching could occur to any of several CH genes.

CONCLUSIONS

One can view B cell development at various levels of resolution. Certainly the central features of this process are shared between mice and men and perhaps among all mammals. Thus the expression of the μ-ΨL complex, which may play a crucial role in Ig V gene assembly, is similar in both species. However, as one delves deeper into the regulation of various aspects of B cell differentiation, significant differences emerge. The fact that isotype switching is apparently regulated differently in mice and humans may be a reflection of the greater complexity of the IgH locus in man. Other species differences, such as the secretion of IgD in humans, are not so easily explained. However, they are probably important, if not yet interpretable, clues as to functional differences.

ACKNOWLEDGEMENTS

We would like to thank Andreas Radbruch for helpful discussion, Maxine Aycock for the artistic figure graphics, and Ann Brockshire for preparing this manuscript. These studies have been supported by grants CA 16673 and CA 13148, awarded by the National Institutes of Health. W. G. K. is a fellow of the Irvington Istitute for Medical Research. M. D. C. is an HHMI investigator. P. D. B. is a Scholar of the Leukemia Society of America.

REFERENCES

1. M. D. Cooper and A. R. Lawton, The development of the immune system, *Sci. Am.* 231:58 (1974).
2. L. Du-Pasquier, J. Schwager and M. F. Flajnik, The immune system of *Xenopus*, *Ann. Rev. Immunol.* 7:251 (1989).
3. M. D. Cooper and P. D. Burrows, B cell differentiation, *in* "Immunoglobulin Genes" T. Honjo, F. W. Alt, and T. H. Rabbitts, eds., Academic Press, London (1989).
4. F. W. Alt, T. K. Blackwell, R. A. DePinho, and M. Reth, Regulation of genome rearrangement events during lymphocyte differentiation, *Immunol. Rev.* 89:5 (1986).
5. U. Storb, Immunoglobulin gene analysis in transgenic mice, *in* "Immunoglobulin Genes," T. Honjo, F. W. Alt and T. H. Rabbitts, eds., Academic Press, London, (1989).
6. S. M. Fu, J. N. Hurley, J. M. McCune, H. G. Kunkel, and R. A. Good, Pre-B cells and other possible precursor lymphoid cell lines derived from patients with X-linked agammaglobulinemia, *J. Exp. Med.* 152:1519 (1980).
7. S. Katamine, M. Otsu, K. Tada, S. Tsuchiya, T. Sato, N. Ishida, T. Honjo, and Y. Ohno, Epstein-Barr virus transforms precursor B cells even before immunoglobulin gene rearrangements, *Nature* 309:368 (1984).
8. T. F. Tedder, L. T. Clement, and M. D. Cooper, Expression of C3d receptors during human B cell differentiation: Immunofluorescence analysis with the HB-5 monoclonal antibody. *J. Immunol.* 133:678 (1984).
9. G. R. Nemerow, M. D. Moore, and N. R. Cooper, Structure and function of the B lymphocyte Epstein-Barr virus/C3d receptor, *Adv. Cancer. Res.* 54:273 (1990).
10. H. Kubagawa, M. D. Cooper, A. J. Carroll, and P. D. Burrows, Light chain gene expression before heavy chain gene rearrangement in pre-B cells transformed by Epstein-Barr virus, *Proc. Natl. Acad. Sci. USA* 86:2356 (1989).
11. S. Tonegawa, Somatic generation of antibody diversity, *Nature* 302:575 (1983).
12. G. D. Yancopoulos and F. W. Alt, Regulation of the assembly and expression of variable-region genes, *Ann. Rev. Immunol.* 4:339 (1986).
13. B. Lange, M. Valtieri, D. Santoli, D. Caracciola, F. Mavilio, I. Gemperlein, C. Griffin, B. Emanuel, J. Finan, P. Nowell, and G. Rovera, Growth factor requirements of childhood acute leukemia: establishment of GM-CSF-dependent cell lines, *Blood* 70:192 (1987).
14. K. Sheibani, A. Wu, J. Ben-Erza, R. Stroup, H. Rappaport, and C. Winberg, Rearrangement of kappa-chain and T-cell receptor beta-chain genes in malignant lymphomas of "T-cell" phenotype, *Am. J. Path.* 129:201 (1987).
15. D. Delia, M. G. Borrello, E. Berti, M. A. Pierotti, D. Biassoni, R. Gianotti, I. Alessi, M. G. Rizzetti, R. Caputo, and G. Della Porta, Clonal immunoglobulin gene rearrangements and normal T-cell receptor, bcl-2, and c-myc genes in primary cutaneous B-cell lymphomas, *Cancer Res.* 49:4901 (1989).
16. M. D. Schlissel and D. Baltimore, Activation of immunoglobulin kappa gene rearrangement correlates with induction of germline kappa gene transcription, *Cell* 58:1001 (1989).
17. E. J. Siden, D. Baltimore, D. Clark, and N. Rosenberg, Immunoglobulin synthesis by lymphoid cells transformed *in vitro* by Abelson Murine leukemic virus, *Cell* 16:389 (1979).
18. J. Sen, N. Rosenberg, and S. J. Burakoft, Expression and ontogeny of CD2 on murine B cells, *J. Immunol.* 144:2925 (1990).
19. H. Kubagawa, P. D. Burrows, C. E. Grossi, J. Mestecky, and M. D. Cooper, Precursor B cells transformed by Epstein-Barr virus undergo sterile plasma cell differentiation: J chain expression without immunoglobulin, *Proc. Natl. Acad. Sci. USA* 85:875 (1988).
20. P. A. Scherle, K. Dorshkind, and O. N. Witte, Clonal lymphoid progenitor cell lines expressing the BCR/ABL oncogene retain full differentiative function, *Proc. Natl. Acad. Sci. USA* 87:1908 (1990).
21. R. Wall. M. Birskin, C. Carter, H. Govan, A. Taylor, and P. Kincade, A labile inhibitor blocks immunoglobulin κ light chain gene transcription in a pre-B leukemic cell line, *Proc. Natl. Acad. Sci. USA* 83:295 (1986).

22. P. A. Baeuerle and D. Baltimore, IκB: a specific inhibitor of the NF-κB transcription factor, *Science* 242:540 (1988).

23. M. J. Lenardo and D. Baltimore, NF-κB: a pleiotropic mediator of inducible and tissue-specific gene control, *Cell* 58:227 (1989).

24. C. Jamieson, F. Mauxion, and R. Sen, Identification of a functional NF-κB binding site in the murine T cell receptor 2 locus, *J. Exp. Med.* 170:1737 (1989).

25. P. D. Burrows, J. F. Kearney, A. R. Lawton, and M. D. Cooper, Pre-B cells: Bone marrow persistence in anti-μ suppressed mice, conversion to B lymphocytes and recovery after destruction by cyclophosphamide, *J. Immunol.* 120:1526 (1978).

26. G. M. Fulop and D. G. Osmond, Regulation of bone marrow lymphocyte production. IV. Cells mediating the stimulation of marrow lymphocyte production by sheep red blood cells: studies in anti-IgM-suppressed mice, athymic mice, and silica-treated mice, *Cell. Immunol.* 75:91 (1983).

27. I. G. Haas and M. R. Wabl, Immunoglobulin heavy chain binding protein, *Nature* 306:387 (1983).

28. L. M. Hendershot, D. Bole, and J. F. Kearney, The role of immunoglobulin heavy chain binding protein, *Immunol. Today* 8:111 (1987).

29. S. Munro and H. R. B. Pelham, An Hsp70-like protein in the ER: identity with the 78 kd glucose-regulated protein and immunoglobulin heavy chain binding protein, *Cell* 46:291 (1986).

30. L. B. Vogler, W. Crist, D. E. Bockman, E. R. Pearl, A. R. Lawton, and M. D. Cooper, Pre-B cell leukemia: a new phenotype of childhood lymphoblastic leukemia, *New Engl. J. Med.* 298:872 (1978).

31. H. W. Findley, Jr., M. D. Cooper, T. H. Kim, C. Alvarodo, and A. H. Ragab, Two new acute lymphoblastic leukemia cell lines with early B cell phenotype, *Blood* 60:1305 (1982).

32. N. Sakaguchi and F. Melchers, λ5, a new light chain related locus selectively expressed in pre-B lymphocytes, *Nature* 324:579 (1986).

33. A. Kudo and F. Melchers, A second gene, Vpre-B, in the λ5 locus of the mouse which appears to be selectively expressed in pre-B lymphocytes, *EMBO J.* 6:2267 (1987).

34. S. Pillai and D. Baltimore, Formation of disulphide-linked μ2ω2 tetramers in pre-B cells by the 18 K ω-immunoglobulin light chain, *Nature* 329:172 (1987).

35. S. Pillai and D. Baltimore, The omega and iota surrogate immunological light chains, *Curr. Top. Microbiol. Immunol.* 137:136 (1988).

36. W. G. Kerr, M. D. Cooper, L. Feng, P. D. Burrows, and L. M. Hendershot, Mu heavy chains can associate with a pseudo-light chain complex (ΨL) in human pre-B cells, *Inter. Immunol.* 1:355 (1989).

37. H. Kubagawa, T. Ohno, P. D. Burrows, and M. D. Cooper, Putative human μ-ΨL complex defined by a monoclonal antibody (MAb) with reactivity to an IgMλ conformational determinant, *7th Inter. Congress of Immunol.* p. 180, A34-13 (1989).

38. S. Bauer, A. Kudo, and F. Melchers, Structure and pre-B lymphocyte expression of the Vpre-B gene in humans and conservation of its structure in other mammalian species, *EMBO J.* 7:111 (1988).

39. F. Mami, P. A. Cazenave, and T. J. Kindt, Conservation of the immunoglobulin Cλ5 gene in the *Mus* genus, *EMBO J.* 7:117 (1988).

40. C. Schiff, M. Milili, and M. Fougereau, Isolation of early immunoglobulin λ-like gene transcripts in human fetal liver, *Eur. J. Immunol.* 19:1873 (1989).

41. G. F. Hollis, R. J. Evans, J. M. Stafford-Hollis, S. J. Korsmeyer, and J. P. McKearn, Immunoglobulin λ light-chain-related genes 14.1 and 16.1 are expressed in pre-B cells and may encode the human immunoglobulin ω light chain protein, *Proc. Natl. Acad. Sci. USA* 86:5552 (1989).

42. T. Ohno, M. D. Cooper, M. C. Sekar, P. D. Burrows, L. M. Hendershot, and H. Kubagawa, Biochemical and functional characterization of the μ heavy chain and the surrogate light chain complex expressed on human pre-B cell lines, *Fed. Proc.* 4:A1846 (1990).

43. N. Nishimoto, H. Kubagawa, T. Ohno, G. L. Gartland, A. K. Stankovic, and M. D. Cooper, Normal pre-B cells can express the μ heavy chain and surrogate light chain complex on their surface, *Fed. Proc.* 4:A2885 (1990).

44. P. W. Kincade, G. Lee, C. E. Pietrangeli, S. Hayashi, and G. M. Gimble, Cells and molecules that regulate B lymphopoiesis in bone marrow, *Ann. Rev. Immunol.* 7:111 (1989).

45. G. Möller, Immunoglobulin D: Structure, synthesis, membrane representation and function, *Immunol. Rev.* 37:1 (1977).

46. F. R. Blattner and P. W. Tucker, The molecular biology of immunoglobulin D, *Nature* 307:417 (1984).

47. E. L. Mather, K. J. Nelson, J. Haimovich, and R. P. Perry, Mode of regulation of immunoglobulin μ- and δ-chain expression varies during B lymphocyte maturation, *Cell* 36:329 (1984).

48. D. Yuan and P. W. Tucker, Transcriptional regulation of the μ-δ heavy chain locus in normal murine B lymphocytes, *J. Exp. Med.* 160:564 (1984).

49. D. S. Rowe and J. L. Fahey, A new class of human immunoglobulins. I. A unique myeloma protein, *J. Exp. Med.* 121:171 (1965).

50. A. Bargelles, G. Corte, E. E. Cosulich, and M. Ferrarini, Presence of serum IgD and IgD-containing plasma cells in the mouse, *Eur. J. Immunol.* 9:490 (1979).

51. W. G. Kerr, L. M. Hendershot, and P. D. Burrows, Regulation of μ- and δ-chain expression in human B-lineage cells, *J. Immunol.* (in press, 1990).

52. D. Yuan and P. W. Tucker, Regulation of IgM and IgD synthesis in B lymphocytes. I. Changes in biosynthesis of mRNA for μ- and δ-chains, *J. Immunol.* 132:1561 (1984).

53. W. A. Kuziel, C. J. Word, D. Yuan, M. B. White, J. F. Mushinski, F. R. Blattner, and P. W. Tucker, The human immunoglobulin Cμ-Cδ locus: regulation of μ and δ RNA expression during B cell development, *Inter. Immunol.* 1:310 (1989).

54. A. Iglesias, M. Lamers, and G. Köhler, Expression of immunoglobulin δ chains causes allelic exclusion in transgenic mice, *Nature* 330:482 (1987).

55. R. F. Coico, G. W. Siskind, and G. J. Thorbecke, Role of IgD and T delta cells in the regulation of the humoral immune response, *Immunol. Rev.* 105:45-67 (1988).

56. W. E. Gathings, A. R. Lawton, and M. D. Cooper, Immunofluorescent studies of the development of pre-B cells, B lymphocytes and immunoglobulin isotype diversity in humans, *Eur. J. Immunol.* 7:804 (1977).

57. E. R. Abney, M. D. Cooper, J. F. Cooper, A. R. Lawton, and R. M. E. Parkhouse, Sequential expression of immunoglobulin on developing mouse B lymphocytes. A systematic survey which suggests a model for the generation of immunoglobulin isotype diversity, *J. Immunol.* 120:2041 (1978).

58. T. Honjo, A. Shimizu, and Y. Yaoita, Constant-region genes of the immunoglobulin heavy chain and the molecular mechanism of class switching, *in* "Immuno-globulin Genes," T. Honjo, F. W. Alt, and T. H. Rabbitts, eds., Academic Press, London (1989).

59. C. Esser and A. Radbruch, Immunoglobulin class switching: molecular and cellular analysis, *Ann. Rev. Immunol.* 8:717 (1990).

60. Y. Yaoita, Y. Kumagai, K. Okumura, and T. Honjo, Expression of lymphocyte surface IgE does not require switch recombination, *Nature* 297:697 (1982).

61. A. P. Perlmutter and W. Gilbert, Antibodies of the secondary response can be expressed without switch recombination in normal mouse B cells, *Proc. Natl. Acad. Sci. USA* 81:7189 (1984).

62. E. A. Weiss, P. W. Tucker, and D. Yuan, The Cμ gene is transcribed in IgG-bearing B lymphocytes, *J. Molec. Cell. Immunol.* 3:69 (1987).

63. Y. W. Chen, C. Word, V. Dev, J. W. Uhr, E. S. Vitetta, and P. W. Tucker, Double isotype production by a neoplastic B cells lines. II. Allelically excluded production of μ and $\delta1$ heavy chians without C$_H$ gene rearrangement, *J. Exp. Med.* 164:562 (1986).

64. T. Miyawaki, J. L. Butler, A. Radbruch, G. L. Gartland, and M. D. Cooper, Isotype commitment of human B cells that are transformed by Epstein-Barr virus, submitted.

65. A. Shimizu, M. C. Nussenzweig, T. R. Mizuta, P. Leder, and T. Honjo, Immunoglobulin

double isotype expression by trans-mRNA in a human immunoglobulin transgenic mouse, *Proc. Natl. Acad. Sci. USA* 86:8020 (1989).

66. G. B. Borzillo, M. D. Cooper, H. Kubagawa, A. Landay, and P. D. Burrows, Isotype switching in human B lymphocyte malignancies occurs by DNA deletion: evidence for nonspecific switch recombination, *J. Immunol.* 139:1326 (1987).

67. G. V. Borzillo, M. D. Cooper, L. F. Bertoli, A. Landay, R. Castleberry, and P. D. Burrows, Lineage and stage specificity of isotype switching in humans, *J. Immunol.* 141:3625 (1988).

68. C. Schultz, J. Petrini, J. Collins, J. L. Claflin, K. A. Denis, P. Gearhart, C. Gritzmacher, T. Manser, M. Shulman, and W. Dunnick, Patterns and extent of isotype-specificity in the murine H chain switch DNA rearrangement, *J. Immmunol.* 144:363 (1990).

69. W. G. Kerr and P. D. Burrows, Stage-specific transcription of human Cγ and Cα loci during human B cell differentiation, submitted.

STROMAL CELL LINES WHICH SUPPORT LYMPHOCYTE GROWTH

II. CHARACTERISTICS OF A SUPPRESSIVE SUBCLONE

P.W. Kincade, K. Medina, C. E. Pietrangeli, S-I. Hayashi, and A. E. Namen

Oklahoma Medical Research Foundation
Oklahoma City, Oklahoma

INTRODUCTION

Cloned stromal cell lines developed by a number of laboratories are revealing complexity in the hemopoietic inductive microenvironment.[1,2] From these, we now know a considerable amount about genes that may be constitutively expressed in stromal cells and those whose expression is elicited by exogenous stimuli. It is also becoming obvious that stromal cells retain differentiation potential, with a striking example being their differentiation to adipocytes.[3] However, no single stromal cell clone supports all the B lineage differentiation events which normally occur within bone marrow and there are indications that stromal cells in culture differ in some ways from their counterparts *in situ*. Most stromal cell lines which have been described were originally selected for their ability to support growth and/or differentiation of a particular type of hemopoietic cell. Our BMS2 clone permits continuous growth of a number of cloned lymphocytes, which were themselves derived from long term bone marrow cultures.[4] Subclones of BMS2 were prepared and analyzed to determine how stable the line is with repeated passage and if any variations in function correspond to expression of known genes. We now describe one subclone with several unique characteristics which antagonizes lymphocyte growth.

MATERIALS AND METHODS

Cell Cultures

The parent BMS2 stromal cell clone was subcloned by limiting dilution in Iscoves modified MEM (Gibco) with 20% FCS (Hyclone lot #1111794 or #1110638). Cells were subcultured by use of trypsin and EDTA as in our previous reports.[4,5] Lipopolysaccharide (LPS) (Difco, 10μgm/ml) and/or recombinant rat γ interferon (Amgen, 1000 units/ml) were used as inductive stimuli by addition at the time of subculture. The methylcellulose co-culture technique was modified from our previously described[4] cloning method as follows. Stromal cell clones were mixed with either NIH 3T3 cells (one experiment) or C1K cells (two experiments) in different ratios. The C1K line was derived from 3T3 cells by retroviral introduction of the bacterial *lacZ* gene.[6] It was similar to 3T3 with respect to growth support and provided a convenient control for differential growth rates of BMS2 and 3T3 cells. Proportions of blue cells in adherent layers developed with X-gal (5-bromo-4-chloro-3-indoyl β-D-galacto-pyranoside) were as would be expected from the ratios of the initial mixtures. A final density of 3×10^4 cells was plated into 35mm dishes with 10% FCS in DMEM and allowed to adhere overnight. The medium was then aspirated and replaced with 5×10^3 lymphocytes (2E8 cells) in methyl cellulose containing medium.

Mechanisms of Lymphocyte Activation and Immune Regulation III
Edited by S. Gupta *et al.*, Plenum Press, New York, 1991

227

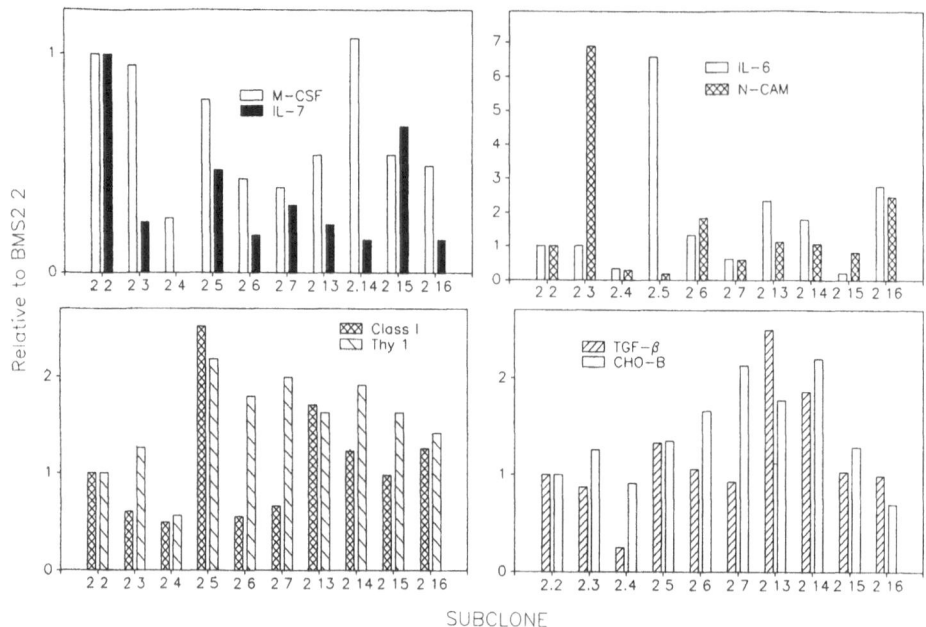

Figure 1. Relative expression of various genes in stromal cell subclones. A Northern blot was prepared with 2 μgm of poly A⁺ RNA per lane from each of the subclones. The cells were allowed to grow to relatively high density before harvesting under what were thought to be similar conditions. The blot was sequentially probed, stripped and re-probed with the indicated cDNA probes. Optimal exposures were then evaluated by laser densitometry and all values normalized with respect to actin to correct for variations in loading. Results are expressed relative to the level of each message in the BMS2.2 subclone. CHO-B is included as a housekeeping gene probe.

Northern Blots

Preparation of total RNA, selection of poly A⁺ RNA, Northern blotting, hybridization and densitometry were as described in our previous publications.[3,5,7] Those reports also contain detailed references for most of the cDNA probes used in this study. Their origins were M-CSF (Dr. P. Ralph); IL-7 (A. Namen); IL-6 (S. Clark); N-CAM (C. Goridis); Class I (G. Waneck); Thy-1 (J. Silver); c-jun (R. Bravo); TGF-β (R. Derynck), CHO-B (R. Wall) and JE (C. Stiles). In addition, we used probes to murine TNF donated by Dr. B. Beutler,[8] murine lymphotoxin from Dr. N. Ruddle;[9] TIS 1 and TIS 7 from Dr. H. Herschman[10] and the PLF1 and PLF2 probes to murine proliferin donated by Dr. D. Linzer.[11]

Viscosity

Conditioned medium was harvested from stromal cell subclones after one week of culture in DMEM, with 10% FCS and 5 x 10⁻⁵M 2-mercaptoethanol. It was centrifuged to remove dead cells and debris and evaluated with a simple glass viscometer purchased from VWR Scientific. The viscometer was standardized before and during use with double distilled water and viscosity measurements compared to fresh medium. The marked viscosity of BMS2 conditioned medium was completely eliminated by treatment at 37°C for 6 hours with 5 units/ml of chondroitin sulfitase ABC (Sigma) or for 30 minutes at 56°C with 10 units/ml of Streptomyces hyaluronidase (ICN Immunobiologicals).

RESULTS AND DISCUSSION

Ten subclones were generated from BMS2 by limiting dilution and tested in a methylcellulose assay with three lymphocyte clones. All except the BMS2.4 subclone supported lymphocyte growth. A possible explanation was initially suggested by Northern blot analysis of RNA prepared from the subclones. In two experiments the 2.4 subclone contained less mRNA corresponding to a number of genes such as those encoding M-CSF, IL-6, MHC Class I and TGF-β. Notably, message corresponding to interleukin 7 (IL-7) was not detectable (Figure 1). This cytokine is thought to be critical to pre-B cell replication and one of the lymphocyte clones (2E8) can grow continuously in the presence of IL-7 (manuscript in preparation).[12,13] However, in subsequent experiments with RNA prepared from BMS 2.4 cultures, interleukin 7 mRNA was clearly visible and we found that the ability to detect it depends on when the cells are evaluated after subculture. These results indicate that the lymphocyte support function is relatively stable in BMS2 subclones. While there can be marked variation in the abundance of various mRNA's in a given experiment, IL-7 expression does not account for the failure of lymphocytes to grow on BMS2.4 cells.

In fibroblastoid cells, the expression of many genes is known to be related to the proliferative state. For example, proliferin is silent in quiescent 3T3 cells but subculture or fresh complete medium activates this member of the prolactin gene family.[11] We found that expression of this and many other genes was markedly affected by passage of BMS2 subclones (Figure 2). In addition to the results shown, levels of cJun, Tis 1, and Tis 7 were sensitive to culture manipulation. Growth was probably not synchronously initiated in the different clones by trypsination and subculture. Accordingly, the kinetics of gene expression often varied from experiment to experiment (not shown). These findings point out an important pitfall when comparing two different cell lines. The results would be very misleading if cytokine mRNA's were compared at only one time point after subculture. We conclude that failure to elaborate IL-7 is not responsible for the inability of BMS2.4 to support lymphocyte growth. The message for IL-7 is just at the limit of detection in many preparations and

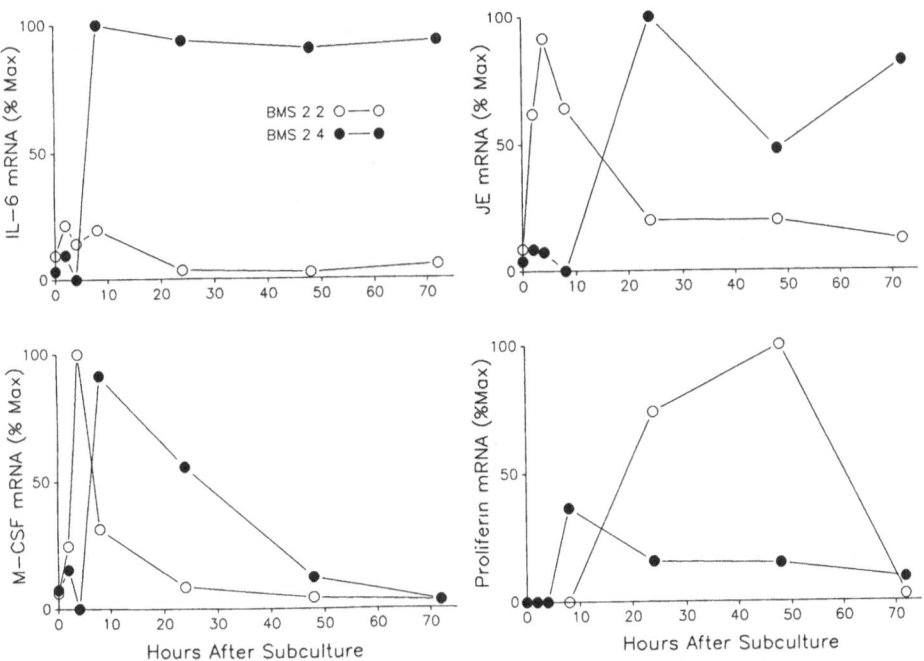

Figure 2. Responsiveness of stromal cell subclones to subculture evaluated by Northern blot analysis. Cells were subcultured after trypsin-EDTA treatment into 100 mm dishes at a density of 2 x 10⁶ cells in 6 ml of DMEM containing 10% FCS. At the indicated times, the adherent cells were extracted and 10 μgm per lane of total RNA was used per lane to prepare Northern blots. Values obtained by densitometry were normalized with the CHO-B housekeeping gene probe and plotted relative to the maximum level of expression obtained in the experiment.

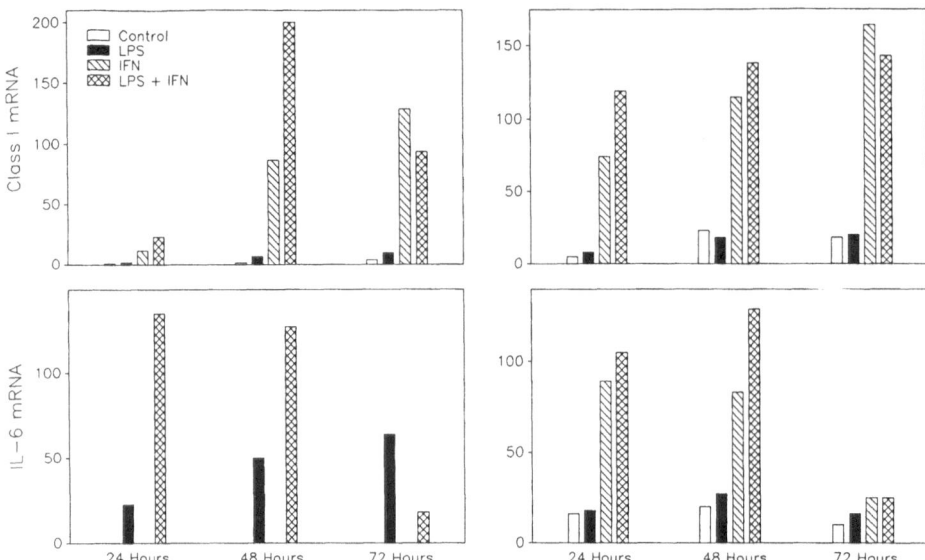

Figure 3. Responsiveness of stromal cell subclones to lipopolysaccharide (LPS) and γ interferon evaluated by Northern blot analysis. Small differences in loading of RNA were adjusted by use of the CHO-B housekeeping gene and normalized to unstimulated BMS2.2 cells at 24 hours after subculture. All results are expressed as fold increases over the mRNA level detected in that sample.

there can be experiment-to-experiment variation in peak expression at various times after subculture.

Previous studies indicate that stromal cells are responsive to a variety of exogenous stimuli, including contact with lymphocyte precursors.[5,14,15] The BMS2.2 and 2.4 subclones were therefore compared after exposure to LPS and/or γ IFN (Figure 3). While both subclones have functional receptors for these agents, it was again important to evaluate the responses at several times after subculture and addition of stimuli. One notable finding was a more rapid response of BMS2.4 cells to interferon in the induction of MHC Class I and IL-6 mRNA's. In contrast, the BMS 2.2 subclone responded more vigorously to LPS, as shown by the up regulation of IL-6 mRNA.

The methylcellulose cloning assay used for the initial characterization of BMS2 permits quantitative asessment of the lymphocyte support function.[4] Mixtures of NIH 3T3 cells and the parent BMS2 line were used to prepare adherent underlayers with limiting numbers of lymphocytes added in the viscous medium (Figure 4). In that case, numbers of lymphocyte clones directly corresponded to the proportion of BMS2 cells that were initially plated. In contrast, similar cocultures with BMS 2.4 cells revealed that they actively suppressed clonal lymphocyte proliferation. Addition of exogenous IL-7 to cultures containing BMS2.4 cells in the adherent layer did not restore the capacity for lymphocyte cloning (not shown).

One stromal cell clone has been reported which selectively suppresses growth of plasma cell tumors via a membrane associated inhibitor.[16] From a limited series of experiments, it appears that BMS2.4 may interfere with proliferation of a different range of hemopoietic cell types. For example, growth of 70Z/3 pre-B lymphoma cells was inhibited (4 experiments), but the growth of neither EL4 T lymphoma cells (3 experiments) nor MPC 11 plasma cells (1 experiment) were affected (data not shown). These findings suggest that the BMS2.4 subclone may selectively inhibit proliferation of certain types of lymphoid cells.

BMS2.4 differed from the parent clone, as well as the other nine subclones, in several

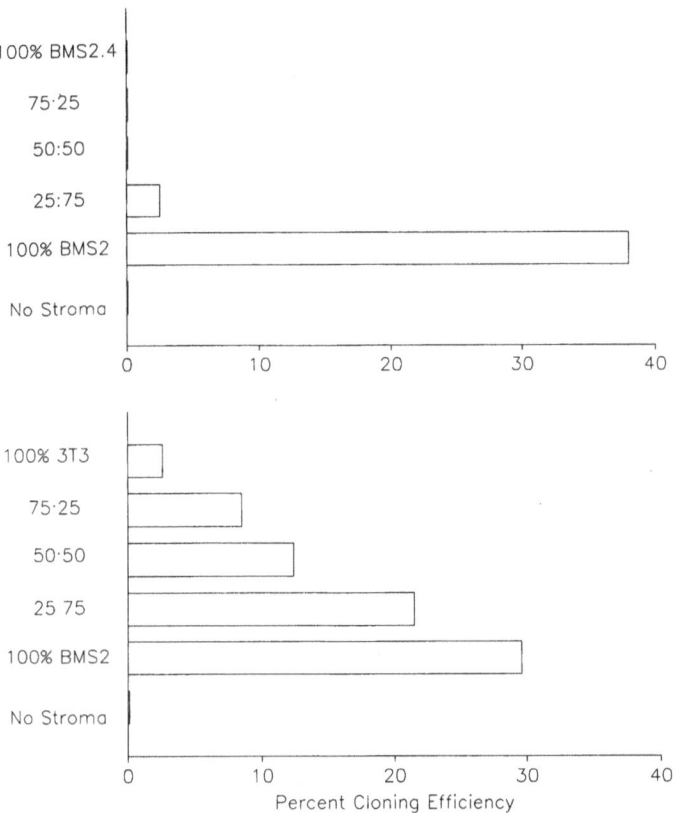

Figure 4. Active suppression of stromal cell dependent lymphocyte proliferation by BMS 2.4 cells. The indicated mixtures
of NIH 3T3 cells and the parent BMS2 line were made and
used to establish adherent layers in 35 mm dishes. A lymphocyte clone (2E8) was plated in methylcellulose over these and
colonies of proliferating cells scored 7 days later. The
lymphocyte clone is able to grow continuously in the presence
of IL-7 alone.

other respects. The other clones caused a notable increase in the viscosity of the medium
measurable as average outflow time (AOT) with a viscometer (Figure 5). This viscosity was
eliminated by treatment of the conditioned medium with either chondroitin sulfitase ABC or
Streptomyces hyaluronidase (not shown). We conclude that most BMS2 clones elaborate
substantial amounts of hyaluronate. The other BMS2 subclones spread considerably after
attachment to culture dishes, but BMS2.4 always retained a spindle like morphology. A
peculiar interaction with the pH indicator, phenol red, is another unique feature of the
BMS2.4 subclone. A bright yellow color consistently developed within hours of the cells
being placed in fresh medium. However, repeated measurements of the medium with pH
electrodes did not reveal values below neutral. Phenol red is not necessarily inert in
cultures and is known to influence estrogen-responsive cells.[17] Our stromal cell subclone
metabolized or interacted in some other way with the dye to result in a striking color
change. It is possible that the morphology of BMS2.4 cells, interaction with phenol red, and
failure to make large amounts of hyaluronate are related and could result from the disregu-
lated expression of a single gene. However, further investigation would be required to test
that hypothesis.

The mechanism through which BMS2.4 suppresses lymphocyte growth is not yet clear.
Preliminary Northern blot analyses did not reveal elevated expression of message for

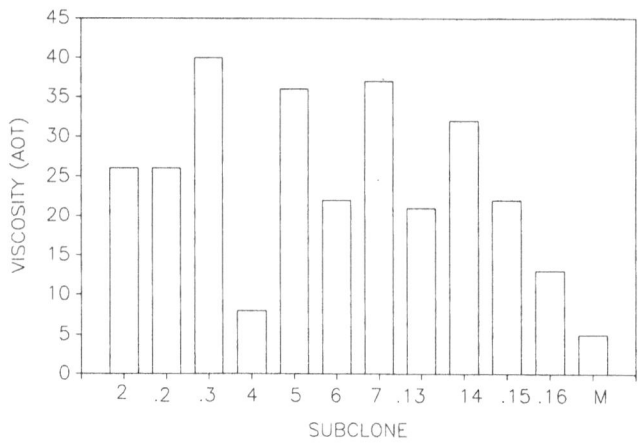

Figure 5. Increased viscosity of medium conditioned by
BMS2 subclones. Conditioned medium from the subclones
was harvested after 7 days of culture and evaluated by
means of a viscometer. Results are shown as average
outflow times (AOT) with fresh medium (M) included as
a control. This is typical of 3 similar experiments.

TGF-β, which is a known antagonist of B lymphopoiesis.[18] Neither tumor necrosis factor
(TNF) nor lymphotoxin (LT) messages were detectable in our stromal cell lines. In some
experiments, mRNA corresponding to IL-6 was very high in BMS2.4 cells (Figure 2) and
active cytokine was detectable in culture supernates with a bioassay (data not shown).
Studies done with the parent cell line indicate that the expression of IL-6 message is
normally low, but easily modulated by exposure of the cells to exogenous stimuli.[5] Disreg-
ulated production of other known cytokines should be evaluated, but this is complicated by
the fact that only some preparations of conditioned medium from BMS2.4 were markedly
suppressive. Regardless of whether this is a variant of the parent BMS2 cells or a contaminating subpopulation in the original clone, the cells may reveal one pattern of gene expression
and function which is possible by microenvironmental elements. It has been proposed that
antagonists play an essential role in lymphohemopoiesis by controlling stem cell self renewal
and regulating the selective recruitment of progenitors in different lineages.[19] A number
of purified recombinant cytokines have been found to inhibit discrete proliferation and differ
entiation events in lymphoid and myeloid progenitors.[18,20-24] This stromal cell subclone
may provide a suitable model for identifying new antagonists and investigating how they are
expressed at a cellular level.

ACKNOWLEDGEMENTS

This work was supported by grants AI-20069 and AI-19884 from the National Institutes
of Health. Dr. Merton Bernfield provided valuable suggestions about hyaluronate detection and
cDNA probes were obtained from many individuals listed in the Materials and Methods.

REFERENCES

1. P. W. Kincade, G. Lee, C. E. Pietrangeli, S-I. Hayashi, and J. M. Gimble, Cells and
       molecules that regulate B lymphopoiesis in bone marrow, *Ann. Rev. Immunol.*
       7:111 (1988).
2. D. Zipori, Cultured stromal cell lines from hemopoietic tissues, *in*: "Handbook
       of the Hemopoietic Microenvironment," M. Tavassoli., ed., Humana Press, Clifton,
       NJ (1989).

3. J. M. Gimble, M-A. Dorheim, Z. Cheng, K. Medina, C-S. Wang, R. Jones, E. Koren, C. E. Pietrangeli, and P. W. Kincade, Adipogenesis in a murine bone marrow stromal cell line capable of supporting B lymphocyte growth and proliferation: biochemical and molecular characterization, *Eur. J. Immunol.* 20:379 (1990).

4. C. E. Pietrangeli, S-I. Hayashi, and P. W. Kincade, Stromal cell lines which support lymphocyte growth: Characterization, sensitivity to radiation and responsiveness to growth factors, *Eur. J. Immunol.* 18:863 (1988).

5. J. M. Gimble, C. E. Pietrangeli, A. Henley, M. A. Korheim, J. Silver, A. E. Namen, M. Takeichi, C. Goridis, and P. W. Kincade, Characterization of murine bone marrow and spleen derived stromal cells: Analysis of leukocyte marker and growth factor mRNA transcript levels, *Blood* 74:303 (1989).

6. J. R. Sanes, J. L. R. Rubenstein, and J-F. Nicols, Use of a recombinant retrovirus to study post-implantation cell lineage in mouse embryos, *EMBO J.* 5:3133 (1986).

7. J. M. Gimble, M. A. Dorheim, Q. Cheng, P. Pekala, S. Enerback, L. Ellingsworth, P. W. Kincade, and C-S. Wang, Response of bone marrow stromal cells to adipogenetic antagonists, *Mol. Cell. Biol.* 57:4587 (1989).

8. D. Caput, B. Beutler, K. Hartog, R. Thayer, S. Brown-Shimer, and A. Cerami, Identification of a common neucleotide sequence in the 3'-untranslated region of mRNA molecules specifying inflammatory mediators, *Proc. Natl. Acad. Sci. USA* 83:1670 (1986).

9. C. B. Li, P. W. Gray, P. F. Lin, K. M. McGrath, F. H. Ruddle, and N. H. Ruddle, Cloning and expression of murine lymphotoxin cDNA, *J. Immunol.* 138:4496 (1987).

10. M. T. Tippetts, B. C. Varnum, R. W. Lim, and H. R. Herschman, Tumor promoter-inducible genes are differentially expressed in the developing mouse, *Mol. Cell. Biol.* 8:4570 (1988).

11. D. I. H. Linzer and D. Nathans, Nucleotide sequence of a growth-related mRNA encoding a member of the prolactin-growth hormone family, *Proc. Natl. Acad. Sci. USA* 81:4255 (1984).

12. A. E. Namen, A. E. Schmierer, C. J. March, R. W. Overell, L. S. Park, D. L. Urdal, and D. Y. Mochizuki, B cell precursor growth-promoting activity. Purification and characterization of a growth factor active on lymphocyte precursors, *J. Exp. Med.* 167:988 (1988).

13. G. Lee, A. E. Namen, S. Gillis, L. R. Ellingsworth, and P. W. Kincade, Normal B cell precursors responsive to recombinant murine IL-7 and inhibition of IL-7 activity by transforming growth factor-B, *J. Immunol.* 142:3875 (1989).

14. D. Rennick, G. Yang, L. Gemmell, and F. Lee, Control of hemopoiesis by a bone marrow stromal cell clone: Lipopolysaccharide- and interleukin-1-inducible production of colony-stimulating factors, *Blood* 69:682 (1987).

15. T. Sudo, M. Ito, Y. Ogawa, M. Iizuka, H. Kodama, T. Kunisada, S.-I. Hayashi, M. Ogawa, K. Sakai, S. Nishikawa, and S.-I. Nishikawa, Interleukin 7 production and function in stromal cell-dependent B cell development, *J. Exp. Med.* 170:333 (1989).

16. D. Zipori, M. Tamir, J. Toledo, and T. Oren, Differentiation stage and lineage-specific inhibitor from the stroma of mouse bone marrow that restricts lymphoma cell growth, *Proc. Natl. Acad. Sci. USA* 83:4547 (1986).

17. Y. Berthois, J. A. Katzenellenbogen, and B. S. Katzenellenbogen, Phenol red in tissue culture media is a weak estrogen: implications concerning the study of estrogen-responsive cell in culture, *Proc. Natl. Acad. Sci. USA* 83:2496 (1986).

18. G. Lee, L. R. Ellingsworth, S. Gillis, R. Wall, and P. W. Kincade, β transforming growth factors are potential regulators of B lymphopoiesis, *J. Exp. Med.* 166:1290 (1986).

19. D. Zipori, M. Kalai, and M. Tamir, Restrictins: stromal cell associated factors that control cell organization in hemopoietic tissues, *Natl. Immun. Cell Growth Regul.* 7:185 (1988).

20. H. E. Broxmeyer, D. E. Williams, L. Lu, S. Cooper, S. L. Anderson, G. S. Beyer, R. Hoffman, and B. Y. Rubin, The suppressive influences of human tumor necrosis factors on bone marrow hematopoietic progenitor cells from normal donors and patients with leukemia: Synergism of tumor necrosis factor and interferon-γ, *J. Immunol.* 136:4487 (1986).

21. D. E. Williams, S. Cooper, and H. E. Broxmeyer, Effects of hematopoietic suppressor molecules on the *in vitro* proliferation of purified murine granulocyte-macrophage progenitor cells, *Cancer Res.* 48:1548 (1988).

22. G. K. Sing, J. R. Keller, L. R. Ellingsworth, and F. W. Ruscetti, Transforming growth factor β selectively inhibits normal and leukemic human bone marrow cell growth *in vitro*, *Blood* 72:1504 (1988).

23. D. Rennick, G. Yang, C. Muller-Sieburg, C. Smith, N. Arai, Y. Takabe, and L. Gemmell, Interleukin 4 (B-cell stimulatory factor 1) can enhance or antagonize the factor-dependent growth of hemopoietic progenitor cells, *Proc. Natl. Acad. Sci. USA* 84:6889 (1987).

24. B. S. Polla, A. Poljak, J. Ohara, W. E. Paul, and L. H. Glimcher, Regulation of class II gene expression: Analysis in B cell stimulatory factor 1-inducible murine pre-B cell lines, *J. Immunol* 137:3332 (1986).

RESTRICTED UTILIZATION OF GERM-LINE VH GENES IN RABBITS: IMPLICATIONS FOR INHERITANCE OF VH ALLOTYPES AND GENERATION OF ANTIBODY DIVERSITY

Katherine L. Knight*, Robert S. Becker*, and Luisa A. DiPietro[+]

*Department of Microbiology
Stritch School of Medicine
Loyola University Chicago
Maywood, Illinois
[+]Department of Pathology
Northwestern University
Chicago, Illinois

SUMMARY

The presence of inherited VH region allotypic specificities, a1, a2 or a3, on nearly all rabbit immunoglobulins has presented a paradox. We know the germline contains hundreds of VH genes, and if we assume that most of these are used in the generation of antibody diversity, then we must ask how have the a allotype-encoding regions been maintained over time? On the other hand, if we assume that only one (or a small number) of these VH gene(s) is (are) used in VDJ gene rearrangements, then, how is antibody diversity generated? To address these questions, we have cloned and determined the nucleotide sequence of the 3'-most germline VH genes from the a1, a2 and a3 chromosomes and shown in each case that the 3'-most H gene, *VH1-a1*, *VH1-a2*, or *VH1-a3*, encodes an a1, a2 or a3 VH region, respectively. Analysis of rearranged VDJ genes from leukemic B cells showed that *VH1* was utilized in these rearrangements. Based on these data, we propose that the allelic inheritance of the VH allotypes is explained by the preferential usage of the VH1 gene in VDJ rearrangements. Support for this hypothesis was obtained from analysis of the mutant rabbit Alicia in which most serum Ig molecules do not have VHa allotypic specificities, but instead have so-called VHa-negative Ig molecules. In this rabbit, *VH1* is not expressed as it has been deleted. Analysis of cDNA clones from spleen of Alicia rabbits suggests that the expressed VHa-negative molecules also are encoded by a single germline VH gene. Thus, we suggest that nearly all rabbit VH regions are encoded by one to two germline VH genes and that antibody diversity is generated primarily by somatic hypermutation and gene conversion.

INTRODUCTION

The presence of VH region allotypic specificities on rabbit immunoglobulins (Ig) has puzzled immunologists for more than 20 years. The VHa allotypic specificities a1, a2 or a3, are found on 80 to 90% of the Ig molecules and are inherited as if controlled by allelic genes.[1-4] Specific amino acid correlates of each of the three VHa allotypes have been identified at several positions within framework regions (FR) FR 1 and 3.[3] In normal rabbits, 10 to 30% of circulating Ig molecules do not carry any VHa allotypic specificity and are designated VHa-negative.[2] Although w, x, and y allotypic specificities have been found on some of the VHa-negative molecules, such molecules have not been studied extensively at the molecular level.[5,6] Partial amino acid sequence analysis of two samples of VHa-negative Ig has been reported, and comparison of these sequences with those of over 30

Mechanisms of Lymphocyte Activation and Immune Regulation III
Edited by S. Gupta *et al.*, Plenum Press, New York, 1991

235

VHa-positive (a1, a2 or a3) molecules has led to a preliminary identification of amino acid residues associated with VHa-negative molecules.[3,7]

There are several possible explanations for the allelic inheritance pattern of the VHa allotypes. First, meiotic recombinations among the VH genes may be suppressed by some unknown mechanism. Alternatively, meiotic recombinations may have occurred in rabbit families such that the VH repertoire of rabbits is shared; in this case, the inheritance pattern of the VHa allotypes may reflect allelic "regulatory" genes. It is also possible that all VHa-positive molecules are derived from one or a small number of closely linked VH genes rather than from the hundreds of VH genes in the germline.

To investigate the molecular nature of VHa-negative molecules and to understand the allelic inheritance pattern of the a1, a2 and a3 allotypes, we have begun to compare the germline and expressed VH gene repertoire of a1, a2 and a3 rabbits as well as of a mutant a2 rabbit, Alicia, that expresses low levels of a2 Ig and yet has normal Ig levels.

RESULTS

Organization of VH, D and JG Genes

A cosmid library was prepared from sperm DNA of an homozygous $\underline{a}^3/\underline{a}^3$ rabbit,[8] and VH encoding clones were identified with a pan-VH probe, p181. The VH probe p181 is derived primarily from FR 2-encoding regions of rabbit VH genes;[9] comparison of the FR 2-encoding sequences of over 30 rabbit VH genes shows this region is highly conserved, and as such, p181 is expected to hybridize with all rabbit VH genes. The VH containing cosmid clones were restriction mapped and analyzed by Southern blot hybridization with the VH probe, p181. Several of the clones contained overlapping regions of DNA and a total of 765 kb of VH region DNA with approximately 100 VH genes was characterized.[8] One group of overlapping cosmid clones spanned 120 kb. This segment of DNA contained a cluster of VH genes at one end and one JH hybridizing fragment at the other end; the JH hybridizing region was separated from the VH genes by 63 kb (Figure 1).[10] Nucleotide sequence analysis of the JH hybridizing region showed that the genomic DNA contains five JH gene segments of which four appear to be functional.[10]

Three D gene families, D1, D2 and D3, have been identified between the VH and JH genes (Figure 1).[10] The four members of the D1 family are identical to each other whereas the two members of the D2 family are similar, but not identical; the D3 family appears to have a single member.

Predominance of Germline VHa-negative Encoding Genes

The approximately 100 germline VH genes in the cosmid clones were analyzed by Southern blots to determine whether they encoded VHa-positive molecules or VHa-negative molecules. Amino acid sequence anlysis of VH regions has identified GLU at position 2 in each of two VHa-negative protein samples; this amino acid is not present in any of approxi-

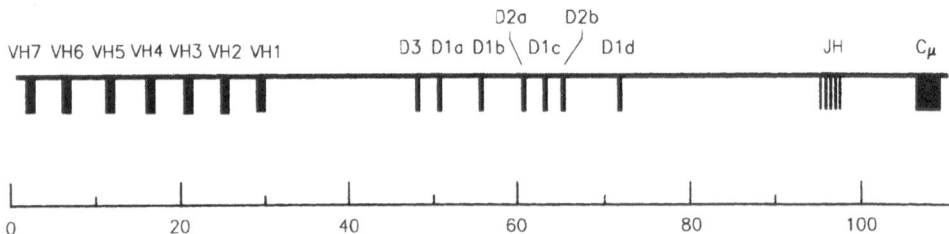

Figure 1. Organization of the rabbit germ-line VH-D-JH region of the a3 chromosome. The VH, D and JH gene segments are indicated by solid boxes. The scale indicates distance in kb.

Figure 2 (sequence alignment of rabbit VH genes; figure is printed sideways on the page):

```
       -19
       M   E   T   G   L   R   W   L   L   L   V   A   V   L   K
5.5    ATG GAG ACT GGG CTG CGC TGG CTT CTC CTG GTC GCT GTG CTC AAA G gtaatgatgagaacgtgggcactgagtcgggagaagagactgagaga
5C3    --- --- --- --- --- --- --- --- --- --- --- --- --- --- --- - ---
4K7    --- --- --- --- --- --- --- --- --- --- --- --- --- --- --- - ---
VH1-a1 --- --- --- --- --- --- --- --- --- --- --- --- --- --- --- - ---

          gacacagagtgtgagtgacagtgtcctgaccatgtctgtgttcag
5.5
5C3
4K7
VH1-a1    ---------------------------------------------

                  -1 FR1                              9                  40
                        3                             *   *   *
              G   V     Q   C   Q   S   V   E   E   S   G   G   R   L   V
5.5           GT  GTC   CAG TGT CAG TCG GTG GAG GAG TCC GGG GGT CGC CTG GTC
5C3           --  ---   --- --- --- --- --- --- --- --- --- --- --- --- ---
4K7           --  ---   --- --- --- --- --- --- --- --- --- --- --- --- ---
VH1-a1        --  ---   --- --- --- --- --- --- --- --- --- --- --- --- ---

       *                   20                  30 CDR1                  FR2
       T   P   G   T   P   L   T   C   T   V   S   G   F   S   L   S   S   Y   A   M   S
5.5    ACG CCT GGG ACA CCC CTG ACA TGC ACA GTC TCT GGA TTC TCC CTC AGT AGT TAT GCA ATG AGC
5C3    --- --- --- --- --- --- --- --- --- --- --- --- --- --- --- --- --- --- --- --- ---
4K7    --- --- --- --- --- --- --- --- --- --- --- --- --- --- --- --- --- --- --- --- ---
VH1-a1 --- --- --- --- --- --- --- --- --- --- --- --- --- --- --- --- --- --- --- --- ---

       FR2                                49 CDR2                  60
       W   V   R   Q   A   P   G   K   G   L   E   W   I   G   I   I   S   S   S   T   Y   Y   A   S   W   A   K   G
5.5    TGG GTC CGC CAG GCT CCA GGG AAG GGG CTG GAA TGG ATC GGA ATC ATT AGT AGT AGT ACC TAT TAC GCG AGC TGG GCG AAA GGC
5C3    --- --- --- --- --- --- --- --- --- --- --- --- --- --- --- --- --- --- --- --- --- --- --- --- --- --- --- ---
4K7    --- --- --- --- --- --- --- --- --- --- --- --- --- --- --- --- --- --- --- --- --- --- --- --- --- --- --- ---
VH1-a1 --- --- --- --- --- --- --- --- --- --- --- --- --- --- --- --- --- --- --- --- --- --- --- --- --- --- --- ---

                                 FR3
       R   F   T       70                    80  82 82A                      90
       R   G   R   T   I   S   K   T   S   T   V   D   L   K   I   T   E   D   T   A   T   Y   F   C   A   R
                                                             *   *       *
5.5    CGA TTC ACC ATC TCC AAA ACC TCG ACC ACG GTG GAT CTG AAA ATC ACC GAG GAC ACG GCC ACC TAT TTC TGT GCC AGA GGG
5C3    --- --- --- --- --- --- --- --- --- --- --- --- --- --- --- --- --- --- --- --- --- --- --- --- --- AGA --A
4K7    --- --- --- --- --- --- --- --- --- --- --- --- --- --- --- --- --- --- --- --- --- --- --- --- --- AGA --T
VH1-a1 --- --- --- --- --- --- --- --- --- --- --- --- --- --- --- --- --- --- --- --- --- --- --- --- --- AGA
                                                                                                          CDR3

                              102 FR4
5.5    GGG TGG TTA TGC TGG TTA TTG TGC TAT TGC TAT \\\ CGG TTG GAT CTC TGG GGC CAG GGC ACC CTG GTC AC
5C3    \\\ -TT -C- --- --- --- --- --- --- --- --- \\\ GAT --- --- -C- --- --- --- --- --- --- --- \\\
4K7    \\\ CA- -AC GAT GAC -AT G-T GAT -A- GGG \\\     --- --- --- -AA --- --- --- --- --- --- --- \\\
```

Figure 2. Nucleotide sequences of three a1-encoding VDJ genes, 5.5, 5C3 and 4K7, cloned from leukemic B cells, and of the germline VH1-a1. VDJ genes were cloned in Charon 28 recombinant libraries constructed with DNA from B cells of leukemic rabbits.[11] The encoded amino acids are indicated by the one letter code. The a1 allotype-associated amino acid residues are indicated by asterisks. The nucleotide sequences of the intron between the first (leader peptide) exon and the VH exon segment are shown in lowercase letters. (\) indicates absence of nucleotide; dashes represent identity. Amino acid number is from[7]; framework regions are designated FR1, FR2, FR3 and FR4 (JH); complementarity determining regions are designated CDR1, CDR2 and (D region).

```
        -19
         M   E   T   G   L   R   W   L   L   L   V   A   V   L   K
4.1     ATG GAG ACT GGG CTG CGC TGG CTT CTC CTG GTC GCT GTG CTC AAA G gtaatgatggggaacgcgggacactgagtctgggagaggatgtgagtgag
5C2     --- --- --- --- --- --- --- --- --- --- --- --- --- --- --- - --------------------------------------------------
4.8     --- --- --- --- --- --- --- --- --- --- --- --- --- --- --- - --------------------------------------------------
5C1     --- --- --- --- --- --- --- --- --- --- --- --- --- --- --- - --------------------------------------------------
VH1-a2  --- --- --- --- --- --- --- --- --- --- --- --- --- --- --- - --------------------------------------------------

                                                                -1 FR1  3       *           *      10       *
                                                     G   V   Q   C   Q   S   V   K   E   S   E   G   G   L   F
4.1     agacacagagagtgtgagtgacagtgtcctgaccatgtcgtctgtgtttgcag GT GTC CAG TGT CAG TCG GTG AAG GAG TCC GAG GGA GGT CTC TTC
5C2     ----------------------------------------------------- -- --- --- --- --- --- --- --- --- --- --- --- --- --- ---
4.8     ----------------------------------------------------- -- --- --- --- --- --- --- --- --- --- --- --- --- --- ---
5C1     ----------------------------------------------------- -- --- --- --- --- --- --- --- --- --- --- --- --- --- ---
VH1-a2  ----------------------------------------------------- -- --- --- --- --- --- --- --- --- --- --- --- --- --- ---

                     *   *          20                                      30 CDR1                FR2             40
         K   P   T   D   T   L   T   L   T   C   T   V   S   G   F   S   L   S   S   N   A   I   S   W   V   R   Q   A
4.1     AAG CCA ACG GAT ACC CTG ACA CTC ACC TGC ACA GTC TCT GGA TTC TCC CTC AGT AGC AAT GCA ATA AGC TGG GTC CGC CAG GCT
5C2     --- --- --- --- --- --- --- --- --- --- --- --- --- --- --- --- --- --- --- --- --- --- --- --- --- --- --- ---
4.8     --- --- --- --- --- --- --- --- --- --- --- --- --- --- --- --- --- --- --- --- --- --- --- --- --- --- --- ---
5C1     --- --- --- --- --- --- --- --- --- --- --- --- --- --- --- --- --- --- --- --- --- --- --- --- --- --- --- ---
VH1-a2  --- --- --- --- --- --- --- --- --- --- --- --- --- --- --- --- --- --- --- --- --- --- --- --- --- --- --- ---

                                    49 CDR2                                 60                          * FR3 *
         P   G   N   G   L   E   W   I   G   A   I   G   S   S   G   S   A   Y   Y   A   S   W   A   K   S   R   S   T
4.1     CCA GGG AAC GGG CTG GAA TGG ATC GGA GCC ATT GGT AGT AGT GGT AGC GCA TAC TAC GCG AGC TGG GCG AAA AGC CGA TCC ACC
5C2     --- --- --- --- --- --- --- --- --- --- --- --- --- --- --- --- --- --- --- --- --- --- --- --- --- --- --- ---
4.8     --- --- --- --- --- --- --- --- --- --- --- --- --- --- --- --- --- --- --- --- --- --- --- --- --- --- --- ---
5C1     --- --- --- --- --- --- --- --- --- --- --- --- --- --- --- --- --- --- --- --- --- --- --- --- --- --- --- ---
VH1-a2  --- --- --- --- --- --- --- --- --- --- --- --- --- --- --- --- --- --- --- --- --- --- --- --- --- --- --- ---

        69   *   *           *                  (V)     82 82A 82B 82C                                  90
         I   T   R   N   T   N   L   N   T   V   T   L   K   M   T   S   L   T   A   A   D   T   A   T   Y   F   C   A
4.1     ATC ACC AGA AAC ACC AAC CTG AAC ACT GTG ACT CTG AAA ATG ACC AGT CTG ACA GCC GCG GAC ACG GCC ACC TAT TTC TGT GCG
5C2     --- --- --- --- --- --- --- --- --- --- --- --- --- --- --- --- --- --- --- --- --- --- --- --- --- --- --- ---
4.8     --- --- --- --- --- --- --- --- --- --- --- --- --- --- --- --- --- --- --- --- --- --- --- --- --- --- --- ---
5C1     --- --- --- --- --- --- --- --- --- --- --- G-- --- --- --- --- --- --- --- --- --- --- --- --- --- --- --- ---
VH1-a2  --- --- --- --- --- --- --- --- --- --- --- --- --- --- --- --- --- --- --- --- --- --- --- --- --- --- --- ---

            CDR3                                                                    102 FR4
         R
4.1     AGA GAT TAT \\\ \\\ GCT GGT TAT GCT GGT TAT GGT TAT GCT AGG AAA TAC TTT AAC TTG TGG GGC CAG GGC ACC CTG GTC AC
5C2     --- --- --- --- --- --- --- --- --- --- --- --- --- --- --- --- --- --- --- --- --- --- --- --- --- --- --- --
4.8     --- -GG GGA TAT TCT TA- --- --- --- --- --- -C- --- --- \\\ GAT GCT \\\ GAT CCC --- --- --- --- --- --- --- --
5C1     --- --- --- --- --- --- --- --- --- --- --- --- --- --- --- --- --- --- --- --- --- --- --- --- --- --- --- --
VH1-a2  --- --
```

Figure 3. Nucleotide sequences of four a2-encoding VDJ genes, 4.1, 5C2, 4.8 and 5C1, cloned from leukemic B cells and of the germline VH gene, *VH1-a2*. The a2 allotype-associated amino acids are indicated by an asterisk. See the legend to Figure 2 for further details.

mately 30 VHa-positive (a1, a2 and a3) protein sequences. Based on these observations, we synthesized oligonucleotides that should distinguish genes encoding VHa-positive molecules from those encoding VHa-negative molecules. Southern blots of Bam HI digested VH cosmid DNA, probed with the VHa-negative and VHa-positive oligomer probes, showed that over 50% of the VH genes hybridized with the VHa-negative oligomer and less than 15% hybridized with the VHa-positive oligomer.[8] These data suggest that more than one-half of the germ-line VH genes encode VHa-negative molecules. This observation is surprising, considering that most serum Ig molecules, 80 to 95%, are VHa-positive and only 5 to 20% are VHa-negative. Thus, it appears that the majority of functional VH regions is derived from a limited number of germline VH genes.

VH and D Gene Utilization in Leukemia B Cells from Eμ-myc Transgenic Rabbits

Transgenic rabbits with the Eμ-*myc* transgene develop polyclonal B cell leukemias at an early age, 14 to 20 days of age.[11] The leukemic B cells appear to be late pre-B or early B cells as they have rearranged both the heavy and light chain V genes but they have little or no surface Ig, and they do not secrete Ig. A total of nine VDJ rearrangements from the leukemic cells were cloned from three rabbits and their nucleotide sequences were determined. Each of the nine VDJ genes had distinct VD or DJ joints and hence must represent different B cell clones. Surprisingly, the nucleotide sequences of the three a1-encoding VDJ genes were identical in their VH segments, FR1, CDR1 (complementarity determining region), FR2, CDR2 and FR3 (Figure 2); similarly, the nucleotide sequences of the four a2 encoding VDJ genes were identical in their VH segments except for a single nucleotide (Figure 3).[12] These data suggested that the same germline VH gene segment had been utilized in each of the VHa2-encoding VDJ genes and similarly, that the same germline VH gene segment had been utilized in each of the VHa1-encoding VDJ genes.

With the exception of VD and DJ joint regions, the nucleotide sequences of the D regions of each clone were identical to germline D1 or D2 sequences,[12] i.e., little or no somatic mutation had occurred in the D region genes. The similarity of the D regions of VDJ rearrangements in leukemic B cells to each other and to germline D regions is in sharp contrast to the dissimilarity of D regions found in cDNA clones from splenic mRNA; in this case, the D regions bear little or no resemblance to the germline D1, D2 or D3 sequences.[13]

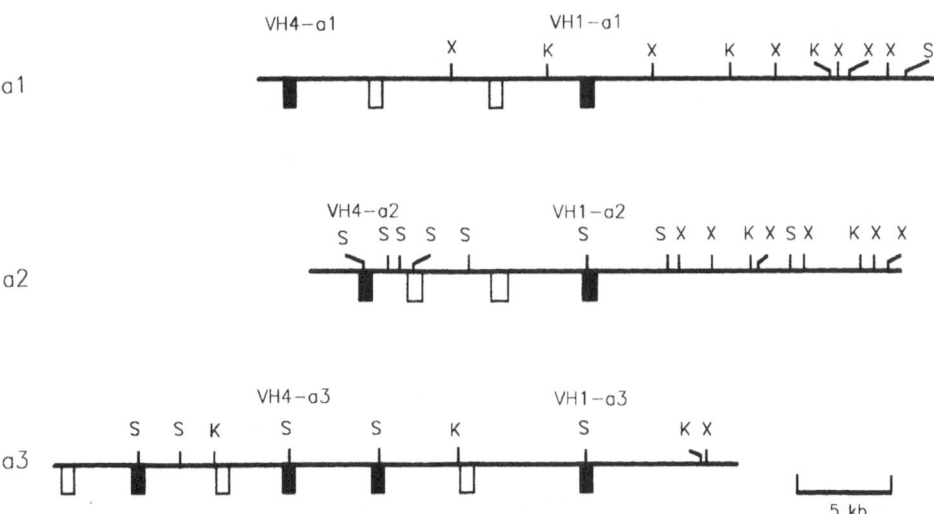

Figure 4. Partial restriction maps of DNA from cosmid clones that contain the 3'-most VH genes from a¹/a¹, a²/a², and a³/a³ rabbits. Solid boxes represent apparently functional VH genes; open boxes represent non-functional VH genes. The orientation of each clone is 5' to 3' from left to right.

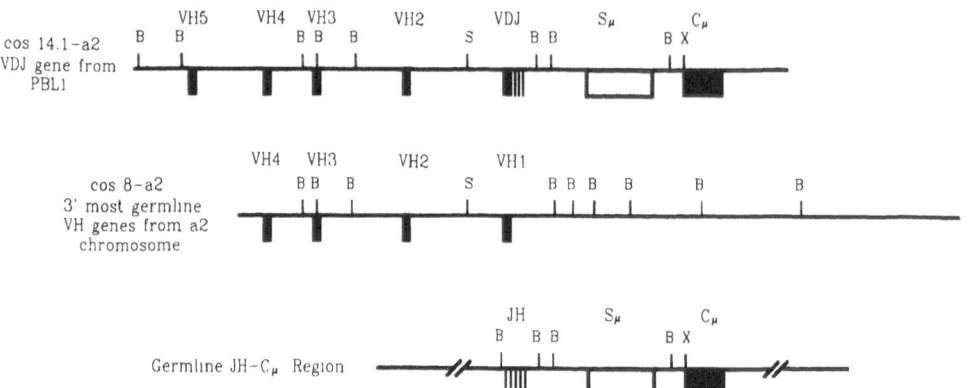

Figure 5. Comparison of the restriction map of the cosmid clone Cos 14.1-a2 containing the VDJ gene from the a2 chromosome of PBL-1 cell line with the restriction map of the cosmid clone Cos 8-a2 containing the 3'-most germline VH genes from the a2 chromosome and with germ-line JH-Cμ region DNA.

Leukemic B Cells Utilize 3' Most VH Gene

Since the D-proximal VH genes in mouse and man have been shown to be preferentially utilized in B cells early in ontogeny,[14,15] we cloned and analyzed the 3'-most VH genes of the rabbit genome to determine if one or more of these genes had been utilized in the VDJ rearrangements in the leukemic B cells. Cosmid libraries were constructed from sperm or liver DNA of homozygous rabbits, a^1/a^1, a^2/a^2 or a^3/a^3, and then screened with a single copy probe obtained 5 kb 3' of *VH1*, the 3'-most VH gene. VH-containing cosmid clones were restriction mapped and the VH genes were identified with the pan-VH probe p181. Comparison of the maps showed that the locations of restriction sites and of VH genes were distinctive between the a1, a2 and a3 chromosomes (Figure 4).[16] The nucleotide sequences of the 3'-most VH gene, *VH1*, from the a1, a2 and a3 chromosomes were determined and all three appeared to be functional.[16] The nucleotide sequence of the VH1 gene from the a1 chromosome, *VH1-a1* was identical to that of the V segment of the leukemic B cell VDJ gene from the a1 chromosome (Figure 2). Similarly, *VH1* from the a2 (Figure 3) and a3 (data not shown) chromosomes, *VH1-a2* and *VH1-a3*, were identical to the V segments of the leuke-mia B cell VDJ genes of the a2 and a3 chromosomes, respectively. These data suggested that *VH1* was the utilized VH gene in the leukemic B cell VDJ rearrangements. To determine if *VH1* was indeed the utilized gene, the VDJ gene was cloned from a cosmid library of DNA from the leukemic B cell line of the a2 allotype, PBL-1,[11] and the region 5' of the VDJ gene was compared to the germline a2 chromosome. The restriction sites 5' of germline *VH1* and 5' of the VDJ gene of PBL-1 were identical, as were the locations of *VH2*, *VH3* and *VH4* (Figure 5). These data indeed confirm that *VH1* was utilized in the VDJ rearrange-ment in PBL-1.

VHa Allotypes Encoded by Allelic VH1

The amino acid sequences encoded by the VH1 gene from the a1, a2 and a3 chromo-some were compared with the amino acid sequences of VH regions of a1, a2 and a3 serum Ig molecules.[7] *VH1* from the a1 chromosome encoded VH regions identical to the sequence of the FR's of pooled a1 serum Ig and it encoded the four a1 allotype-associated amino acids at positions 10, 13, 84 and 85 (Figure 6A). Similarly, VH1 from the a2 and a3 chromosomes encode VH FR's identical to the partial amino acid sequences of FR's of pooled a2 and a3 serum Ig, respectively (Figures 6B and 6C). *VH1-a2* encoded 10 of the 11 predicted a2 allotype-associated amino acids at positions 5, 8, 12, 16, 17, 65, 67, 70, 71, 74 and 75 and *VH1-a3* encoded the one potential a3 allotype-associated amino acid, ASP, at position 10.[3] Thus the allelic genes *VH1-a1*, *VH1-a2* and *VH1-a3* encode prototypic a1, a2 and a3 VH regions, respectively. This finding, coupled with the fact that the leukemic B cells utilize *VH1* in their VDJ rearrangements, strongly suggests that the allelic inheritance

```
A
        FR1  3                        9    *          *                       20
POOL-a1      S  V,L  E   E   S   G   G   R   L   V   T   P   G   T   P   L   T   L   T
VH1-a1   Q   -   V   -   -   -   -   -   -   -   -   -   -   -   -   -   -   -   -

                                      30  CDR1                 FR2             40
POOL-a1  C   T  V,A  S   G   F   S   L   S   S   Y   A,D  #  G,S  W   V   R   Q   A   P
VH1-a1   -   -   A   -   -   -   -   -   A   M   S   -   -   -   -   -   -   -   -   -

                                 49  CDR2                                      60
POOL-a1  G   K   G   L   E  T,W  I   G  FYI I,V  #                   Y   Y   A   TSN
VH1-a1   -   -   -   -   -   W   -   -   I   I   S   S   S   G   S   T   -   -   -   S

                FR3              70                                  80
POOL-a1  W   A   K   G   R   F   T   I   S   K   T   S   T   T   V   D   L   K  ILM  T
VH1-a1   -   -   -   -   -   -   -   -   -   -   -   -   -   -   -   -   -   -   I   -

            82A      *   *              90
POOL-a1  S  P,L  T   T   E   D   T   A   T   Y   F   C   A   R
VH1-a1   -   P   -   -   -   -   -   -   -   -   -   -   -   -


B
        FR1  3       *            *      10        *           *   *          20
POOL-a2      S   V   K   E   S   E   G   G   L   F   K   P   T   D   T   L   T   L   T
VH1-a2   Q   -   -   -   -   -   -   -   -   -   -   -   -   -   -   -   -   -   -   -

                                 30  CDR1              FR2          40
POOL-a2  C   T  V,A  S   G   F               S   S   Y   G   #   S   W   V   R       40
VH1-a2   -   -   V   -   -   -   S   L   -   -   N   A   I   -   -   -   -   Q   A   P

                                 49  CDR2                                      60
POOL-a2
VH1-a2   G   N   G   L   E   W   I   G   A   I   G   S   S   G   S   A   Y   Y   A   S

            *   FR3  *      69   *   *           *                 80
POOL-a2          S   T   I   T   R               T                     L
VH1-a2   W   A   K   S   R   -   -   -   -   -   N   T   N   L   N   T   V   T   L   K

         82 82A 82B 82C                      90
POOL-a2      T   S   L   T   A   A   D   T   A   T   Y   F   C   A   R
VH1-a2   M   -   -   -   -   -   -   -   -   -   -   -   -   -   -


C
        FR1  3                   10                              20
POOL-a3      S   L   E   E   S   G   G   D   L   V   K   P   G   A   S   L   T   L   T
VH1-a3   Q   -   -   -   -   -   -   -   -   -   -   -   -   -   -   -   -   -   -   -

                                 30                                  40
POOL-a3  C  T,K  A   S   G   F   N  G,A  S   S  F,Y  Y   #               Q   A
VH1-a3   -   T   -   -   -   -   S   F   -   -   S   -   Y   M   C   W   V   R   -   -

                                      50
POOL-a3  P   G   K
VH1-a3   -   -   -   G   L   E   W   I   A   C   I   Y   A   G   S   S   G   S   T   Y

            60                          70
POOL-a3          A   K   G   R   F   T   I   S   K
VH1-a3   Y   A   S   W   -   -   -   -   -   -   -   -   -   T   S   S   T   T   V   T

            80                          90
POOL-a3      #   T   S   L   T   A   A   D   T   A   T   Y   F   C   A   R
VH1-a3   L   Q   M   -   -   -   -   -   -   -   -   -   -   -   -   -   -
```

Figure 6. Comparison of the translated amino acid sequence of (A) *VH1-a1*, (B) *VH1-a2*, and (C) *VH1-a3*, with the partial amino acid sequence of pooled a1, a2 and a3 Ig, respectively.[7] The allotype associated amino acid residues[3] are marked by an asterisk. Dashes represent identity; # represents more than three amino acids found in pooled Ig; gaps within the pool a1, a2 and a3 sequences represent residues that were not identified in the protein sequence.[7] The amino acid numbering is from [7].

pattern of the VHa allotypes is the result of utilization of *VH1* in VDJ rearrangements. Since 80 to 95% of normal serum Ig have the VHa allotypes a1, a2 or a3, we suggest that *VH1* is preferentially utilized in B cell VDJ arrangements.

If *VH1* is the preferentially utilized VH gene in VDJ gene arrangements, then we reasoned that the Alicia rabbit, which has the a2 chromosome but little or no a2 Ig,[17] could have a mutation in *VH1*. Consequently, we constructed a cosmid library from germline DNA of an homozygous *ali/ali* Alicia rabbit and cloned and analyzed the 3'-most VH genes. Comparison of the restriction map of the Alicia DNA containing the 3'-most VH genes with the restriction map of similar clones from the normal a2 chromosome showed *ali* DNA to have a 10kb deletion, encompassing both *VH1* and *VH2*. Thus, the loss of expression of the VHa2 allotype in the Alicia rabbit correlates with the loss of *VH1* expression and supports the view that VHa allotype expression is due to preferential utilization of *VH1*.

Preferential VH Gene Utilization in Alicia Rabbits

The 3'-most VH gene in Alicia rabbits is *VH3*; this gene is non-functional as the 5' end is truncated and it has no identifiable leader exon or RNA splice site.[16] The VH gene immediately 5' to *VH3* is *VH4*, which, based on nucleotide sequence data, appears functional.[16] Alicia rabbits, however, do not seem to utilize *VH4* extensively in VDJ rearrangements as *VH4* appears to encode molecules with a2 allotopes and less than 30% of Alicia Ig has a2 allotopes. To investigate the diversity of VH regions of Alicia Ig, heavy chain encoding cDNA clones were prepared from adult spleen poly A$^+$ RNA and were analyzed:[18] the nucleotide sequences of the VH regions of seven clones were determined. One clone, 3.2 (Figure 7), appears to encode the VHa2 allotype as it encodes 9 of the 11 potential VHa2 allotype-associated amino acids in FR's 1 and 3; the other six clones probably do not encode VHa2 molecules as they encode only 1 of the 11 VHa2 allotype-associated amino acids. Since serum Ig of the Alicia rabbit, from which the cDNA library was prepared, contained more than 97% VHa-negative Ig, it is likely that these six cDNA clones encode VHa-negative molecules.

Strikingly, the six VHa-negative cDNA clones encoded VH regions remarkably similar to each other but different from those of most VH regions previously reported. These clones each encode an insertion of four amino acids after position 10 in FR1 (Figure 7); such an insertion has been observed in only 2 of 54 germline and rearranged VH genes.[18] The findings that the six VHa-negative encoding clones encode similar molecules and that the sequence of these is unique, indicate that these VH regions may be derived from one or a small number of very similar germline VH genes. Thus, the VHa-negative Ig molecules of the Alicia rabbit, like the VHa-positive molecules of normal rabbits, seem to be derived predominantly from one germline gene.

DISCUSSION

Our evidence for preferential usage of *VH1* in rabbit is obtained from studies of VDJ rearrangements in B cell leukemias and from studies of the mutant rabbit Alicia. The preferential utilization of the 3'-most VH gene family has been observed in VDJ rearrangements in mouse and human B-lineage cells but this is usually limited to B cells taken early in ontogeny.[14,15] The utilization of *VH1* in VDJ rearrangements in the rabbit leukemic B cells is consistent with this finding in that the leukemic B cells were obtained in 2 to 3 week old rabbits. Also, studies of the Alicia rabbit indicate that *VH1* is preferentially utilized in normal adult rabbits, i.e., deletion of *VH1* in the Alicia rabbit resulted in loss of expression of VHa2 Ig.

The basic organization of rabbit immunoglobulin VH, D and JH genes is similar to that of other mammals[19] and of chicken[20] in that groups of D gene segments are clustered 3' to the cluster(s) of VH gene segments and 5' to the JH gene segments. The VH and JH genes are separated by 65 to 80 kb in mammals[10,19] and a functional VDJ gene results from somatic rearrangements of the VH, D and JH gene segments. Several hundred VH gene segments are found in the germline of mammals, and in mouse, many of these genes are utilized in VDJ rearrangements. In contrast, rabbit appears to preferentially utilize the 3'-most VH gene *VH1*, in VDJ rearrangements. The observation is particularly striking since the germline contains multiple, VH gene segments that appear functional. While VDJ rearrangements in chicken B cells also utilize only one VH gene segment, the chicken germline contains only a single functional VH gene segment.

The predominant utilization of *VH1* explains the inheritance pattern of the a1, a2 and a3 allotypes. VH1 genes from the a1, a2 and a3 chromosomes encode prototypic a1, a2 and

```
        FR1  3
        Q   Q   L   E   Q   S   G   G   G   A   G   G   G   L   V   K   P   G   G   S   L   E
10.2 (γ) CAG CAG CTG GAG CAG TCC GGA GGA GGA GCC GGA GGA GGC CTG GTC AAG CCT GGG GGA TCC CTG GAA
3.1 (μ)  --- --- --- --- --- --- --- --- --- --- --- --- --- --- --- --- --- --- --- --- --- ---
3.3 (μ)  --- --- --- --- --- --- --- --- --- --- --- --- --- --- --- --- --- --- --- --- --- ---
5.2 (γ)  --- --- --- --- --- --- --- --- --- --- --- --- --- --- --- --- --- --- --- --- --- ---
8.1 (μ)      --- --- --- --- --- --- --- --- --- --- --- --- -C- --- --- --- --- --- --- --- ---
9.3 (μ)  --- --- --- --- --- --- --- --- --- --- --- --- --- -C- --- --- --- --- --- --- --- AC-
3.2 (γ)  --- TCA G-- A-- G-- --- -AG --- --T ... ... ... ... --C T-- --- --A ACC -AC C-T --- AC-

        20                              30  CDR1        32A         FR2              40
        L   C   C   K   A   S   G   F   S   L   S   S   N       Y   M   C   W   V   R   Q   A   P   G   K
        CTC TGC TGC AAA GCC TCT GGA TTC TCC CTC AGT AGC AAC ... TAC ATG TGC TGG GTC CGC CAG GCT CCA GGG AAG
        --- --- --- --- --- --- --- --- GA- T-- --- --- T-- ... -GG --- --- --- --- --- --- --- --- --- ---
        --- --- --- --- --- --- --- --- --- --- --- --- -G- TAC -GG --A -A- --- --- --- --- --- --- --- ---
        --- --- --- --- --- --- --- --- --- --- --- --- T-- ... -GG --A --- --- --- --- --- --- --- --- ---
        --- --- --- --- --- --- --- --- A-- T-- --- --- T-- ... --- T-- --- --- --- --- --- --- --- --- ---
        --- -C- --- --- --- --- --- --- GA- T-- --- --- T-- ... --- --- A--- --- --- --- --C A-- --- --- ---
        --- AC- --- -C- -T- --- --- --- --- --- --- --- T-T ... GGA --A AC- --- --- --- --- --- --- --- --C

        CDR2              52A 52B                     60                          FR3
        G   L   E   W   I   G   C   I   Y   A   G   S   S   D   S   T   Y   Y   A   S   W   V   N   G   R
        GGG CTG GAG TGG ATC GGA TGC ATT TAT GCT GGT AGT AGT GAT AGC ACT TAC TAT GCG AGC TGG GTG AAT GGC CGA
        --- --- --- --- --- --- --- --- --- --- --- --- --- -G- --- --A --- --C --- --- --- --- --- --- ---
        --- --- --- --- --- --- --- --- --- --- --- --A --- -G- --- --A --- --- GC- --- --- --- --- --- ---
        --- --- --- --- --- --- AT- --- G-A -G- -T- --- --- -G- --- --- --- --- --A -A- --- --- --- --- ---
        --- --- --- --- --- --- --- --- --- --- --- --- --- -G- --- --A --- --C --- --- --- --- --- --- ---
        --- --- --- --- --- --- AC- --- --- A-- ... ... --- TA- TA- -G- --- --A --- --C --- -C- --- -GC G-A AAG
        --- --- --- --- --- --- AC- --- A-- ... ... TA- TA- -G- --- --A --- --C --- -C- --- -GC G-A AAG ---

        70                              80      82A     B   C
        F   T   L   S   R   D   I   D   Q   S   T   G   C   L   Q   L   N   S   L   T   A   A   D   T   A
        TTC ACT CTC TCC AGA GAC ATC GAC CAG AGC ACA GGT TGC CTA CAA CTG AAC AGT CTG ACA GCC GCG GAC ACG GCC
        --- --- --- --- --- --- --- --- --- --- --- --- --- --- --- --- --- --- -T- --- --- --- --- --- ---
        --- --- --- --- --- --- --- --- --- --- --- --- -C- --- --- --- --- --- --- --- --- --- --- --- ---
        --- --- --- --- --- --- --- --- AC- --- --- --- --A --- --- --- --- --- -T- --- --- --- --- --- ---
        --- --- --- --- --- --- --- --- --- --- --- --- -C- --- --- --- --- --- -T- --- --- --- --- --- ---
        -C- --C A-- A-- --- A-- -C- A-- G-- -A- --G -TG ACT --G A-- A-- -C- --- --- --- --- --- --- --- ---

        90              94  DH                          D1
        M   Y   Y   C   A   R   E   G   Y   S   Y   G   D   Y   G   D   S   I   H   F   Y   G   M   D   P
        ATG TAT TAT TGT GCG AGA GAG GGG TAT AGC TAC GGT GAC TAT GGT GAT TCC ATA CAT TTC TAC GGC ATG GAC CCC
        --- --- --C --- --- --- AGT A-T GG- TAT --- T-C CC- --T- AAC ATC ... ... ... ... ... ... ... ... ...
        --- --- --C --- --- --- AGT C-T GG- --T AC- T-G --T ATG --A T-C -TT -AC ATC ... ... ... ... ... ...
        --- --- --C --- --- --- --T T-- GC- G-T AGT A-- TCT --- ACC CC- -TT -AC ATC ... ... ... ... ... ...
        --- --- --C --- --- --- AGT ACT G-- GT- -TT -AC T-- -C- AAC ATC ... ... ... ... ... ... ... ... ...
        --- --- --C --- --- --- -G- -AC --- -TT -CT --- T-T GG- AA- ACG -A- T-T A-C A-- ... ... ... ... ...
        -CC --- -TC --- --- --- -GA T-- -TA TTG G-- ATC ... ... ... ... ... ... ... ... ... ... ... ... ...

        JH
        W   G   P   G   T   L   V   T
        TGG GGC CCA GGG ACC CTC GTC ACC
        --- --- --- --- --- --G --- ---
        --- --- --T --C --- --G --- ---
        --- --- --- --C --- --G --- ---
        --- --- --- --C --- --G --- ---
        --- --- --- --C --- --G --- ---
        --- --- --- --C --- --G --- C--
```

Figure 7. Nucleic acid and deduced amino acid sequences of seven cDNA VH regions
from an Alicia rabbit. The one letter amino acid code is used and represents the
sequence encoded by clone 10.2. The μ and γ identity of each cDNA clone is indicated. Dashes represent identity to the prototype sequence of clone 10.2; dots represent no nucleotide at that position. Amino acid number is per [7]. Nucleotide changes
that result in an altered amino acid are indicated in bold.

a3 VH regions, respectively, and as these genes are allelic, their preferential utilization
will result in the allelic behavior of the VHa allotypes.

In normal rabbits, a small percentage (5 to 20%) of Ig molecules are VHa-negative.
Based on the similarity of the VHa-negative encoding cDNA clones derived from splenic
RNA of the Alicia rabbit, we suggest that most of the Alicia VHa-negative molecules, like
normal a1, a2 and a3 molecules, are derived primarily from one VH gene. At present we do
not know if the Alicia VHa-negative molecules are of the x, y, and/or z subgroup but we propose that molecules of each subgroup are encoded by a separate but small group of clustered
VH genes and that this (these) gene(s) is (are) inherited in a Mendelian fashion and preferentially utilized in VDJ rearrangements. Such restricted usage of VH genes in VDJ rearrangements would require extensive somatic diversification to generate sufficient antibody
diversity. Such diversification may result from hypermutation and gene conversion.

REFERENCES

1. S. Dray, G. O. Young, and L. Gerald, Immunochemical identification and genetics of rabbit γ-globulin allotypes, *J. Immunol.* 91:403 (1963).

2. S. Dray, G. O. Young, and A. Nisonoff, Distribution of allotypic specificities among rabbit γ-globulin molecules genetically defined at two loci, *Nature (Lond.)* 199:52 (1963).

3. R. G. Mage, K. E. Bernstein, N. McCartney-Francis, C. B. Alexander, G. O. Young-Cooper, E. A. Padlan, and G. H. Cohen, The structural and genetic basis for expression of normal and latent V_Ha allotypes of the rabbit, *Molec. Immunol.* 21:1067 (1984).

4. T. J. Kindt, Rabbit immunoglobulin allotypes, *Adv. Immunol.* 21:35 (1975).

5. B. S. Kim and S. Dray, Identification and genetic control of allotypic specificities on two variable region subgroups of rabbit immunoglobulin heavy chains, *Eur. J. Immunol.* 2:509 (1972).

6. K. H. Roux, A fourth heavy chain variable region subgroup, w, with 2 variants defined by an induced auto-antiserum in the rabbit, *J. Immunol.* 127:626 (1981).

7. E. A. Kabat, T. T. Wu, M. Reid-Miller, H. M. Perry, and K. S. Gottesman, "Sequences of Proteins of Immunologic Interest," U.S. Department of Health and Human Services, Public Health Service, National Institutes of Health, 4th Edition (1987).

8. S. J. Currier, J. L. Gallarda, and K. L. Knight, Partial molecular genetic map of the rabbit VH chromosomal region, *J. Immunol.* 140:1651 (1988).

9. J. L. Gallarda, K. S. Gleason, and K. L. Knight, Organization of rabbit immunoglobulin genes I. Structure and multiplicity of germ-line VH genes, *J. Immunol.* 135:4222 (1985).

10. R. S. Becker, S. Zhai, S. J. Currier, and K. L. Knight, Ig VH, DH, and JH germ-line gene segments linked by overlapping cosmid clones of rabbit DNA, *J. Immunol.* 142:1351 (1989).

11. K. L. Knight, H. Spieker-Polet, D. S. Kazdin, and V. T. Oi, Transgenic rabbits with lymphocytic leukemia induced by the c-myc oncogene fused with the immunoglobulin heavy chain enhancer, *Proc. Natl. Acad. Sci. USA* 85:3130 (1988).

12. R. S. Becker, M. Suter, and K. L. Knight, Restricted utilization of VH and DH genes in leukemic rabbit B cells, *Eur. J. Immunol.* 20:397 (1990).

13. L. A. DiPietro and K. L. Knight, Restricted utilization of germ-line VH genes and diversity of D regions in rabbit splenic Ig mRNA, *J. Immunol.* 144:1969 (1990).

14. G. D. Yancopoulos, S. V. Desiderio, M. Paskind, J. F. Kearney, D. Baltimore, and F. Alt, Preferential utilization of the most J_H-proximal V_H gene segments in pre-B cell lines, *Nature* 311:727 (1984).

15. H. W. Schroeder, Jr., J. L. Hillson, and R. M. Perlmutter, Early restriction of the human antibody repertoire, *Science* 283:791 (1987).

16. K. L. Knight and R. S. Becker, Molecular basis of the allelic inheritance of rabbit immunoglobulin VH allotypes: Implications for the generation of antibody diversity, *Cell* 60:963 (1990).

17. A. S. Kelus and S. Weiss, Mutation affecting the expression of immunoglobulin variable regions in the rabbit, *Proc. Natl. Acad. Sci. USA* 83:4883 (1986).

18. L. A. DiPietro, J. A. Short, S. Zhai, A. S. Kelus, J. Meier, and K. L. Knight, Limited number of immunoglobulin VH regions expressed in the mutant rabbit "Alicia", *Eur. J. Immunol.* (in press, 1990).

19. H. W. Schroeder, Jr., M. A. Walter, M. H. Hofker, A. Ebens, K. Willems Van Dijk, L. C. Liao, D. W. Cox, E. C. B. Milner, and R. M. Perlmutter, Physical linkage of a human immunoglobulin heavy chain variable region gene segment to diversity and joining region elements, *Proc. Natl. Acad. Sci. USA* 85:8196 (1988).

20. C.-A. Reynaud, A. Dahan, V. Anquez, and J.-C. Weill, Somatic hyperconversion diversifies the single V_H gene of the chicken with a high incidence in the D region, *Cell* 59:171 (1989).

CONTROL OF IMMUNOGLOBULIN HEAVY CHAIN CONSTANT REGION GENE

EXPRESSION

*^Suzanne C. Li, *^Paul Rothman, #Mark Boothby, *Pierre Ferrier,
#Laurie Glimcher, and *Frederick W. Alt

*The Howard Hughes Medical Institute and Departments of Biochemistry and
 Microbiology
^Department of Medicine
 College of Physicians and Surgeons of Columbia University
 New York City, New York
#Department of Cancer Biology
 Harvard School of Public Health, and Department of Medicine
 Harvard Medical School
 Boston, Massachusetts

INTRODUCTION

Control of VDJ Assembly

The N-terminus of immunoglobulin (Ig) heavy (H) and light chains is highly variable and determines antigen-binding specificity. The C-terminus of H chains has a constant amino acid sequence that is responsible for determining effector activities such as the localization of the immunoglobulin molecule and the type of secondary pathways activated. The variable region of Ig genes is encoded by multiple germline elements that are assembled into complete V(D)J variable region genes during precursor (pre)-B cell differentiation. Both Ig H and L chain variable region gene segments, as well as related gene segments that encode T cell receptor (TCR) variable regions, have been shown to be assembled by a common enzymatic ctivity referred to as VDJ recombinase. The ability of a single VDJ recombinase to assemble particular gene segments in appropriate cell types and stages within lymphoid lineages is proposed to be effected by modulating the accessibility of substrate gene segments to the common VDJ recombinase. This accessibility has been correlated with transcription of target unrearranged gene segments,[1] both with respect to endogenous genes as well as in transfected recombination substrates.[2,3]

Recently, we have directly tested the hypothesis that known Ig transcriptional control elements may also be involved in regulating VDJ recombination. For these experiments, we constructed a mini-locus that contained unrearranged TCRβ V, D, and J segments linked to the Ig μ heavy chain constant region by a segment of DNA that either contained or lacked the Ig heavy chain transcriptional enhancer element (Eμ). This construct was introduced into the germline of transgenic mice and assayed for its ability to rearrange in normal T and B lineage cells or in A-MuLV transformed precursor T and B cell lines derived from the transgenics. The question was whether the Eμ element could override the relative T cell specificity of TCRβ gene segment rearrangement and force rearrangement of these gene segments in B lineage cells. As the V to DJ rearrangement step appears to be the more tightly regulated step in VDJ assembly, we also asked whether the Eμ element would equally affect the TCRβ D to J and V to DJ rearrangements within the construct.

Mechanisms of Lymphocyte Activation and Immune Regulation III
Edited by S. Gupta *et al.*, Plenum Press, New York, 1991

245

Remarkably, we found that transgenic constructs containing the Eμ transcriptional control element underwent TCRβ D to J rearrangement in nearly 100% of developing T and B cells; those that lacked the enhancer did not rearrange in either lineage.[4] However, we also demonstrated that complete assembly of the TCRβ elements in the construct (Vβ to DJβ joining) occurred at high frequency only in T lineage cells. In correlation with this rearrangement pattern, we found that the unrearranged Vβ segment in the construct was detectably transcribed in pre-T (i.e., thymus) but not pre-B (i.e., bone marrow) rich tissues.[4] Therefore, we have clearly demonstrated that a segment of DNA that contained the Eμ element can have a direct and dominant role in influencing VDJ recombination in these transgenic substrates. However, the T cell specificity of the complete VDJ joining appears to be provided by elements associated with the Vβ gene segment that correlate in activity with the germline Vβ promoter. These results suggested that combinations of stage-specific enhancer and/or promoter elements may play a role in the normal stage-, lineage-, and allele-specificity of the VDJ recombination process. To date, however, it is not clear exactly how such elements influence recombination; transcription of germline variable region gene segments may directly target them for rearrangement by VDJ recombinase or may simply reflect other alterations in these loci that simultaneously leads to transcriptional and recombinational availability.[1]

HEAVY CHAIN CLASS-SWITCHING

Murine C_H genes are located downstream of the variable region locus. The 8 murine C_H genes are found within a 200 kb region. Differentiating B lymphocytes first produce μ H chains which associate with Ig light chains to form the IgM surface receptor. Upon interaction with cognate antigen or stimulation with polyclonal activators, B cells secrete large amounts of Ig and may change the C_H region of the antibody produced. The latter process is termed H chain class-switching and allows a clonal lineage of B cells to produce an antibody that retains variable region specificity in association with a different C_H gene (effector function). The mechanism of class-switching in B lineage cells is through a recombination event that juxtaposes a downstream C_H gene to the expressed V(D)J gene usually accompanied by deletion of intervening sequences including the previously expressed C_H gene. This recombination process involves areas of repetitive sequences termed switch (S) regions that lie upstream of C_H genes. The S regions contain repetitive sequence elements and can vary from 1 to 10 kb in length. Switching recombination can occur over a broad portion of a given S region. It was originally thought that class-switching might simply represent a normal illegitimate recombination event catalyzed by enzymes present in all cells. However, recent class-switch recombination substrate experiments have provided evidence that at least some components of the class-switch recombination activity are expressed specifically in B lineage cells.[5]

TRANSCRIPTION OF GERMLINE C_H GENES

We and others have defined transcription units that initiate upstream of class-switch recombination target sequences of germline H chain genes including $\gamma1$, $\gamma2b$, $\gamma3$, ϵ, and α.[6-12] These germline transcription units initiate upstream of the given S region, proceed downstream through the S region, and terminate at the normal sites downstream of the given constant region gene. All characterized primary germline C_H gene transcripts are spliced so that a small upstream exon (I region) that runs from about 100 to 500 bp downstream of the promoter site is spliced onto the downstream constant region exons. The I regions of characterized germline transcripts do not appear capable of initiating the synthesis of a protein in frame with the attached C region. Because of the apparent lack of coding capacity of these germline transcripts, they are also referred to as "sterile" transcripts. Potential functions of these transcripts will be considered below.

Treatment of normal B lineage cells or A-MuLV transformed cell lines with the polyclonal activator LPS induces transcription from the germline $\gamma2b$ and $\gamma3$ promoters followed by induction of switch recombination to these genes (Figure 1).[13] Conversely, addition of the T cell factor interleukin-4 (IL-4) simultaneously with LPS inhibits germline $\gamma2b$ and $\gamma3$ transcription and switching in both normal spleen cells and pre-B cell lines, but now induces germline $\gamma1$ and ϵ transcription followed by switching to those C_H genes (Figure 1).

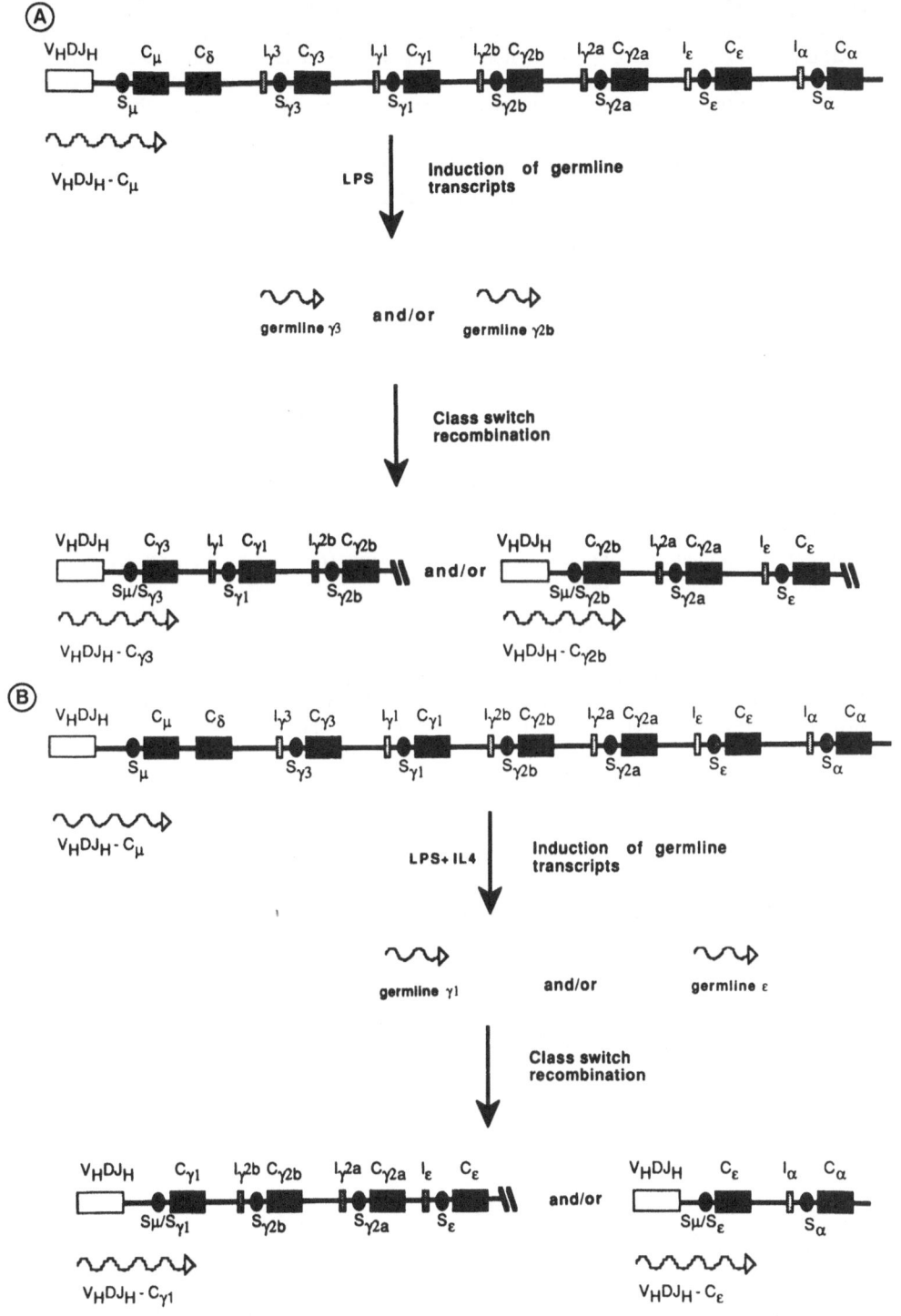

Figure 1. Model for factor-directed class switching. Open boxes represent rearranged Ig variable region genes, dark boxes represent C_H genes, ovals represent switch regions and dotted boxes represent germline exons. A. LPS induced switching to $\gamma 2b$ and/or $\gamma 3$. B. LPS + IL4 induced switching to $\gamma 1$ and/or ϵ. Other details are in the text.

Together, these findings are consistent with the notion that class-switching also may be controlled by modulating access of different C_H genes to a common switch recombinase, with accessibility being modulated by external agents through a mechanism that is correlated with transcripton of target gene sequences.[14,15] Thus, class-switch recombination activity may be targeted to its appropriate substrate by the same general types of mechanisms that target VDJ recombinase activity.

LYMPHOKINES MAY ALTER THE PRE-COMMITMENT

We find that most pre-B cell lines either spontaneously switch to γ2b or can be induced to do so by treatment with LPS. Likewise, treatment of pre-B cells with LPS also induces switching to γ3. In both cases, the induction of switching is correlated with induction of the corresponding germline γ2b and γ3 transcripts. It is notable that γ2b and γ3 are the major isotypes expressed by LPS induced normal splenic B cells and are also the dominant isotypes expressed in T cell independent immune responses.[16-18] This and other data support the notion that differentiating B lineage cells inherit a pre-commitment to switch to γ2b and γ3; a manifestation of this commitment is the LPS induction of γ2b and γ3 germline transcripts in these cells. We propose that interaction with lymphokines such as IL-4 can change this pre-commitment, and upon simultaneous exposure to LPS (or antigen) will lead to germline transcription of a different C_H gene followed by switching to that gene. This change in commitment appears possible even in pre-B cells.[14,15,19] In the latter context, we have clearly demonstrated that, while LPS alone induces only germline γ2b and γ3 transcripts, LPS plus IL-4 treatment can induce germline ε transcripts followed by class-switching to ε in A-MuLV transformed cell lines.[10]

Molecular Control of Germline Transcription

The molecular basis of the pre-commitment to switch to certain isotypes has not been clearly defined. Currently, we are further characterizing the regions and events involved in controlling expression of germline C_H transcripts. In particular we have focused on the identification of the LPS and IL-4 responsive elements around the germline γ2b and ε promoters. As an initial means to identify such regions, we have prepared a number of constructs spanning the genomic γ2b and ε regions in which the germline constant region genes have been mutated so that they can be distinguished from the endogenous genes by S1 nuclease protection assays (Figure 2). These constructs were then transfected into the A-MuLV trans-

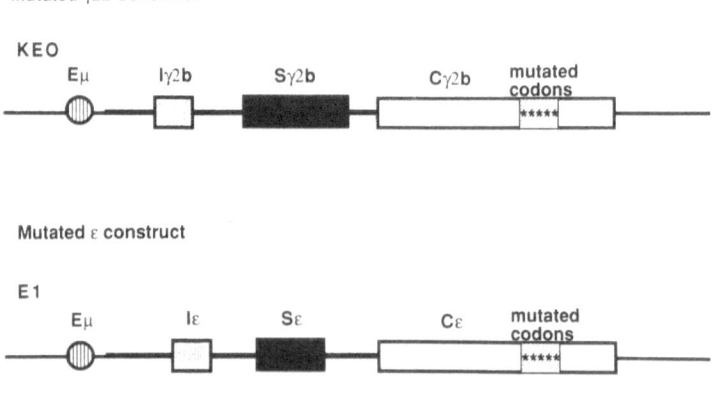

Figure 2. Diagram of mutated germline γ2b and ε constructs. Each construct extends from approximately 3 kb upstream of the germline exon to approximately 3 kb downstream of the C_H gene. The C_H gene contains a 24 bp mutation which allows the construct to be distinguished from the endogenous locus at the DNA and RNA level. Each construct also contains the Ig heavy chain transcriptional enhancer element (Eμ).

formed pre-B cell line 18-8A20 (A20) that is known to be LPS and IL-4 responsive with respect to germline C_H expression and class-switching. Initial transfection experiments utilized constructs that consisted of genomic clones spanning from approximately 3 kb upstream of the C_H exons.

We failed to detect expression of the γ2b construct and only detected a low level of expression of the ε construct after transfection into A20 cells (data not shown). To further test expression of these clones, we linked each of these constructs to the IgH transcriptional enhancer element (Eμ) described above. Linkage to the Eμ element drove constitutive expression of the γ2b clone; this constitutive expression was further augmented by treatment of cells with LPS and inhibited by treatment with IL-4 simultaneously with LPS (data not shown). Although the precise structure of the transcripts arising from the transfected germline γ2b clone have not yet been defined in detail, these findings indicate that IL-4 (and LPS) responsive elements are contained within the γ2b and/or Eμ genomic sequences employed.

Linkage of the Eμ element also led to constitutive expression of the genomic ε clone; however, S1 assays demonstrated that the resultant transcripts did not correspond to spliced germline transcripts (IεCε) in structure (Figure 3). Significantly, treatment with LPS resulted in

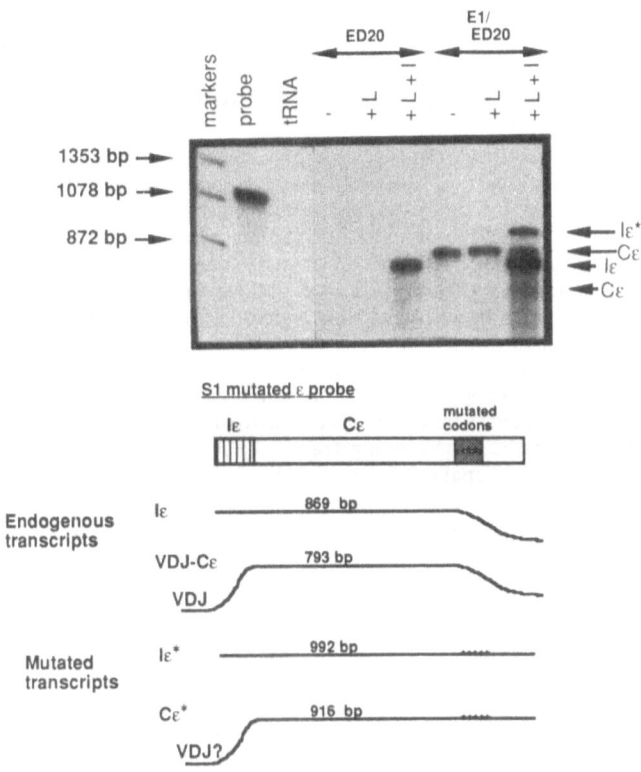

Figure 3. S1 mapping analysis of the endogenous and mutated epsilon transcripts in A20 cells transfected with E1. The probe used can distinguish between endogenous and mutated germ line epsilon transcripts (Iε and Iε*), as well as endogenous and mutated constant region epsilon transcripts (Cε and Cε*). The Cε transcripts presumably represent endogenous switched transcripts (VDJ-Cε).

a slight induction of correctly spliced IεCε transcripts from the transfected ε genomic clone and addition of IL-4 with LPS dramatically augmented this induction (Figure 3). Therefore, addition of the Eμ element permits LPS and IL-4 inducible transcription of the ε genomic clone.

We currently do not understand why the construct is expressed poorly or not at all without the associated Eμ element. Conceivably, the sequences employed lack a positive enhancer element for which the Eμ substitutes. Alternatively, it is conceivable that these constructs contain negative elements that down-regulate expression of the transfected clone. Preliminary CAT assay experiments are consistent with the latter possibility. In these experiments a 1 kb fragment of DNA starting 700 bp upstream of the ε promoter and ending just downstream of Iε was linked to a promoterless CAT gene. CAT expression above background was barely expressed after transfection of this construct into cells; however, treatment of cells harboring the construct with LPS and IL-4 resulted in a 10-fold induction of CAT activity indicating that LPS and IL-4 responsive elements are contained within a minimal region surrounding the Iε promoter. A similar DNA fragment spanning the surrounding the Iε promoter region was tested in gel shift assays for its ability to bind nuclear factors present in untreated or IL-4 treated spleen cells. Several specific IL-4 induced bands were detected and this binding was completed by a 35 bp oligonucleotide representing an IL-4 inducible factor binding site (BRE-2) upstream of the class II gene promoter.[20] Footprinting analyses, however, demonstrated that, despite the competition, the Iε promoter element that binds the Il-4 inducible factor is a nucleotide sequence distinct from that of BRE-1. Additional work is in progress to identify the gene encoding this factor to further assess the role of transcription in directing heavy chain class-switching.

ROLE OF GERMLINE TRANSCRIPTION IN SWITCHING

The precise role of germline transcription with respect to directed class-switching remains to be elucidated. All germline transcripts initiate 5' of the switch region consistent with the possibility that transcription through a targeted region may be directly involved in promoting accessibility. However, there are several additional, not necessarily mutually exclusive, possibilities. Transcription may be a by-product of other events (perhaps binding of specific factors) involved in promoting recombinational accessibility. In additional, germline transcripts themselves may play a role in this process. The general structures of the germline C_H transcripts are very similar; each consists of a 5' exon derived from germline sequences upstream of the corresponding S region that is spliced to the C_H1 of the immediate downstream C_H gene. These structural similarities suggest an analogous function, although the different transcripts contain no significant sequence homologies. The upstream germline exons of all C_H transcripts have stop codons in the open reading frame of the C_H region to which they are spliced; therefore, they do not appear capable of encoding large proteins. Germline RNA transcripts themselves may promote specific recombination events through interaction with S region DNA or could conceivably be used in putative trans-RNA splicing mechanisms for H chain class switching.[14,15] We have employed the marked germline constant region constructs linked to the Eμ element (described above) as well as other similar constructs in a variety of cell-transfection and transgenic mouse experiments to further elucidate these possibilities.

ACKNOWLEDGEMENTS

This work was supported by the Howard Hughes Medical Institute and NIH grants AI-200047 and CA-40427 to F. W. A., and by NIH and Leukemia Society grants to L. G.. S. C. L. is supported by an Arthritis Foundation Loeb Postdoctoral Fellowship. M. B. is supported by a grant from the Arthritis Foundation and NIH grant GM-42550. P. R. is supported by an NIH Physician Scientist Award (DK01336).

REFERENCES

1. T. K. Blackwell and F. W. Alt, Mechanism and developmental program of immunoglobulin gene rearrangement in mammals, *Ann. Rev. Genetics* 23:605 (1989).
2. G. D. Yancopoulos and F. W. Alt, Developmentally controlled and tissue-specific expression of unrearranged VH gene segments, *Cell* 40:271 (1985).

3. T. K. Blackwell, M. W. Moore, G. D. Yancopoulos, H. Suh, S. Lutzker, and F. W. Alt, Recombination between immunoglobulin variable region gene segments is enhanced by transcription, *Nature* 324:585 (1986).

4. P. Ferrier, B. Krippl, T. K. Blackwell, A. J. W. Furley, H. Suh, A. Winoto, W. E. Cook, L. Hood, F. Constantini, and F. W. Alt, Separate elements control DH and VDJ rearrangement in a transgenic recombination substrate, *EMBO* 6:117 (1990).

5. D. E. Ott, F. W. Alt, and K. B. Marcu, Immunoglobulin heavy chain switch region recombination within a retroviral vector in murine pre-B cells, *EMBO* 6:577 (1987).

6. M. T. Berton, J. W. Uhr, and E. S. Vitteta, Synthesis of germline $\gamma 1$ immunoglobulin heavy-chain transcripts in resting B cells; Induction by interleukin 4 and inhibition by interferon γ *Proc. Natl. Acad. Sci. USA* 86:2829 (1989).

7. C. Esser and A. Radbruch, Rapid induction of transcription of unrearranged S$\gamma 1$ switch regions in activated murine B cells by interleukin 4, *EMBO* 8:483 (1989).·

8. S. Lutzker and F. W. Alt, Structure and expression of germ line immunoglobulin $\gamma 2b$ transcripts, *MCB* 8:1849 (1988).

9. P. Rothman, S. Lutzker, B. Gorham, V. Stewart, R. Coffman, and F. W. Alt, Structure and expression of germline immunoglobulin $\gamma 3$ heavy chain gene transcripts: Implication for mitogen and lymphokine directed class-switching, *Internat. Immunol.* 2:621 (1990.

10. P. Rothman, Y.-Y. Chen, S. Lutzker, S. C. Li, V. Stewart, R. Coffman, and F. W. Alt, Structure and expression of germ line immunoglobulin heavy-chain ϵ transcripts: Interleukin-4 plus lipopolysaccharide-directed switching to Cϵ, *MCB* (1990, in press).

11. G. Radcliffe, Y.-C. Lin, M. Julius, K. B. Marcu. and J. Stavnezer, Structure of germ line immunoglobulin α heavy-chain RNA and its location of polysomes, *MCB* 10:382 (1990).

12. P. Sideras, T.-R. Mizuta, H. Kanamori, N. Suzuki, M. Okamoto, K. Kuze, H. Ohno, S. Doi, S. Fukuhara, M. S. Hassan, L. Hammarstrom, E. Smith, A. Shimizu, and T. Honjo, Production of sterile transcripts of Cγ genes in an IgM-producing human neoplastic B cell line that switches to IgG-producing cells, *Int. Imm.* 1:631 (1989).

13. S. Lutzker, P. Rothman, R. Pollock, R. Coffman, and F. Alt, Mitogen- and IL-4-regulated expression of germline Ig $\gamma 2b$ transcripts: Evidence for directed heavy chain class switching, *Cell* 53:177 (1988).

14. S. Lutzker and F. W. Alt, Immunoglobulin heavy chain class switching. In "Mobile DNA," D. D. Berg and M. M. Howe, eds., American Society for Microbiology, Washington (1989).

15. P. Rothman, S. C. Li, and F. W. Alt, The Molecular events in heavy chain class-switching, *Seminars in Immunology* 1:65 (1989).

16. Y. V. Rosenberg, Isotype-specific T cell regulation of immunoglobulin expression, *Immunol Rev.* 67:33 (1982).

17. E. Severinson, S. Bergstedt-Lindqvist, W. van der Loo, and C. Fernandez, Characterization of the IgG response induced by polyclonal B cell activators, *Immunol. Rev.* 67:73 (1982).

18. D. Yuan and E. Vitteta, Structural studies of cell surface and secreted IgG in LPS-stimulated murine B cells, *Mol. Immunol.* 20:367 (1983).

19. W. Paul, Pleiotropy and redundancy: T cell-derived lymphokine in the immune response, *Cell* 57:521 (1989).

20. M. Boothby, E. Gravallese, H.-C. Liou, and L. H. Glimcher, A DNA binding protein regulated by IL-4 and by differentiation in B cells, *Science* 242:1559 (1988).

INDEX

CONTRIBUTORS

JAMES P. ALLISON--Cancer Research Laboratory, University of California, Berkeley, California 94720

FREDERICK ALT--Department of Biochemistry, Columbia University College of Physicians and Surgeons, New York, New York 10032

RICHARD ARMITAGE--Immunex Corporation, Seattle, Washington 98101

ELIZABETH BARBER--Dana-Farber Cancer Institute, Boston, Massachusetts 02215

MARCIA BLACKMAN--Howard Hughes Medical Institute, National Jewish Center, Denver, Colorado 80206

PETER BURROWS--Howard Hughes Medical Institute, University of Alabama, Birmingham, Alabama 35294

MAX COOPER--Howard Hughes Medical Institute, University of Alabama, Birmingham, Alabama 35294

LOUIS DuPASQUIER--Basel Institute for Immunology, Basel, Switzerland

F. A. FLETCHER--Howard Hughes Medical Institute, Baylor College of Medicine, Houston, Texas 77030

B. J. FOWLKES--Laboratory of Cellular and Microbial Immunity, National Institute of Allergy and Infectious Diseases, National Institutes of Health, Bethesda, Maryland 20892

SUDHIR GUPTA--Department of Medicine, University of California, Irvine, California 92717

LEONARD A. HERZENBERG--Department of Genetics, Stanford University School of Medicine, Stanford, California 94305-5120

TASUKU HONJO--Kyoto University, Kyoto, Japan

TOSHIYUKI HORI--DNAX Research Institute of Molecular and Cellular Biology, Palo Alto, California 94304-1104

MARC K. JENKINS--Department of Microbiology, University of Minnesota, Minneapolis, Minnesota 55455-0312

PAUL W. KINCADE--Department of Immunobiology and Cancer, Oklahoma Medical Research Foundation, Oklahoma City, Oklahoma 73104

KATHERINE L. KNIGHT--Department of Microbiology, Loyola University Stritch School of Medicine, Maywood, Illinois 60153

NICOLE LeDOUARIN--Institut d'Embryologie Cellulaire et Moleculaire du C. N. R. S. et du College de France, Nogent-sur-Marne, France

GARY W. LITMAN--Department of Molecular Genetics, Tampa Bay Research Institute, St. Petersburg, Florida 33716

YANG LIU--Howard Hughes Medical Institute, Yale University School of Medicine, New Haven, Connecticut 06510

DENNIS YUNG-DUK LOH--Howard Hughes Institute, Washington University School of Medicine, St. Louis, Missouri 63110

IAN MacNEIL--DNAX Research Institute, Palo Alto, California 94304

TAK W. MAK--Departments of Medicine, Biophysics, and Immunology, University of Toronto, Toronto, Ontario, Canada M4X 1K9

FRITZ MELCHERS--Basel Institute for Immunology, Basel, Switzerland

WILLIAM PAUL--Laboratory of Immunology, National Institute of Allergy and Infectious Diseases, National Institutes of Health, Bethesda, Maryland 20892

MICHAEL RETH--MPI für Immunologie, Freiburg, Germany

ELLEN V. ROTHENBERG--Department of Biology, California Institute of Technology, Pasadena, California 91125

JONATHAN SPRENT--Department of Immunology, Research Institute of Scripps Clinic, La Jolla, California 92037

SUSUMU TONEGAWA--Department of Biology, Massachusetts Institute of Technology, Cambridge, Massachusetts 02139

HARALD von BOEHMER--Basel Institute for Immunology, Basel, Switzerland